ADVANCES IN TRANSPORTATION CYBERSECURITY AND RESILIENCY

ADVANCES IN TRANSPORTATION CYBERSECURITY AND RESILIENCY

Editors

Satish Ukkusuri
Purdue University, USA

Mashrur Chowdhury
Clemson University, USA

Bhavani Thuraisingham
The University of Texas at Dallas, USA

World Scientific

NEW JERSEY • LONDON • SINGAPORE • BEIJING • SHANGHAI • HONG KONG • TAIPEI • CHENNAI • TOKYO

Published by

World Scientific Publishing Co. Pte. Ltd.
5 Toh Tuck Link, Singapore 596224
USA office: 27 Warren Street, Suite 401-402, Hackensack, NJ 07601
UK office: 57 Shelton Street, Covent Garden, London WC2H 9HE

Library of Congress Cataloging-in-Publication Data
Names: Ukkusuri, Satish V. S. K., 1978– editor | Chowdhury, Mashrur A. editor |
Thuraisingham, Bhavani M. editor
Title: Advances in transportation cybersecurity and resiliency / editors,
Satish Ukkusuri, Purdue University, USA, Mashrur Chowdhury, Clemson University, USA,
Bhavani Thuraisingham, The University of Texas at Dallas, USA.
Description: New Jersey : World Scientific, [2027] | Includes bibliographical references and index.
Identifiers: LCCN 2026010531 | ISBN 9789819831531 hardcover |
ISBN 9789819831548 ebook for institutions | ISBN 9789819831555 ebook for individuals
Subjects: LCSH: Intelligent transportation systems--Security measures |
Intelligent transportation systems--Data processing | Artificial intelligence--Security measures |
Computer security | Computer networks--Security measures
Classification: LCC TE228.3 .A394 2027
LC record available at https://lccn.loc.gov/2026010531

British Library Cataloguing-in-Publication Data
A catalogue record for this book is available from the British Library.

For any available supplementary material, please visit
https://www.worldscientific.com/worldscibooks/10.1142/14831#t=suppl

Desk Editors: Eshak Nabi Akbar Ali/Veronica Lee

Typeset by Stallion Press
Email: enquiries@stallionpress.com

https://doi.org/10.1142/9789819831548_fmatter

About the Editors

Satish Ukkusuri is a Hubert and Audrey Kleasen Professor in the Lyles School of Civil and Construction Engineering at Purdue University. He also serves as the Director of the Urban Mobility Networks and Intelligence (UMNI) Lab and Director of the Purdue Transportation Cybersecurity and Resilience Center. His research expertise lies in interdisciplinary transportation networks, with his current interests in data-driven mobility solutions, disaster management, resilience of interdependent networks, connected and autonomous traffic systems, transportation cybersecurity, and smart logistics. Professor Ukkusuri is an ASCE Fellow, Fulbright Fellow, and Senior Member of IEEE and INFORMS, previously serving as a University Faculty Scholar from 2017 to 2022. He has been recognized by the National Academies of Engineering and Science through the JAFOE (2016) and Arab American Frontiers (2017) programs and is a recipient of the CUTC/ARTBA Faculty Award. He has published more than 350 peer-reviewed papers in top-tier transportation journals and interdisciplinary journals such as PNAS, Nature and Nature Communications. He is the co-Editor-in-Chief of both the Springer journal, *Data Science for Transportation*, and the *ACM Journal of Autonomous Transportation*.

Mashrur "Ronnie" Chowdhury is the Eugene Douglas Mays Chair of Transportation in the Glenn Department of Civil Engineering at Clemson University. He serves as the Founding Director of the USDOT-sponsored National Center for Transportation Cybersecurity and Resiliency (TraCR), as well as the USDOT-sponsored Center for Connected Multimodal Mobility (C2M2). He is also the Director of the Complex Systems, Analytics, and Visualization Institute (CSAVI) and a co-Associate Director of the USDOT-sponsored Center for Regional and Rural Connected Communities (CR2C2).

Professor Chowdhury obtained his MS in Transportation from Morgan State University (1991) and his PhD in Civil Engineering from the University of Virginia (1995). Following his industry career at Bellow-McGee Inc. and Iteris Inc., he served as an Assistant Professor of Civil Engineering at the University of Dayton before joining Clemson University.

Professor Chowdhury's current research focuses on the evolving realms of sensing, communications, computing, cybersecurity, and cyber-resiliency, all with the goal of establishing a secure and resilient IoT environment for smart cities and regions. Alongside his team, he delves into the myriad vulnerabilities present in the nation's transportation systems, public infrastructures, and cyber-physical-social systems, developing robust strategies to mitigate these threats and ensure long-term system resilience.

Bhavani Thuraisingham is currently serving as the Founders Chair Professor of Computer Science at the University of Texas at Dallas, where she served as the Founding Executive Director of the Cyber Security Research and Education Institute from 2004 to 2021. A Fellow of the ACM, IEEE, AAAS, and NAI, her research focuses on the integration of cybersecurity with AI and data science. Her 45-year career spans industry (Honeywell), federal research (MITRE), the government (NSF) sector, and academia. She is the recipient of several prestigious honors, including the IEEE Computer Society's 1997 Edward J. McCluskey Technical Achievement Award, the ACM SIGSAC 2010 Outstanding Contributions Award, and the IEEE CS 2023 Taylor L Booth Education Award.

https://doi.org/10.1142/9789819831548_fmatter

Introduction

The global transportation ecosystem is undergoing a significant transformation. Connected vehicles, automated driving systems, networked traffic infrastructure, drones operating in complex airspaces, and cooperative mobility services now form the backbone of modern transportation. These systems promise unprecedented gains in safety, efficiency, sustainability, and accessibility. Yet they also bring a profound risks and vulnerabilities: the boundary between cyber risk and physical risk is dissolving. In transportation, cybersecurity is no longer an adjunct concern—it has become a primary determinant of operational safety, public trust, national security, and economic stability.

Across federal agencies, state and local governments, transportation authorities, and research institutions, there is a growing recognition that traditional IT-centric cybersecurity frameworks are insufficient for cyber-physical mobility systems. These systems blend software, sensing, machine learning, control algorithms, communications networks, and human-machine interaction into tightly coupled, safety-critical environments. A single exploited vulnerability—whether in a sensor, an onboard model, a V2X interface, a roadway signal, or a software supply chain—can cascade into large-scale disruptions, safety hazards, or systemic failures. That reality demands new cyber-transportation models, new testing paradigms, new regulatory perspectives, and new collaborations between government, academia, and industry.

This book, *Advances in Transportation Cybersecurity*, is written to address that need. It synthesizes state-of-the-art scientific research, emerging technological trends, and the evolving legal and policy landscape into a unified treatment aimed at researchers, federal agencies, policymakers, system designers, and practitioners. The chapters that follow span conceptual foundations, analytical tools, experimental platforms, AI-native security methods, and forward-looking opportunities such as quantum computing and international governance. While diverse in scope, they share a common motivation: **ensuring that the transportation systems society depends on remain safe, trustworthy, and resilient under adversarial conditions.**

Cyber-Physical Mobility: A New Category of National Infrastructure

Transportation is now deeply intertwined with digital infrastructure. Roadways and intersections host connected sensors; vehicles exchange data with cloud services; transit networks rely on predictive analytics; and drones navigate through algorithmically managed airspace. The result is a distributed, federated, continuous system whose reliability depends on the integrity and authenticity of computation and communication at every level.

For national security communities, one of the most significant shifts is that the attack surfaces once limited to enterprise IT now include:

- perception and machine-learning pipelines,
- V2X systems and roadside units,
- traffic management centers,
- software update channels spanning multiple global vendors,
- learning-based decision systems,
- drones and other unmanned systems operating with real-time autonomy.

This complexity has outpaced regulatory structures and siloed standards, which historically evolved separately for aviation, automotive, rail, maritime, and municipal traffic systems. The chapters in this book argue for a vertically integrated approach: one that links cybersecurity requirements to safety cases, aligns them with engineering constraints, and embeds verification, monitoring, and recovery mechanisms into every layer of the transportation stack.

The book begins with foundational chapters that provides an overview of the cyber-physical threat landscape for mobility systems. Understanding adversarial behavior is essential, but in transportation, the implications go beyond data breaches or system outages. Threats may compromise sensor accuracy, distort environmental perception, affect routing guidance, or induce unsafe maneuvers in autonomous systems. The early chapters present taxonomies of threats, attack surfaces, and system vulnerabilities informed by real-world incidents and experimental evidence.

A key theme that emerges is the inadequacy of traditional perimeter-based defense models. Transportation systems are inherently open and interconnected; communication cannot be tightly restricted without degrading performance and safety. Thus, the book emphasizes **zero-trust architectures**, **data provenance and endorsement**, **continuous verification**, and **behavior-based monitoring** as core strategies. Regulatory agencies will find in these chapters a conceptual bridge from IT norms to cyber-physical safety requirements.

For policymakers and infrastructure owners, one of the central questions is: *How can we assess risk before failures occur?*

Chapters in the second part introduce mathematically grounded approaches such as reachability analysis, stealthy attack modeling, and resilience metrics for multi-vehicle systems and networked traffic flows. These tools provide a quantitative basis for determining worst-case attack impacts, identifying critical vulnerabilities, and evaluating mitigation strategies. Such methods are especially valuable for government agencies responsible for safety certification, rulemaking, and national risk assessments.

The chapters on urban traffic resilience and CAV data endorsement extend this analysis to real-world constraints. They highlight how adversarial manipulation might reroute vehicles, degrade urban mobility, or corrupt trust in shared datasets. These insights are essential for agencies evaluating regional transportation planning, smart-city deployments, or public-private data-sharing agreements.

Testbeds, Experimentation, and Validation

Cybersecurity in transportation cannot rely solely on theoretical guarantees. Because cyber-physical systems exhibit emergent and context-dependent behaviors, **testing environments must integrate simulation, hardware, communications, and realistic mobility patterns**. The book's third section presents an end-to-end co-simulation framework designed to support reproducible research, red-team evaluations, adversarial testing of AI models, and system-level verification.

Government research labs, state DOTs, and regulated manufacturers will find this section particularly relevant. It provides concrete architectural guidance for building testbeds that meet regulatory expectations and produce evidence suitable for safety case frameworks and compliance auditing. The chapter on FLUTE advances this idea further through federated Q-learning within a zero-trust framework, illustrating how distributed learning systems can operate securely and scalably across heterogeneous jurisdictions.

As machine learning becomes central to perception, decision-making, and prediction in transportation systems, the cybersecurity landscape shifts from static vulnerabilities to dynamic, data-driven risks. The fourth part of the book explores generative AI, embodied intelligence in drones, and AI-enhanced threat modeling.

The chapter on drones provides a powerful case study: autonomous systems operating in open environments must make safety-critical decisions under uncertainty and under adversarial influence. Embodied AI systems cannot rely solely on preprogrammed rules; they require adaptive reasoning supported by robust, secure, and constrained decision mechanisms.

The chapters on generative AI focus on two related concerns. First, how AI models themselves can be hardened against data poisoning, prompt manipulation, and adversarial examples. Second, how generative models can enhance security operations, from anomaly detection to predictive situational awareness. These chapters are aligned with ongoing federal efforts to develop AI assurance frameworks and provide scientifically grounded strategies for integrating AI into cybersecurity workflows responsibly.

The final section of the book examines the broader strategic environment. Quantum computing presents both potential risks—such as breaking current cryptographic standards—and opportunities for new defensive capabilities. For policymakers, this chapter provides high-level timelines, technology readiness assessments, and transition planning considerations.

The concluding chapter addresses the legal and regulatory landscape. Transportation cybersecurity is governed by a mosaic of sector-specific and cross-sector frameworks spanning the United States, the European Union, and Asia-Pacific regions. As systems grow more interconnected, harmonization becomes essential to ensure interoperability, data consistency, mutual trust, and coordinated incident response. The chapter surveys existing regulations, identifies gaps, and proposes directions for future governance, including certification regimes, liability models, and public-private data-sharing policies.

This book is designed to advance knowledge in the transportation cybersecurity landscape and support informed decision-making across several stakeholders:

- **Federal agencies** seeking science-based frameworks for rulemaking, safety certification, and infrastructure modernization.
- **Researchers** exploring the frontiers of autonomous mobility, cyber-physical security, and AI-enabled resilience.
- **State and regional authorities** designing secure smart-city deployments and connected infrastructure.
- **Industry partners** involved in vehicle manufacturing, software development, sensor design, and mobility services.
- **Legal and policy professionals** shaping governance models for emerging transportation technologies.

By integrating foundational theory, practical engineering, AI-native methods, and policy analysis, the book offers a comprehensive resource for securing the transportation systems that underpin modern society. The challenges are formidable, but with rigorous scientific tools, robust engineering practices, and coordinated governance, secure and resilient transportation is not only possible—it is achievable at scale.

https://doi.org/10.1142/9789819831548_fmatter

Contents

Part I

Foundations and Evolving Threat Landscape

https://doi.org/10.1142/9789819831548_0001

Chapter 1

Understanding Cyber Threats in Cyber-Physical Mobility Systems

Omar Faruqe Hamim*, Zengxiang Lei*, Eunhan Ka*, and Satish V. Ukkusuri*,†‡

**Lyles School of Civil and Construction Engineering*
Purdue University, West Lafayette, IN, USA
†*Department of Computer Science (Courtesy)*
Purdue University, West Lafayette, IN, USA
‡*sukkusur@purdue.edu*

Abstract
As road transportation systems become increasingly intelligent, they rely heavily on software-driven sensing, automated decision-making, and digital communication with infrastructure. This transformation introduces significant cybersecurity vulnerabilities alongside its benefits in mobility, efficiency, and safety. This chapter synthesizes peer-reviewed literature (2021–2025) on cyber threats to transportation as a cyber-physical system, organizing attacks into three functional categories. Sensor-based attacks exploit perception technologies including GPS, cameras, radar, and lidar by injecting false environmental data to mislead vehicle interpretation of road geometry, traffic participants, or navigation cues. These attacks can cascade into erroneous routing or safety interventions without triggering internal diagnostics. Maneuver-based attacks intervene at the motion planning or control level, using unauthorized message injections, software tampering, or spoofed vehicle-to-vehicle communication to provoke dangerous behaviors such as sudden braking or lane departures, often undetected by conventional fault monitors. Infrastructure-level attacks target roadside units, traffic control systems, charging stations, and cloud-based services, enabling adversaries to

simultaneously disrupt large numbers of vehicles through falsified broadcasts or compromised back-end servers. Real-world incidents confirm these threats are no longer theoretical. We leverage natural language processing to analyze over 600 significant transportation cyber incidents recorded between 2005 and 2025, identifying prevalent attack types, their cyber-physical impacts, and how attack methods have evolved over time. The chapter concludes by reviewing countermeasures across sensing, control, and infrastructure layers, including redundancy, anomaly detection, encryption, and secure authentication, while highlighting critical defense gaps at system interfaces. Cybersecurity in transportation must ensure not just breach prevention, but continuous safe operation under adversarial conditions.

Keywords: Cyber-physical mobility systems, Autonomous Vehicle (AV), Connected Vehicle (CV), Electric Vehicle (EV).

1.1 Introduction

As road transportation systems become increasingly intelligent, they rely heavily on software-driven sensing, automated decision-making, and digital communication with infrastructure (Ali *et al.*, 2023; Huang *et al.*, 2023; Sumalee and Ho, 2018; Ukkusuri *et al.*, 2025). This transformation, while promising unprecedented gains in mobility, efficiency, and safety (Underwood, 2014), also introduces a new frontier of cybersecurity vulnerabilities (Durlik *et al.*, 2024; Schlenoff *et al.*, 2024; Ward and Wooderson, 2021). The complex interplay between automation, communication technologies, and infrastructure demands a rethinking of how we assess and ensure the resilience of transportation systems against cyber threats (Aljohani and Almutairi, 2024; Kim *et al.*, 2021; Mwanje *et al.*, 2024; Shen *et al.*, 2022).

This chapter synthesizes current knowledge on the cybersecurity risks faced by transportation as a cyber-physical system. Drawing from a systematic review of peer-reviewed research spanning 2021 to mid-2025, we group the literature into three primary categories of attack: sensor-based, maneuver-oriented, and infrastructure-level, and conclude with a discussion of transportation as a cyber-physical system with distinctive resilience challenges, as illustrated in Fig. 1.1.

Sensor-based attacks exploit weaknesses in perception technologies that interpret the physical environment. These include attacks on positioning signals, visual inputs, and remote sensing devices such as cameras, radar, and LiDAR (Yan *et al.*, 2022; Zeng *et al.*, 2017). By distorting or injecting false environmental data, adversaries can manipulate how vehicles interpret road geometry, traffic participants, or navigational cues. Practical examples include spoofed satellite signals that mislead location estimations or adversarial patterns that deceive object recognition modules (Pavlitska *et al.*, 2023; Song *et al.*, 2023). Because these systems often fuse data from multiple sources, a successful sensor-level attack can cascade into erroneous route choices, lane shifts, or safety interventions, often without being flagged by internal diagnostics (Shen *et al.*, 2022).

Maneuver-oriented attacks intervene at the level of motion planning or control execution. Rather than targeting perception, these attacks directly influence how the vehicle behaves. Techniques include unauthorized message injection on internal control networks (Miller and Valasek, 2015), tampering with real-time decision software, or mimicking legitimate communication from neighboring vehicles to provoke unnecessary maneuvers such as sudden braking or lane departures (Nie *et al.*, 2017). Such attacks may remain undetected by conventional fault monitors, especially when they originate from trusted interfaces or mimic normal but erratic driving patterns (Kim *et al.*, 2021). The challenge here lies in distinguishing adversarially induced maneuvers from legitimate safety-oriented responses, since the vehicle is intentionally designed to exhibit conservative behavior under uncertain or abnormal conditions.

Infrastructure-level attacks target components beyond the vehicle, including roadside communication units, networked traffic control devices, energy supply systems, and cloud-based routing or maintenance services (Garofalaki *et al.*, 2022; Ghosal and Conti, 2020; Lamssaggad *et al.*, 2021). In these cases, attackers manipulate shared data or public interfaces to disrupt large numbers of vehicles simultaneously. Common vectors include falsified congestion broadcasts (Li *et al.*, 2018), compromised charging stations (Muhammad *et al.*, 2023), and manipulation of digital signage or signal timing (Chen *et al.*, 2018). Since many mobility systems rely on centralized coordination and remote updates, vulnerabilities in back-end servers or communication protocols can have far-reaching physical and operational consequences, ranging from localized confusion to city-wide disruptions in flow and safety (Bao *et al.*, 2018).

To support effective countermeasures, we also review emerging strategies to improve resilience across sensing platforms, control layers, and infrastructure interfaces. Technical measures such as sensor redundancy, encryption, and secure communication protocols form the baseline of defense (International Organization for Standardization, 2011). At a higher level, context-aware anomaly detection and behavior verification can help identify abnormal system behavior in real time (Durlik *et al.*, 2024; Shen *et al.*, 2022). Infrastructure-focused solutions include robust authentication, fault isolation, and hardened gateways for energy and fleet management systems (National Highway Traffic Safety Administration, 2022; Ward and Wooderson, 2021). By organizing attack literature around the functional targets within intelligent mobility systems, this chapter provides a clear and actionable framework for understanding emerging threats. It also highlights gaps in current defense strategies, particularly at the interfaces between perception, control, and infrastructure. Through this lens, cybersecurity is not only about preventing breaches, it is about ensuring that the entire transportation system continues to operate safely and predictably, even in the face of adversarial interference.

The remainder of this chapter is structured as follows. Section 1.2 outlines the methodology for literature collection, filtering, and synthesis. Sections 1.3–1.5 analyze cyberattacks according to the categories in Fig. 1.1, with discussions organized by system functions and attack paths. Section 1.6 reviews major cyber incidents from 2005 to 2025 and their impacts on stakeholders. Section 1.7 highlights key findings, policy and industry implications, and unresolved challenges. Finally, Section 1.8 provides concluding remarks and directions for future research.

1.2 Data Collection and Methodology

This study employs a systematic literature review (SLR) approach (Kitchenham *et al.*, 2009; Petersen *et al.*, 2015) to analyze cybersecurity threats in cyber-physical mobility systems, emphasizing autonomous vehicles (AV), connected vehicles (CV), and electric vehicles (EV). The review identifies different forms of cyberattacks,

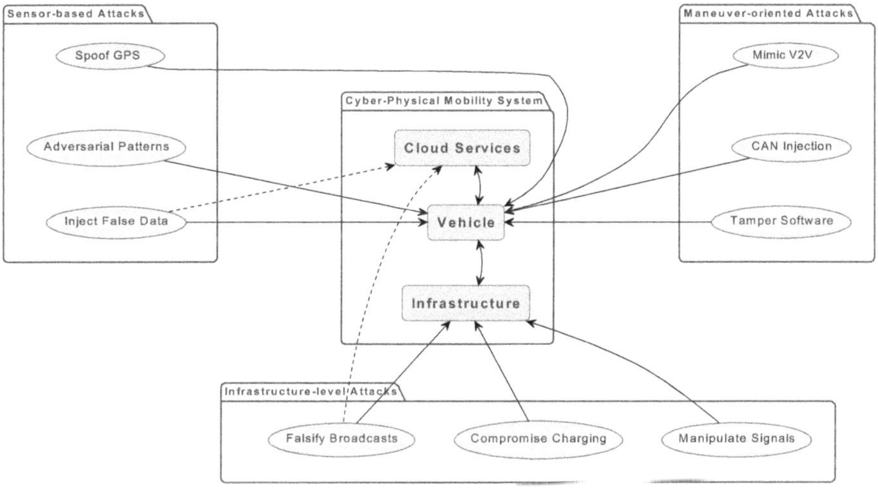

Fig. 1.1. Cyber threats in cyber-physical mobility systems. The threats are grouped based on the target components.

groups the literature into three primary categories of attack that are sensor-based, maneuver-oriented, and infrastructure-level, and examines system vulnerabilities that adversaries may exploit. Following established SLR protocols, the process was divided into four main phases: (1) defining research questions, (2) designing a search strategy, (3) applying screening and quality assessment criteria to the collected literature, and (4) integrating findings to map attack types to a structured workflow of next-generation transportation applications (see Fig. 1.1). This framework ensures comprehensive coverage of current research and provides a critical synthesis of cybersecurity challenges in road transport systems.

1.2.1 *Research Questions*

This systematic literature review is guided by the following research questions:

- **RQ1:** What cyberattacks have been reported in the literature for AVs, CVs, and EVs?
- **RQ2:** How can these attacks be classified into sensor-based, maneuver-oriented, and infrastructure-level categories?

- **RQ3:** Which system components and vulnerabilities are targeted by each attack category across the cyber-physical mobility workflow?

These research questions provide the foundation for the literature search and synthesis conducted in this study.

1.2.2 *Literature Search Strategy*

Following the research questions defined earlier, a keyword-based search was conducted in the Scopus database, targeting AVs, CVs, and EVs individually. Table 1.1 summarizes the selected keywords along with the number of publications identified within the period 2021–2025. This five-year window was chosen as cybersecurity research on AVs, CVs, and EVs shows notable growth starting in 2021. The search was executed on 12 September 2025.

1.2.3 *Screening and Literature Evaluation*

All documents obtained from the initial search underwent a two-stage screening process: first by reviewing titles, abstracts, and conclusions, and then by conducting a full-text assessment to confirm alignment

Table 1.1. Summary of literature search keywords and count of retrieved documents.

Emerging Technology	Details on Keywords Based Search	Count of Retrieved Scientific Documents
AV	(“autonomous vehicle”) AND (“cyberattack” OR “cybersecurity” OR “cyber threat” OR “cyber risk” OR “cyber vulnerabilities”)	1110
CV	(“connected vehicle”) AND (“cyberattack” OR “cybersecurity” OR “cyber threat” OR “cyber risk” OR “cyber vulnerabilities”)	840
EV	(“electric vehicle”) AND (“cyberattack” OR “cybersecurity” OR “cyber threat” OR “cyber risk” OR “cyber vulnerabilities”)	494

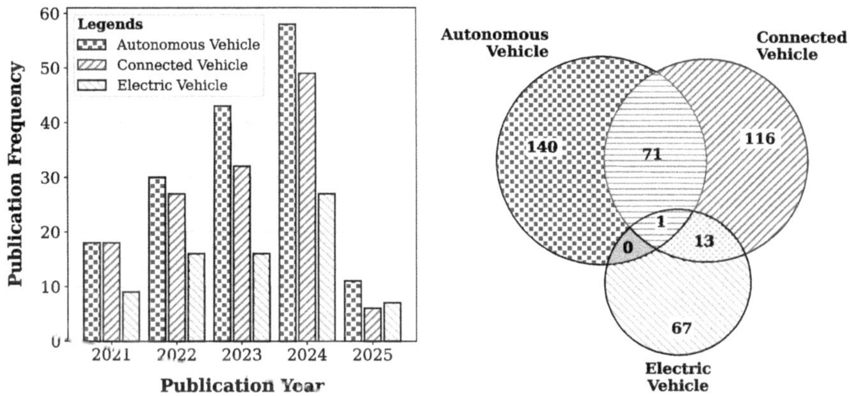

(a) (b)

Fig. 1.2. Summary of Scopus literature search results showing temporal trends and emerging technological application overlaps. (a) Yearly distribution of relevant publications (2021–2025). (b) Overlap of research papers on cyber-physical mobility systems.

with this study's aims and scope. After filtering, 212 publications on AVs, 201 on CVs, and 81 on EVs were retained.

To capture publication dynamics, a histogram was generated to show the annual distribution of research studies related to AVs, CVs, and EVs between 2021 and late 2025 (Fig. 1.2(a)), illustrating the increasing research attention in each domain. In Fig. 1.2(b), a Venn diagram shows how different categories coincide, indicating the extent of cross-cutting studies spanning multiple technologies. These visualizations together highlight both the temporal growth and the interdisciplinary character of cybersecurity research in the domain of road transportation.

1.2.4 *Synthesizing Attacks*

We categorize the attacks into the three functional groups introduced in Section 1.1: sensor-based, maneuver-oriented, and infrastructure-level. The detailed definitions of each category are provided below.

- **Sensor-based.** These attacks **compromise perception and state estimation by tampering with the raw signals that AVs rely on to interpret their surroundings.** By corrupting or forging sensor inputs, adversaries can induce persistent errors that propagate downstream, misleading the planning and control modules in ways that are often difficult to detect in real time. Examples include remote manipulation of sensor data streams, such as falsifying LiDAR point clouds, injecting noise into radar returns, or altering camera feeds, which undermine the integrity of the perception pipeline. GPS spoofing is another well-documented case, where adversaries broadcast counterfeit satellite signals that distort localization estimates, misguiding feature extraction and trajectory prediction and causing centralized decision-making algorithms to issue unsafe commands. Replay attacks represent yet another form, where previously recorded but outdated sensor or communication data are injected into the system, creating inconsistencies between perceived and actual traffic states that can lead to hazardous interactions with surrounding vehicles and infrastructure. In this chapter, replay is treated as sensor-based when it targets perception/state or vehicle-to-everything (V2X)-derived awareness data, while replay of actuator/control commands on in-vehicle networks is categorized as maneuver-oriented.

- **Maneuver-oriented.** These attacks **directly manipulate motion planning, coordination, or actuation, often including interactions with human drivers**. While some exploit communication channels as a delivery mechanism, their primary impact lies in disrupting behavior generation and control. This category includes in-vehicle network intrusions on the Controller Area Network (CAN) bus, such as message injection, fuzzing (fuzzy) attacks, and replay of actuator/control commands. For example, relaying attacks can distort feature extraction and misguide distributed algorithms such as those used in vehicle platooning, resulting in flawed decisions and unsafe robotic actuation. Data falsification attacks similarly interfere with cooperative perception or control signals, degrading human–machine interactions and leading to unsafe maneuvers in mixed traffic environments. Physical manipulation represents a more direct form, where adversaries interfere with vehicle hardware (e.g., disabling braking systems or steering components), immediately compromising actuation and creating hazardous outcomes.

- **Infrastructure-level.** These attacks **compromise data, models, and services that extend beyond a single vehicle, targeting training pipelines, over-the-air (OTA) updates, organizational interfaces, and even the physical roadway environment**. Their scope often spans multiple vehicles simultaneously, amplifying the potential impact. Examples include data poisoning, where adversaries manipulate historical datasets to mislead predictive models and cause centralized decision-making algorithms to fail. Backdoor attacks represent another form, in which tampered training data or models embed hidden triggers that later compromise vehicle control. Man-in-the-Middle (MitM) attacks exploit communication channels by intercepting and altering information exchanged among vehicles or between vehicles and infrastructure, leading to faulty coordination and unsafe actuation. Social-engineering methods such as phishing can also disrupt human–machine collaboration, tricking operators or control centers into unsafe actions. Finally, adversaries may physically alter the environment, for example by creating slippery road conditions or manipulating traffic signs, to mislead both perception systems and human drivers, resulting in hazardous traffic scenarios.

This category-based synthesis clarifies where defensive efforts should be prioritized. Sensor-based attacks underscore the need for redundancy and plausibility checks in perception systems. Maneuver-oriented attacks emphasize the importance of behavior verification and secure control interfaces. Infrastructure-level attacks point to the necessity of protecting data pipelines, ensuring authenticated communication, and hardening both roadside and cloud services. A summary of all attack types is provided in Table 1.2.

1.2.5 *Real-World Cyber Incident Analysis*

We assembled a dataset of publicly reported cyber incidents between 2005 and 2025 (Center for Strategic and International Studies, 2024). Each record included the incident timeline (month and year) and a detailed text description. Descriptions were normalized by lowercasing and tokenizing, while common stopwords were removed to reduce noise. Since the dataset covered incidents across diverse sectors, a keyword-based filter was applied to isolate cases relevant to transportation. The filtering list was extensive, covering road traffic

Table 1.2. Cyberattacks reviewed in this paper by category.

Category	Attack Type	Affected Metrics	Related Works
Sensor-based Attack	Remote Sensor Manipulation	Safety, Efficiency	Jin *et al.* (2019)
	Routing	Safety, Data Integrity	Purohit *et al.* (2017) and Kumar *et al.* (2021)
	Eavesdropping	Safety, Privacy, Security	Bajracharya (2021) and Saltaformaggio *et al.* (2016)
	Replaying	Efficiency, Security	Iqbal *et al.* (2022)
	Location Trailing	Safety, Privacy	Alam *et al.* (2021)
	Close Proximity & CAN	Safety, Efficiency, Privacy	Rouf *et al.* (2010), Murvay and Groza (2018), and Greene *et al.* (2020)
	GPS Spoofing	Safety, Efficiency	Vitale *et al.* (2021), Yang *et al.* (2023), and Guo *et al.* (2021)
Maneuver-oriented Attack	Data Falsification	Safety, Efficiency	Cui *et al.* (2022) and Mia and Amini (2024)
	Relaying	Efficiency, Reliability	Conti *et al.* (2022)
	Repudiation	Economy, Safety, Trust	Huang *et al.* (2018)
	ECUs Software Flashing	Safety, Efficiency	Khemissa and Urien (2022)

(*Continued*)

Table 1.2. (*Continued*)

Category	Attack Type	Affected Metrics	Related Works
Infrastructure-level Attack	Infrastructure	Efficiency, Safety	La Fontaine *et al.* (2022)
	Sybil	Efficiency, Trust, Stability	Fraiji *et al.* (2018)
	Malware Injection	Safety, Efficiency, Economy	Halder *et al.* (2020); Xuefeng and Wei (2022)
	Tampering	Safety, Efficiency, Trust	Gennaro *et al.* (2004)
	Impersonation	Safety, Security, Reliability	Appathurai *et al.* (2019)
	Man-in-the-Middle (MitM)	Safety, Efficiency, Economy	Hwang *et al.* (2008)
	Phishing/Social Engineering	Safety, Economy, Data Integrity	Soykan *et al.* (2021)
	Free-Ride	Economy, Trust	Brighente *et al.* (2023); Jiang *et al.* (2019)
	Denial of Service (DoS)	Safety, Efficiency, Mobility	Groza and Murvay (2018); Mohd *et al.* (2018); Tian *et al.* (2019); Yeasmin and Haque (2023)
	Jamming & Interference	Safety, Efficiency	Fraiji *et al.* (2018); Grover *et al.* (2014)
	Overpower	Safety	Brighente *et al.* (2023)
	Side-Channels	Safety, Economy, Confidentiality	Chandwani *et al.* (2020); Gottumukkala *et al.* (2019); La Cour *et al.* (2021)

management systems (e.g., traffic signals, tolling, intelligent transportation systems), vehicles and subsystems (e.g., CAN bus, Electronic Control Unit (ECU), autonomous driving, EV charging), and logistics and supply chain operations. Incidents referencing any of these terms were retained for further analysis.

Each transport-related incident was assigned a single attack category using curated keyword dictionaries. Multi-word phrases were tokenized, allowing for flexible matching even when tokens did not appear contiguously. A scoring function evaluated token overlap, proportion of matched words, and phrase length to ensure only the most relevant category was selected, avoiding multiple labels per incident.

Each incident was further classified along two orthogonal dimensions:

(1) **Impact Target (who was affected):** Government, Defense, Private Sector, Critical Infrastructure.
(2) **Impact Type (what occurred):** Data Compromise, Operational Disruption, Financial Loss, Reputation Damage, National Security Threat.

Keyword-based dictionaries were used for each dimension, and each description was mapped to the single most relevant target and type. This separation of "sector affected" and "impact outcome" ensured clearer analytical distinctions. All preprocessing, filtering, and classification were implemented in Python using regular expressions and token-level matching. Multi-word phrases were matched flexibly, and thresholds (e.g., requiring at least 50% token overlap) balanced inclusiveness with precision. For each incident, one label was retained per dimension (i.e., attack type, impact target, and impact type), ensuring mutual exclusivity and interpretability. With the review corpus and the incident-labeling process established, the next sections synthesize attacks by the three functional categories in Fig. 1.1, moving from perception compromise (Section 1.3) to control/actuation manipulation (Section 1.4) and then to shared infrastructure threats (Section 1.5).

1.3 Sensor-Based Attacks

Sensor-based attacks manipulate the signals and measurements that convert the physical world into digital inputs. They target the earliest stages of a vehicle's data processing pipeline, where errors in localization, perception, or map matching can cascade through feature extraction and prediction modules and ultimately influence planning and actuation. Since they compromise information at its source, these attacks are particularly dangerous such as advanced sensor fusion and filtering may make malicious inputs appear plausible, concealing the compromise until it has already propagated into unsafe decisions.

1.3.1 *Remote Sensor Manipulation*

Remote sensor manipulation involves falsifying or altering data streams produced by vehicle sensors such as LiDAR, radar, cameras, or even vehicle-to-sensor (V2S) communication links. By injecting falsified signals or interfering with transmissions, adversaries distort a vehicle's situational awareness and increase the likelihood of unsafe maneuvers. For example, falsified LiDAR readings may cause an AV to interpret an open lane as obstructed, while manipulated radar returns could hide an approaching vehicle. Koscher *et al.* (2010) demonstrated that vulnerabilities in automotive sensor and communication systems can be exploited to remotely influence critical vehicle operations, underscoring the importance of secure sensor interfaces. Jin *et al.* (2019) further proposed an adaptive control framework designed to maintain stability in CV platoons even when some sensor inputs are compromised.

1.3.2 *Eavesdropping*

Eavesdropping attacks exploit vulnerabilities in vehicular communication channels, including Vehicle-to-Vehicle (V2V), Vehicle-to-Infrastructure (V2I), Vehicle-to-Network (V2N), and Vehicle-to-Grid (V2G) links. By intercepting data streams without altering them, adversaries compromise confidentiality, gaining unauthorized access to sensitive information such as vehicle location, charging records, or user credentials. Bajracharya (2021) studied the performance of secure communications in the Internet of Vehicles under combined eavesdropping and jamming conditions, showing that additional transmission resources are needed to maintain secrecy rates. In EV ecosystems, eavesdropping can expose sensitive charging data and personal information by intercepting communications between charging stations, mobile applications, or back-end servers (Saltaformaggio *et al.*, 2016). Such passive compromises often serve as a precursor to more active intrusions, including spoofing or session hijacking.

1.3.3 *Routing*

Routing attacks manipulate how data packets are forwarded in vehicular networks, diverting them through malicious nodes or servers. By tampering with routing protocols, attackers compromise the reliability of conveyed data and may inject harmful messages. Purohit *et al.* (2017) addressed the vulnerability of Ad hoc On-Demand Distance Vector (AODV) routing to blackhole attacks and proposed Secure Vehicular On-Demand Distance Routing (SVODR) to mitigate them. Similarly, Kumar *et al.* (2021) developed a cryptography-enhanced protocol that improves resilience against routing exploits. Such threats degrade communication reliability and can mislead higher-level decision-making that depends on timely, accurate data exchange.

1.3.4 *Replay Attacks*

Replay attacks involve capturing authentic sensor or communication data and retransmitting it later to trick systems into accepting stale or malicious inputs as valid. In practice, replay attacks can exploit signals from wireless key fobs, CAN bus messages, or V2V/V2I communications, enabling attackers to unlock vehicles, manipulate cooperative driving data, or disrupt control flows. For clarity, replay of perception/state or Vehicle-to-Everything (V2X)-derived awareness data is discussed as sensor-based, whereas replay of in-vehicle actuator/control commands is treated as a maneuver-oriented CAN intrusion (see Section 1.4). Miller and Valasek (2015) vividly demonstrated the real-world impact of replay-based exploits on production vehicles, emphasizing the urgency of stronger encryption and tailored intrusion detection systems (IDS). Greene *et al.* (2020) showed how conventional rolling-code keyless entry systems remain vulnerable to replay by recording and rebroadcasting unlock signals, and proposed timestamp-enhanced rolling codes as a defense. Iqbal *et al.* (2022) modeled replay scenarios in vehicular ad hoc networks (VANETs), demonstrating how injected stale information disrupts VANET traffic and proposing anomaly detection methods for mitigation.

1.3.5 *Location Trailing*

Location trailing is a privacy threat in which attackers track vehicles by exploiting location data streams transmitted over V2N or V2S links. Persistent monitoring of position information endangers user privacy and can facilitate targeted attacks. Alam *et al.* (2021) proposed detection methods combining Centroid Localization with convex hull algorithms to detect and address location-trailing attacks in vehicular networks.

1.3.6 *Close Proximity and CAN*

Close-proximity exploits leverage vulnerabilities in in-vehicle networks such as the CAN bus. Attackers intercept or inject signals through short-range communication interfaces such as Bluetooth, tire pressure monitoring systems (TPMS), or remote keyless entry (RKE) systems, enabling unauthorized control over vehicle functions. Rouf *et al.* (2010) showed that TPMS messages lack authentication, making them susceptible to spoofing. Murvay and Groza (2018) analyzed flaws in the SAE J1939 protocol and proposed authentication mechanisms to improve bus security. Greene *et al.* (2020) demonstrated replay-based exploits against RKE systems and suggested timestamp-enhanced rolling codes as a countermeasure.

1.3.7 *GPS Spoofing*

GPS spoofing attacks broadcast counterfeit satellite signals that mislead vehicles into computing false position or timing information. Such falsified localization can propagate through routing, prediction, and control modules, ultimately resulting in unsafe navigation or deliberate rerouting into hazardous areas. Checkoway *et al.* (2011) provided an early comprehensive demonstration of vehicles' susceptibility to spoofing and network-level exploits, highlighting the need for robust detection and secure routing protocols. More recently, Vitale *et al.* (2021) developed the CARAMEL framework to detect GPS spoofing incidents using multi-radio access technologies, edge computing, and public key infrastructure, while Yang *et al.* (2023) proposed trajectory-based anomaly detection methods for spoofing in autonomous driving systems.

1.3.8 *Comparison and Limitations*

The literature on sensor security typically classifies mitigation strategies into three distinct themes: cryptographic authentication, cross-sensor plausibility verification, and data-driven anomaly detection (Dibaei *et al.*, 2019). Authentication provides robust guarantees regarding signal integrity; however, the requirement for specific hardware support often precludes retrofitting legacy systems. Conversely, while plausibility checks are lightweight and interpretable, they are often insufficient against sophisticated, coordinated attacks that evade static residual thresholds (Giraldo *et al.*, 2018). Learning-based approaches demonstrate superior efficacy at capturing complex dynamics but struggle to generalize when exposed to distribution shifts in the environment. Furthermore, the prevalence of simulation-based validation in current research leaves a gap in understanding false-alarm behavior in unstructured, real-world driving environments.

1.4 Maneuver-Oriented Attacks

Maneuver-oriented attacks directly affect motion planning, cooperative decision making, and low-level control. They operate at the stage where perception results are translated into decisions and then into actuator commands, altering vehicle behavior even when raw sensor data remains correct. Typical mechanisms include unauthorized message injection on in-vehicle networks (e.g., CAN message injection, fuzzing, or command replay), tampering with planning or control software, and deceptive V2V or V2I messages that trigger abrupt braking, unsafe headways, or unexpected lane changes. Because the resulting behaviors may resemble cautious or erratic human driving, intent attribution and rule-based fault detection are particularly challenging.

1.4.1 *Data Falsification*

Data falsification attacks undermine trust in the information that supports decision making. By injecting, modifying, or poisoning data, adversaries can cause AVs or CVs to misinterpret their surroundings or execute unsafe maneuvers. Such attacks may target real-time control messages, cooperative perception data exchanged over V2V, V2I, or V2N links, or training datasets for machine learning (ML) models. For example, falsified sensor data could cause unnecessary swerving, increasing collision risks. Cui *et al.* (2022) proposed a clean-label poisoning method that significantly degraded classification accuracy in traffic sign recognition systems, while Mia and Amini (2024) analyzed data poisoning against object detection algorithms in AVs and developed defenses for computer vision models. Woo *et al.* (2014) also demonstrated that authentication and message modification vulnerabilities can compromise automotive networks, highlighting the need for rigorous cryptographic protocols and secure logging. Collectively, data falsification can compromise both immediate vehicle safety and the long-term reliability of learned models, threatening efficiency and trust across intelligent transportation systems. Prior research has modeled falsification attacks using high-fidelity simulations and control-theoretic frameworks. For example, fabricated obstacles or false messages have been injected in multi-vehicle driving simulations to mimic ghost objects and spurious V2V data, allowing evaluation of AV control responses under deceptive inputs (Zhang *et al.*, 2024). Attack intensity is parameterized by the fraction of messages or sensor readings corrupted and the timing or duration of falsification events, ranging from sporadic false alarms to persistent feed manipulation. Impact is assessed via safety and performance metrics such as collision occurrences or near-miss counts, increases in vehicle delay and travel time, and degradation of throughput or stability relative to an attack-free baseline (Zhang *et al.*, 2024).

1.4.2 *Relaying*

Relaying attacks position the adversary between two legitimate entities, intercepting and forwarding communication while potentially modifying timing or content. This enables the attacker to impersonate trusted parties without raising suspicion. Relaying threatens distributed algorithms such as vehicle platooning, cooperative driving, and EV charging coordination. Conti *et al.* (2022) showed how relaying attacks against EV charging protocols can enable unauthorized access to controller components. By forwarding, delaying, or altering authentication messages, adversaries can mislead decision-making processes, causing incorrect vehicle spacing or unsafe cooperative maneuvers. Because communication flows appear valid, relaying attacks significantly degrade efficiency and reliability.

1.4.3 *Repudiation*

Repudiation attacks exploit weaknesses in accountability and data verification, allowing malicious actors to deny actions or transactions they performed. This is particularly relevant in EV charging and Electric Vehicle Charging Station (EVCS) systems, where adversaries may tamper with logs, forge digital signatures, spoof identities, or manipulate billing records. Huang *et al.* (2018) highlighted how transaction tampering and session spoofing undermine trust in EV ecosystems. By disrupting verification processes, repudiation enables fraudulent resource consumption and erodes economic integrity. These attacks threaten not only user trust but also the accountability mechanisms essential for large-scale deployment of connected mobility systems.

1.4.4 *ECUs Software Flashing*

ECUs (Engine Control Units) are critical components that govern vehicle actuation, including braking, steering, and energy management. ECU software flashing attacks install malicious firmware either via OTA update mechanisms or through direct physical access. Once compromised, ECUs may mismanage safety-critical functions or embed persistent backdoors. For instance, tampered firmware could manipulate an AV's fuel or energy management to degrade performance or even cause breakdowns. Khemissa and Urien (2022) proposed a centralized ECU security architecture using lightweight cryptographic key agreements to secure communication between ECUs and manufacturer servers. Still, malicious flashing remains a potent threat because it directly undermines safety and reliability at the core of vehicle control systems. Researchers have replicated ECU flashing attacks on both hardware-in-the-loop testbeds and actual vehicles to evaluate their effects. Early experimental studies demonstrated that an attacker with physical or remote access can reprogram critical ECUs by exploiting diagnostic ports, infotainment vulnerabilities, or wireless update channels, thereby gaining unauthorized control of braking, acceleration, and other functions (Koscher *et al.*, 2010). Attack scenarios are parameterized by access vector, targeted ECU subsystem (such as powertrain or steering controller), and whether the malicious firmware remains stealthy or causes overt malfunctions. The impact is then quantified using safety and performance indicators, such as unintended braking or acceleration events induced, changes in engine power output or energy efficiency under attack, and the duration or difficulty of detecting and recovering from the compromised firmware (Koscher *et al.*, 2010).

1.4.5 *Comparison and Limitations*

Maneuver-centric defense mechanisms typically include CAN IDS, cryptographic message authentication, and runtime behavioral monitoring to enforce safety constraints. While CAN-based methodologies offer backward compatibility with legacy platforms, they often require a trade-off between detection sensitivity and false-positive rates, particularly under dynamic conditions such as aggressive driving maneuvers, software updates, or hardware reconfigurations (Lokman *et al.*, 2019). Cryptographic authentication effectively mitigates injection and replay attacks; however, its implementation is often constrained by the heterogeneity of ECUs and stringent real-time latency budgets. Furthermore, while runtime monitors serve as a critical fail-safe mechanism, their reliance on rigorously defined safety envelopes can lead to overly conservative actuation, potentially compromising mobility efficiency (Shalev-Shwartz *et al.*, 2017).

1.5 Infrastructure-Level Attacks

Infrastructure-level attacks target shared digital and physical assets that support many vehicles simultaneously. They span the entire workflow by corrupting training datasets and models, manipulating communication channels, injecting malicious software, exploiting organizational interfaces, and altering the physical roadway environment. Since they compromise common infrastructure, their effects extend beyond a single vehicle, often causing synchronized misbehavior across fleets, degraded traffic flow, and widespread safety risks.

1.5.1 *Infrastructure*

Infrastructure-focused attacks compromise external transportation systems that CVs rely upon, such as traffic management servers or routing platforms. La Fontaine *et al.* (2022) demonstrated how manipulating metropolitan routing infrastructures can reroute vehicles onto inefficient or unsafe paths, creating congestion and systemic safety risks. These attacks exploit the broader mobility environment rather than in-vehicle systems, amplifying their effects across many users. To quantify these threats, researchers have modeled infrastructure attacks using network-level traffic simulations and optimization-based formulations, manipulating traffic signal timing or routing guidance to represent adversarial interference (Laszka *et al.*, 2016; Perrine *et al.*, 2019). Attack intensity is often defined by the number of compromised intersections or the fraction of vehicles rerouted by false information. Evaluations typically report metrics such as total system travel time, average speed, and congestion extent; for example, disabling just a few traffic signals in a city network was shown to increase total delay by over 430% and effectively negate the benefits of optimized traffic control (Perrine *et al.*, 2019).

1.5.2 *Sybil*

Sybil attacks introduce large numbers of fake identities into vehicular or charging networks, overwhelming decision-making systems and disrupting normal operations. In EV contexts, they may distort charging priorities, interrupt services, or gain illegitimate access to resources. Fraiji *et al.* (2018) analyzed how Sybil attacks degrade efficiency and stability in vehicular systems. By flooding systems with false participants, Sybil attacks undermine trust, erode privacy, and destabilize cooperative protocols. Sybil attacks are commonly simulated by introducing numerous ghost vehicles or identities in a traffic or charging network model, with attack magnitude set by the proportion of Sybil nodes among legitimate ones (Söderhäll *et al.*, 2025). Such studies measure impacts on network performance and safety using indicators like throughput, travel time, and collision risk. Results show that even a small fraction of Sybil entities can significantly disrupt normal operations; for instance, injecting Sybils comprising only 3% of participants in a navigation system increased average travel times by roughly 20% due to phantom congestion (Söderhäll *et al.*, 2025).

1.5.3 *Malware Injection*

Malware injection embeds malicious code into software, firmware, or applications that support AVs, CVs, or EVs. Halder *et al.* (2020) documented how malware can compromise charging infrastructure and EV operations, while Xuefeng and Wei (2022) highlighted risks from malicious firmware updates in EV ecosystems. Common vectors include compromised charging station software, backdoors in updates, ransomware in mobile apps, or firmware modifications that alter energy management. Such attacks jeopardize safety, disrupt services, and impose economic costs across the mobility ecosystem. To study malware propagation, researchers have employed co-simulation testbeds and graph based models mapping how infected EVCSs and vehicles spread malicious code across the network (Poudel *et al.*, 2024). Attack scenarios vary the initial number of compromised chargers or EVs and the update frequency of malware payloads to represent different penetration rates. Key metrics include the fraction of devices infected over time, downtime of services, and any physical damage (such as battery degradation or grid disturbances). For example, one study optimized a malware injection strategy targeting a few strategic charging stations and demonstrated accelerated spread to a majority of EVs in a city, illustrating the system-wide risk posed by unmitigated malware attacks (Poudel *et al.*, 2024).

1.5.4 *Tampering*

Tampering involves direct manipulation of infrastructure components or supply chains. Examples include altering charging station hardware, falsifying operational data, or introducing malicious updates into distribution pipelines. Gennaro *et al.* (2004) examined tampering risks in cryptographic and communication systems that underpin vehicular trust models. By corrupting hardware or firmware, tampering attacks reduce efficiency, erode trust, and increase safety hazards. Researchers have examined tampering via supply-chain attacks and direct hardware interference, often using security testbeds to introduce counterfeit components or modified firmware into charging stations or vehicle controllers (Zhou *et al.*, 2023). Attack configurations are defined by the scope of the compromise (e.g., percentage of units tampered or specific modules altered) and timing (persistent vs. one-time tampering). Impacts are evaluated on metrics like system reliability, error rates, and safety margins.

1.5.5 *Impersonation*

Impersonation attacks exploit authentication weaknesses by disguising malicious components as legitimate entities. Appathurai *et al.* (2019) studied impersonation and data injection risks in IoV (Internet of Vehicles) systems and proposed a Washable Physical Unclonable Function (WPUF) authentication method to counteract them. Successful impersonation bypasses standard integrity checks, leading to unauthorized access, compromised vehicle functions, and reduced dependability.

1.5.6 *Man-in-the-Middle (MitM)*

MitM attacks intercept and alter communications between vehicles, infrastructure, or back-end services. Hwang *et al.* (2008) analyzed MitM risks in vehicular networks, showing how attackers can exploit Wi-Fi, cellular, or short-range protocols to inject or modify data. These attacks maintain the appearance of legitimate communication while secretly altering its content, resulting in unsafe actuation, financial fraud, or service disruptions. MitM attacks are often analyzed through network simulations (e.g., using OMNeT++/Veins), where a malicious relay node intercepts and selectively modifies messages exchanged between vehicles and infrastructure (Ahmad *et al.*, 2018). The attacker's strategy can be tuned (forwarding all messages vs. dropping or delaying certain packets) to evaluate different threat levels. Impact metrics include packet delivery ratio, end-to-end latency, and the incidence of cascaded failures in cooperative maneuvers. Simulations indicate that MitM strategies such as message tampering or timing manipulation lead to high packet loss and increased delay, which in turn degrade application performance (e.g., collision warning systems) and can precipitate unsafe behaviors in traffic flow (Ahmad *et al.*, 2018).

1.5.7 *Phishing/Social Engineering*

Phishing and social engineering attacks exploit human operators rather than technical systems. Soykan *et al.* (2021) showed how deceptive portals and communications can trick EV users or operators into revealing credentials or installing malware. Since many vehicle and charging services depend on mobile apps and online accounts, these attacks can lead to account takeovers, unauthorized transactions, and cascading risks to safety and reliability. To evaluate social engineering risks, researchers have employed scenario-based simulations and user studies where false information is disseminated via phishing emails or fake apps to influence EV user behavior (Acharya *et al.*, 2022; Jahangir *et al.*, 2024). Attack campaigns are parameterized by their reach (number of targeted users) and content (e.g., bogus pricing signals or urgent charging notifications). The consequences are measured in terms of compromised accounts, fraudulent transactions conducted, or induced load changes on the grid. Notably, one study demonstrated that sending coordinated fake price alerts could prompt many EV owners to charge simultaneously at peak times, causing an artificial load surge that threatens grid stability and increases charging costs for users (Acharya *et al.*, 2022).

1.5.8 *Free-Ride*

Free-ride attacks exploit weaknesses in authentication and billing systems to consume charging or mobility resources without payment. Brighente *et al.* (2023) and Jiang *et al.* (2019) documented how such attacks target EV charging infrastructure, enabling fraudulent use through piggybacking, credential theft, or session hijacking. Such attacks undermine economic integrity and weaken trust in large-scale EV deployments. Free-ride attacks are typically investigated by analyzing authentication loopholes and billing system weaknesses in charging networks. Researchers model these attacks by allowing adversarial agents to bypass payment verification and then simulate the resulting unauthorized charging sessions (Babu *et al.*, 2022). The intensity is represented by the fraction of charging events that are fraudulent or the duration of free consumption. Impact assessment focuses on revenue loss, fairness (legitimate users being turned away or delayed), and trust erosion in the service. Simulations of token-based payment systems under attack have shown significant revenue reduction and network overload when an attacker can double-spend credits or piggyback on another vehicle's session, underscoring the need for secure payment protocols and real-time monitoring (Babu *et al.*, 2022).

1.5.9 *Denial of Service (DoS)*

DoS attacks overwhelm computational or communication resources, preventing vehicles or infrastructure from operating correctly. Mohd *et al.* (2018) simulated distributed DoS attacks in Connected and Autonomous Vehicles (CAVs) using OMNET++, while Yeasmin and Haque (2023) proposed blockchain-based defenses for OTA updates. Additional studies have shown how DoS can disable EVCSs or overload the CAN bus (Groza and Murvay, 2018; Tian *et al.*, 2019). These attacks degrade service availability, compromise safety, and disrupt mobility. DoS threats are explored both through communication network models and via physical experiments (e.g., the abrupt disconnection of multiple EVs from chargers, Gupta *et al.*, 2024; Johnson *et al.*, 2022). Attack strength is quantified by the attack rate (packets per second) or the number of nodes launching the attack. Researchers evaluate service availability metrics (e.g., the probability of message delivery or station uptime) and secondary effects on safety and operations. For instance, a recent hardware-in-the-loop demonstration of a distributed denial-of-charging attack showed that simultaneously halting several high-power charging stations can not only prevent vehicles from recharging but also induce power grid frequency spikes from sudden load drop, risking wide-area outages (Gupta *et al.*, 2024).

1.5.10 *Jamming and Interference*

Jamming targets wireless communication channels by transmitting disruptive signals. Grover *et al.* (2014) described how jamming undermines vehicular networks by blocking essential communication, while Fraiji *et al.* (2018) analyzed the resulting inefficiency and instability. In EV systems, jamming can interrupt charging authorization or disrupt V2G communication, reducing both safety and efficiency. Jamming attacks are examined using both analytical models and empirical testing of wireless channels. Researchers simulate continuous or intermittent radio interference that overwhelms Dedicated Short-Range Communications (DSRC)/Cellular Vehicle-to-Everything (C-V2X) signals, adjusting jammer power and duty cycle to represent different attack regimes (Punal *et al.*, 2014). Performance degradation is measured using metrics such as packet reception rate, communication range reduction, and missed safety messages.

1.5.11 *Overpower*

Overpower attacks deliberately push electrical systems beyond safe operating thresholds, creating risks for both vehicles and the power grid. Brighente *et al.* (2023) showed how overpowering exploits can damage EV charging systems or destabilize energy distribution. By overwhelming the electrical infrastructure, these attacks compromise both transportation safety and energy security. Overpower attacks are typically studied through integrated vehicle-grid simulations, where an adversary manipulates charging control systems to draw or inject power beyond safe operational limits (Acharya *et al.*, 2024). Attack scenarios may involve compromising the charge control software to ignore current limits or coordinating a fleet of EV chargers to push maximum load into the grid simultaneously. The impacts are evaluated on technical metrics like voltage/frequency deviations, transformer loading, and battery temperatures, as well as service reliability (charger outages or fires). Studies warn that such attacks could blow protective devices and induce grid instability; for example, a coordinated overload of fast chargers in an urban network could lead to over-frequency events or line overloads, potentially causing cascading failures in the power grid (Acharya *et al.*, 2024).

1.5.12 *Side-Channels*

Side-channel attacks exploit unintentional physical emissions, such as electromagnetic signals or power consumption patterns, to extract sensitive information. La Cour *et al.* (2021) demonstrated side-channel risks in cryptographic operations, while Chandwani *et al.* (2020) and Gottumukkala *et al.* (2019) analyzed how attackers can infer EV charging behavior or cryptographic keys from leaked signals. These attacks bypass traditional protections by observing hardware behavior directly, threatening confidentiality and reliability.

1.5.13 *Comparison and Limitations*

Infrastructure-centric studies mainly emphasize secure communication protocols, network segmentation, access control, and the robustness of OTA and data pipelines. While these controls effectively mitigate fleet-wide exposure, current evaluations often rely on static trust boundaries and rarely quantify the propagation of cyber disruptions into dynamic traffic operations or grid stability. This limitation is particularly acute in EV ecosystems, where the tight coupling between mobility and power systems renders coordinated cross-domain attacks physically plausible; however, integrated impact models and shared benchmarks for such coupled scenarios remain scarce (Razzaque *et al.*, 2025).

1.6 Learnings from Cyberattacks in the Real-World

We examined a dataset of publicly reported cyber incidents between 2005 and 2025, filtered to focus exclusively on transportation-related systems. Each incident was categorized by attack type, impact target, and impact type, as described in Section 1.2.5. The analysis highlights how the nature and consequences of cyberattacks in transportation have evolved over time. Based on the analysis, the most commonly observed attack types on transportation systems are summarized in Table 1.3.

1.6.1 *Attack Types*

Figure 1.3 shows the temporal distribution of attack types. Infrastructure-level attacks dominate the landscape across the study period, underscoring the persistent targeting of roadside units, traffic control devices, and charging infrastructure. Sensor-based attacks

Table 1.3. Commonly observed cyberattack types on transportation systems.

Category	Attack Types
Sensor-based	Routing manipulation, GPS spoofing
Maneuver-oriented	Data falsification
Infrastructure-level	Infrastructure attacks, malware injection, MitM, phishing/social engineering, DoS

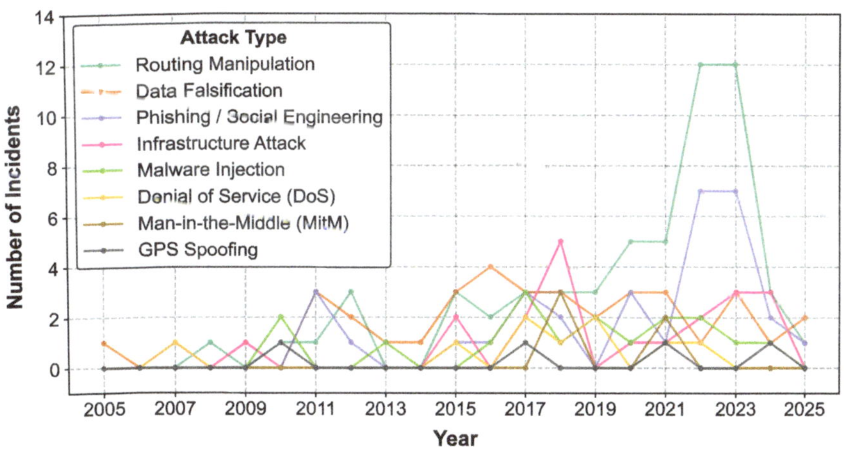

Fig. 1.3. Trends in the cyberattack types affecting transportation systems (2005–2025).

appear less frequently but show an upward trajectory in the last decade, reflecting increasing adversarial focus on perception technologies such as GPS and LiDAR. Maneuver-oriented attacks are rarer overall, but distinct spikes correspond to years when vehicle-level exploits were publicly demonstrated. Together, these patterns reveal a widening threat surface as transportation systems become more digitally integrated. In recent years, routing manipulation and phishing/social engineering have become more prominent, while DoS attacks have declined in frequency.

1.6.2 *Impact Targets*

Figure 1.4 illustrates which sectors were most affected. Government agencies and critical infrastructure entities emerge as the most

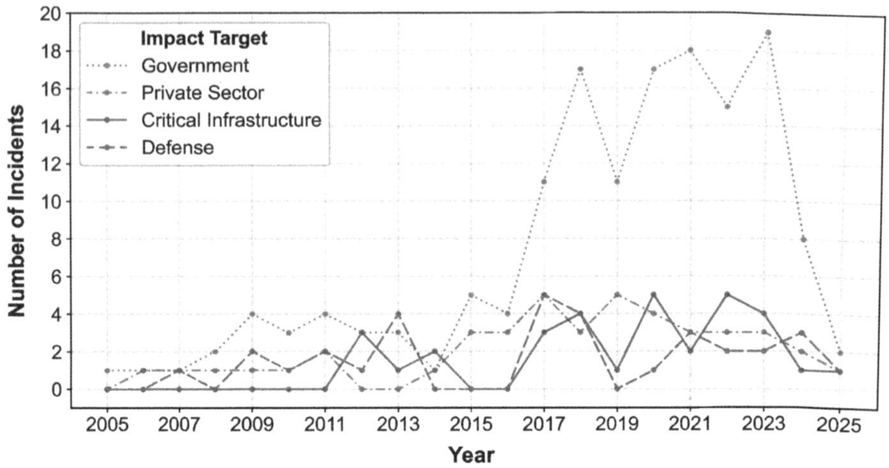

Fig. 1.4. Trends in the impact targets of cyberattacks on transportation systems (2005–2025).

frequent targets, reflecting the strategic value of transportation systems for national security and public safety. Attacks on private sector actors, including logistics firms and fleet operators, show a steady increase in recent years. Defense-related incidents appear less frequently but cluster in periods of heightened geopolitical activity, suggesting targeted campaigns against military transportation assets. Recently, the government sector has been targeted by cyber-attackers more than other sectors.

1.6.3 *Impact Types*

Figure 1.5 reports the relative share of impact types per year. Operational disruption consistently accounts for the largest share, highlighting adversaries' emphasis on interrupting the availability of transportation services. Data compromise is also prominent, reflecting theft of routing data, passenger records, or operational logs. Financially motivated incidents, while smaller in proportion, have risen recently, coinciding with ransomware campaigns against transit agencies and toll operators. Reputation damage and national security threats appear less often but are highly consequential when they occur, especially in cases of exposed systemic vulnerabilities or compromised defense logistics. Lately, data compromise and

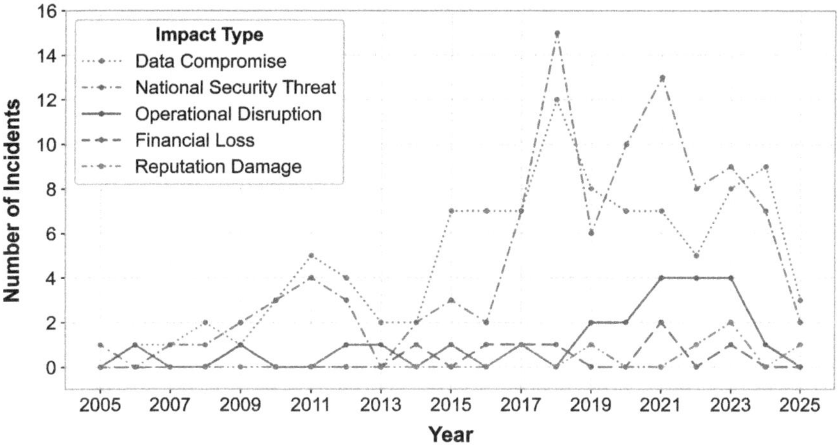

Fig. 1.5. Trends in the impact types of cyberattacks on transportation systems (2005–2025).

national security have been the most vulnerable compared to other impact types.

1.7 Discussions

This discussion reframes the findings using the three functional categories introduced in this chapter: sensor-based, maneuver-oriented, and infrastructure-level. The goal is to explain how different attacks compromise transportation systems, to identify where defenses should concentrate, and to propose category-specific and cross-category recommendations. The analysis emphasizes component-level effects, cascading impacts on traffic operations and charging services, and the role of organizational processes such as software update pipelines and identity management.

1.7.1 *Key Findings by Category*

Sensor-Based. Remote sensor manipulation, GPS spoofing, and replay attacks exploit weak trust in raw or near-raw signals and in derived states such as position and velocity. Multi-sensor fusion can mask small but coordinated perturbations, making detection difficult and shifting the risk to planning and control. When adversaries sustain low-amplitude errors that remain within nominal bounds, perception modules feed plausible but wrong inputs to downstream logic. The main insight is that perception must couple redundancy with semantic and physics checks rather than rely on signal-level agreement alone.

Maneuver-Oriented. Relaying, data falsification, and physical manipulation of control software or ECUs target cooperation protocols and decision execution. These attacks can induce unsafe braking, unsafe headways, or lane departures even when perception is correct. The risk is amplified in cooperative driving and platooning, where small timing or content distortions propagate across agents. Runtime behavior verification and command authentication are therefore essential, together with controller monitors that enforce safe sets and trigger graceful fallback.

Infrastructure-Level. Data poisoning, backdoor, MitM, phishing, and environment or infrastructure manipulation extend the blast radius beyond a single platform. They act through model and data pipelines, identity and access management, OTA update services, roadside and cloud services, and the power interface for charging. These attacks degrade availability and economic fairness, and they can coordinate misbehavior across fleets or neighborhoods. The central insight is that provenance, trust, and segmentation are as critical in shared services as in vehicle controls.

1.7.2 *Cyberattacks and Their Impacts on the Social Space*

The social dimension in cyber-physical-social systems (CPSS) remains underdeveloped despite substantial advances (Yilma *et al.*, 2021; Zeng *et al.*, 2020). This leaves a gap in integrated models that connect cyber, physical, and social spaces. Recent studies emphasize the integration of human and social factors, acknowledging that individuals serve as users, active sensors, and agents within the system. A central aim of CPSS is to leverage social and behavioral data analytics to proactively deliver user-centered services (Wang *et al.*, 2017). In the context of cyber-physical mobility systems, prior reviews have largely concentrated on technical vulnerabilities in the cyber-physical domain without focusing on the potential cyber-social consequences (Chandwani *et al.*, 2020; Euceda *et al.*, 2023; Johnson *et al.*, 2022; Tappeta *et al.*, 2022). Cyberattacks across sensor-based, maneuver-oriented, and infrastructure-level categories, however, extend beyond immediate technical impacts to create cascading effects in social and behavioral domains.

Sensor-based attacks target the perception layer by manipulating raw data or signals collected from the environment. Examples include GPS spoofing, replay attacks, and remote sensor manipulation. These attacks distort situational awareness, leading to navigation errors or traffic disruptions. Beyond the cyber-physical domain, they compromise privacy, facilitate fraud through misinformation, and diminish user confidence in system reliability.

Maneuver-oriented attacks undermine decision-making and vehicle coordination processes. Attacks such as CAN intrusion (e.g., message injection, fuzzing, command replay), relaying, data falsification, and ECU/software flashing can directly manipulate control execution or cooperative decision making, leading to unsafe braking, unsafe headways, misrouting, and operational instability. More broadly, the impacts include financial exploitation, loss of personal data privacy, and reduced trust in connected and automated technologies, which can slow public adoption.

Infrastructure-level attacks focus on external systems and resources that support AV, CV, and EV operations. These include DoS, jamming, malware injection, and overpower attacks, which degrade performance, cause service outages, or induce safety-critical failures. This category also includes identity- and user-facing attacks such as impersonation, Sybil, and phishing/social engineering, which compromise trust and access to shared services. While the immediate effects appear as operational disruptions in vehicles or charging stations, the cyber-social implications involve diminished trust in mobility services, altered travel behavior, and reputational or financial harm to operators and infrastructure providers.

In summary, categorizing cyberattacks into sensor-based, maneuver-oriented, and infrastructure-level domains reveals both distinct technical vulnerabilities and overlapping social consequences. Although the mechanisms differ, their long-term effects converge on reduced trust, compromised privacy, financial losses, and reluctance toward the adoption of cyber-physical mobility systems. Therefore, a CPSS framework that integrates cybersecurity measures across all categories, while explicitly accounting for resilience, transparency, and societal acceptance, needs to be utilized to address these multi-dimensional risks.

1.7.3 *Existing AI/ML Techniques and Emerging Research Avenues*

Existing AI/ML Techniques. Modern transportation systems employ artificial intelligence (AI) and ML at multiple levels to bolster cybersecurity. For example, at the sensor and network level, ML-based anomaly detection techniques continuously monitor sensor outputs and in-vehicle network traffic for signs of malicious deviations in real time (Purohit and Govindarasu, 2022). Recent work has deployed hybrid approaches that combine rule-based logic with ML classifiers on CAN bus data, achieving intrusion detection accuracies above 90% while respecting strict latency requirements (Purohit and Govindarasu, 2022). In practice, automotive IDS using such algorithms have successfully detected over 90% of known attack scenarios in CVs. For example, Garrett's IDS is one example that demonstrates high detection rates for various CV attack patterns (Garrett Motion, 2023).

For higher-level decision-making processes, reinforcement learning (RL) has been applied to improve vehicle control resilience under adversarial conditions. RL-based controllers can be trained against simulated attackers, ensuring they continue to make safe driving decisions (such as obstacle avoidance or safe lane changes) even when sensors or dynamics are perturbed (He *et al.*, 2024). In addition, He *et al.* (2024) developed a robust RL framework that incorporates an online learned adversary model to handle worst-case disturbances in sensor inputs and vehicle dynamics, thereby enhancing an AV's ability to maintain safe operation amid attacks.

In both vehicle and infrastructure settings, secure federated learning (FL) enables collaborative training of defensive models for anomaly detection, traffic prediction, and infrastructure resilience without requiring participants to transmit raw data to a central server (Imteaj *et al.*, 2021a, 2021b). A recent survey indicates that federated learning has been applied in CAV fleets to applications ranging from cooperative perception to motion planning, with deployments often integrating security mechanisms, including secure aggregation and trusted execution environments, to mitigate threats such as model poisoning (Chellapandi *et al.*, 2023). This paper surveys a broad range of AI and ML techniques relevant to transportation cybersecurity. However, our central focus is on how cyberattacks affect road transportation systems rather than on the algorithms themselves. For detailed discussions of specific defense mechanisms based on AI and ML, see comprehensive reviews such as by Ozkan-Okay *et al.* (2024).

Emerging Research Avenues. Looking forward, researchers are exploring advanced AI paradigms to address the limits of current data-driven defenses and to ensure security across a vehicle's life cycle. A promising direction is *neurosymbolic AI*, which integrates the pattern recognition capabilities of neural networks with the structured logical reasoning of symbolic methods. These approaches aim to equip AVs with a layer of common sense reasoning in addition to their learned models (Intelligent Transportation Society of America, 2025). For instance, a neuro-symbolic system could apply logical rules (such as object permanence) to infer the continued presence of a pedestrian who temporarily disappears from sensors behind an obstacle, thereby avoiding mistakes that might confuse a purely neural model and improving safety against corner-case attacks (Intelligent Transportation Society of America, 2025). Early work has integrated symbolic logic into deep RL frameworks to enable safer policy learning. Agents guided by logical safety constraints avoid unsafe actions in simulation and often converge faster than conventional RL methods (Sharifi *et al.*, 2023).

In parallel, there is increasing emphasis on *explainable AI* (XAI) for transportation systems. As AVs, CVs, and EVs come to rely on complex "black-box" ML models, there is increasing demand for techniques that make these models' decisions more interpretable to humans (Kuznietsov *et al.*, 2024). Emerging XAI approaches include inherently interpretable neural network architectures and post-hoc explanation techniques that help engineers and policymakers understand the reasoning behind model decisions. For example, XAI can explain why a perception module misclassified an object or why an IDS issued an alert in network traffic (Kuznietsov *et al.*, 2024).

Recent research increasingly integrates AI-driven intrusion detection with digital twin-based simulation environments to enhance detection performance and system-level cybersecurity resilience in transportation systems. Digital twins of vehicles, charging infrastructure, and traffic networks provide synchronized virtual representations of physical assets, enabling cyberattacks to be emulated, stress-tested, and analyzed under realistic operational conditions. When coupled with ML–based IDS, digital twins support proactive threat assessment, adaptive model calibration, and scenario-based evaluation of defensive strategies, particularly for rare or high-impact attack scenarios that are difficult to observe in real-world deployments. This integration improves early anomaly detection while reducing false positives by contextualizing alerts within expected system dynamics and control behaviors, complementing existing ML-based IDS approaches for vehicular networks (Garrett Motion, 2023; Purohit and Govindarasu, 2022). In parallel, secure FL has emerged as a scalable mechanism for collaboratively training defensive models across distributed vehicles and infrastructure without centralized data sharing (Imteaj *et al.*, 2021a, 2021b). Beyond privacy preservation, FL frameworks increasingly incorporate robustness-enhancing mechanisms such as secure aggregation and defenses against model poisoning, making them well suited for large, heterogeneous transportation networks (Chellapandi *et al.*, 2023; Ozkan-Okay *et al.*, 2024). Together, the convergence of AI-enabled intrusion detection, FL, and digital twin simulation represents a promising direction for developing adaptive, privacy-aware, and continuously improving cybersecurity defenses across the life cycle of cyber-physical mobility systems.

In summary, these emerging paradigms have the potential to significantly enhance the security and trustworthiness of future cyber-physical mobility systems. By embedding logical reasoning and interpretability into their AI components, fleets may become more capable of handling complex or novel attack scenarios while also providing transparent explanations of their defensive actions. This combination of robustness and transparency is expected to strengthen user trust and safety in next-generation transportation networks.

1.7.4 *Category-Based Design Recommendations*

Table 1.4 summarizes practical controls that follow directly from the evidence in Sections 1.3–1.5. The items are prioritized to minimize

Table 1.4. Category-based design recommendations for cyberattacks.

Category	**Key Recommendations**
Sensor-based (Jin *et al.*, 2019; Pavlitska *et al.*, 2023; Rouf *et al.*, 2010; Yan *et al.*, 2022; Zeng *et al.*, 2017)	• Use diverse and independent sensing with cross checks that combine kinematic constraints, map consistency, and environmental plausibility. • Sign and time stamp sensor outputs and intermediate state, and verify freshness and provenance at fusion points. • Deploy residual tests that are aware of vehicle dynamics to bound drift from spoofing and replay while limiting false alarms. • Design perception modules to expose calibrated uncertainty so that planning can down-weight uncertain inputs and select safe actions.
Maneuver-oriented (Checkoway *et al.*, 2011; Koscher *et al.*, 2010; Miller and Valasek, 2015; Sarker *et al.*, 2020; Song *et al.*, 2023)	• Authenticate control and cooperation messages end-to-end and bind them to identities with revocation and replay protection. • Enforce runtime behavior verification with control barrier or invariant checks and trigger graceful fallback when violations are detected. • Protect ECUs and planning software with signed updates, measured boot, and secure storage of keys. • Stress test planning and control in closed loop with adversarial timing and content perturbations to verify stability margins.

(*Continued*)

Table 1.4. (*Continued*)

Category	Key Recommendations
Infrastructure-level (Ahmad *et al.*, 2018; Brighente *et al.*, 2023; Laszka *et al.*, 2016; Punal *et al.*, 2014; Soykan *et al.*, 2021)	• Require signed datasets, model cards, and build provenance for training and map updates, with canary samples and watermarks to expose poisoning or backdoors. • Segment roadside, charging, and cloud services with least privilege access, rate limits, and anomaly detection tuned for fleet scale patterns. • Secure OTA pipelines with multi-party approval, rollback, staged roll-out, and continuous attestation on devices. • Harden radio and power interfaces with jamming detection, adaptive channel selection, surge protection, and fail-safe operation under degraded service.
Cross-category (Dibaei *et al.*, 2019; International Organization for Standardization, 2011; Koopman and Wagner, 2016; Sun *et al.*, 2021)	• Maintain an identity and key life cycle for vehicles, chargers, and services with fast revocation and privacy-preserving credentials. • Define system-wide safe modes that preserve traffic flow and charging fairness under loss of accuracy or connectivity. • Integrate digital twins that couple road traffic and power grids to evaluate cascading risks and rehearse incident response.

the risk of unsafe actions, mitigate cascading effects, and facilitate post-incident diagnosis.

1.7.5 *Implications of Real-World Cyberattack Trends*

The analysis of real-world incidents highlights that cyberattacks on transportation systems are not only increasing in frequency but also diversifying in form and consequence. The predominance of infrastructure-level attacks reflects adversaries' recognition of the systemic leverage gained by targeting roadside units, charging stations, and traffic management systems—components that, if disrupted, have immediate and widespread effects. The increasing prominence of sensor-based and maneuver-oriented attacks suggests a shift toward exploiting vulnerabilities in emerging technologies that enable automation and connectivity, underscoring the risks inherent in the adoption of advanced perception and control systems. From an impact perspective, the concentration of incidents on government agencies and critical infrastructure reinforces the geopolitical salience of transportation as a national security asset, while the growing share of private sector cases illustrates how digitalization of logistics, fleets, and EV services expands the threat surface. The dominance of operational disruption as an impact type further reveals that attackers often aim to maximize visibility and immediate disruption, whereas the increasing footprint of financial loss, particularly through ransomware, signals a parallel trend toward monetization of attacks. These findings emphasize the urgent need for multi-layered defense strategies that address both legacy infrastructure and emerging cyber-physical interfaces, with coordinated involvement from public agencies, private operators, and regulators.

1.8 Conclusions

This survey synthesizes current knowledge on cybersecurity vulnerabilities in the context of road transportation as a cyber-physical system. Based on 408 peer-reviewed studies published between 2021 and late-2025, we group cyber threats into three categories: sensor-based, maneuver-oriented, and infrastructure-level. Sensor-based attacks exploit weaknesses in perception technologies such as positioning signals, cameras, radar, and LiDAR; adversarial inputs or spoofed data can distort environmental interpretation, leading to cascading errors in navigation or safety responses. Maneuver-oriented attacks intervene directly in planning and control by injecting messages or tampering with decision modules, thereby provoking unsafe maneuvers such as abrupt braking, erratic lane shifts, or unintended accelerations that may bypass conventional fault detection mechanisms.

Infrastructure-level attacks extend beyond individual vehicles to target roadside units, traffic management systems, charging interfaces, or cloud-based services, where falsified congestion broadcasts, manipulated digital signage, or compromised energy coordination can trigger widespread disruptions from local traffic slowdowns to city-wide gridlock. Our analysis highlights that most studies focus on vulnerabilities in perception and data exchange, while far fewer investigate the broader operational impacts of successful exploits. Real-world demonstrations, including remote takeovers of steering and braking functions, confirm that these risks are practical and not merely theoretical. Yet the literature often stops short of quantifying consequences such as traffic disruption, grid instability, or loss of system reliability. This disconnect underscores the need for future research to explicitly link technical attack vectors to physical and operational outcomes, thereby supporting a more realistic understanding of cybersecurity risks in transportation systems.

Bibliography

Acharya, S., Khan, H. A. U., Karri, R., and Dvorkin, Y. (2024). MaDEVIoT: Cyberattacks on EV charging can disrupt power grid operation, in *2024 IEEE Power & Energy Society Innovative Smart Grid Technologies Conference (ISGT)* (IEEE), pp. 1–5.

Acharya, S., Mieth, R., Karri, R., and Dvorkin, Y. (2022). False data injection attacks on data markets for electric vehicle charging stations, *Advances in Applied Energy* **7**, p. 100098.

Ahmad, F., Adnane, A., Franqueira, V. N., Kurugollu, F., and Liu, L. (2018). Man-in-the-middle attacks in vehicular ad-hoc networks: Evaluating the impact of attackers' strategies, *Sensors* **18**, 11, p. 4040.

Alam, M. S., Acharya, A., and Oluoch, J. (2021). A novel technique for mapping jammed areas in connected and autonomous vehicles (CAVs), in *2021 IEEE International Conference on Autonomic Computing and Self-Organizing Systems Companion (ACSOS-C)* (IEEE, DC, USA), pp. 111–117.

Ali, W. A., Fanti, M. P., Roccotelli, M., and Ranieri, L. (2023). A review of digital twin technology for electric and autonomous vehicles, *Applied Sciences* **13**, 10, p. 5871.

Aljohani, T. and Almutairi, A. (2024). A comprehensive survey of cyberattacks on EVs: Research domains, attacks, defensive mechanisms, and verification methods, *Defence Technology* **42**, pp. 31–58, doi:https://doi.org/10.1016/j.dt.2024.06.009, https://www.sciencedirect.com/science/article/pii/S221491472400151X.

Appathurai, A., Manogaran, G., and Chilamkurti, N. (2019). Trusted FPGA-based transport traffic inject, impersonate (I2) attacks beaconing in the internet of vehicles, *IET Networks* **8**, 3, pp. 169–178.

Babu, P. R., Palaniswamy, B., Reddy, A. G., Odelu, V., and Kim, H. S. (2022). A survey on security challenges and protocols of electric vehicle dynamic charging system, *Security and Privacy* **5**, 3, p. e210.

Bajracharya, C. (2021). Performance evaluation for secure communications in mobile internet of vehicles with joint reactive jamming and eavesdropping attacks, *IEEE Transactions on Intelligent Transportation Systems* **23**, 11, pp. 22563–22570.

Bao, K., Valev, H., Wagner, M., and Schmeck, H. (2018). A threat analysis of the vehicle-to-grid charging protocol ISO 15118, *Computer Science-Research and Development* **33**, 1, pp. 3–12.

Brighente, A., Conti, M., Donadel, D., Poovendran, R., Turrin, F., and Zhou, J. (2023). Electric vehicles security and privacy: Challenges, solutions, and future needs, arXiv:2301.04587 [cs.CR].

Center for Strategic and International Studies (2024). Significant cyber incidents, https://www.csis.org/programs/strategic-technologies-program/significant-cyber-incidents, accessed: April 18, 2025.

Chandwani, A., Dey, S., and Mallik, A. (2020). Cybersecurity of onboard charging systems for electric vehicles—review, challenges and countermeasures, *IEEE Access* **8**, pp. 226982–226998.

Checkoway, S., McCoy, D., Kantor, B., Anderson, D., Shacham, H., Savage, S., Koscher, K., Czeskis, A., Roesner, F., and Kohno, T. (2011). Comprehensive experimental analyses of automotive attack surfaces, in *Proceedings of the 20th USENIX Conference on Security*, SEC'11 (USENIX Association, USA), p. 6.

Chellapandi, V. P., Yuan, L., Brinton, C. G., Żak, S. H., and Wang, Z. (2023). Federated learning for connected and automated vehicles: A survey of existing approaches and challenges, *IEEE Transactions on Intelligent Vehicles* **9**, 1, pp. 119–137.

Chen, Q. A., Yin, Y., Feng, Y., Mao, Z. M., and Liu, H. X. (2018). Exposing congestion attack on emerging connected vehicle based traffic signal control, in *NDSS* (NDSS, San Diego, CA, USA), pp. 1–15.

Conti, M., Donadel, D., Poovendran, R., and Turrin, F. (2022). EVExchange: A relay attack on electric vehicle charging system, in *European Symposium on Research in Computer Security* (Springer, Cham), pp. 488–508.

Cui, C., Du, H., Jia, Z., He, Y., Yang, Y., and Jin, M. (2022). Data poisoning attack using hybrid particle swarm optimization in connected and autonomous vehicles, in *2022 IEEE Asia-Pacific Conference on Computer Science and Data Engineering (CSDE)* (IEEE, Gold Coast, Australia), pp. 1–5.

Dibaei, M., Zheng, X., Jiang, K., Maric, S., Abbas, R., Liu, S., Zhang, Y., Deng, Y., Wen, S., Zhang, J., *et al.* (2019). An overview of attacks and defences on intelligent connected vehicles, *arXiv preprint* arXiv:1907.07455.

Durlik, I., Miller, T., Kostecka, E., Zwierzewicz, Z., and Łobodzińska, A. (2024). Cybersecurity in autonomous vehicles—Are we ready for the challenge? *Electronics* **13**, 13, p. 2654.

Euceda, G. R., Akundi, A., and Luna, S. (2023). Cybersecurity challenges in electric vehicles: An initial literature review and research agenda, in *2023 IEEE International Systems Conference (SysCon)* (IEEE, Vancouver, BC, Canada), pp. 1–8.

Fraiji, Y., Azzouz, L. B., Trojet, W., and Saidane, L. A. (2018). Cyber security issues of internet of electric vehicles, in *2018 IEEE Wireless Communications and Networking Conference (WCNC)* (IEEE, Barcelona, Spain), pp. 1–6.

Garofalaki, Z., Kosmanos, D., Moschoyiannis, S., Kallergis, D., and Douligeris, C. (2022). Electric vehicle charging: A survey on the security issues and challenges of the open charge point protocol (OCPP), *IEEE Communications Surveys & Tutorials* **24**, 3, pp. 1504–1533.

Garrett Motion (2023). Garrett's AwardWinning Intrusion Detection System (IDS) Receives Acclaim for Increasing Safety in Intelligent Connected Vehicles, https://en.eeworld.com.cn/news/qcdz/eic627006.html, accessed: May 07, 2025.

Gennaro, R., Lysyanskaya, A., Malkin, T., Micali, S., and Rabin, T. (2004). Algorithmic tamper-proof (ATP) security: Theoretical foundations for security against hardware tampering, in M. Naor (ed.), *Theory of Cryptography* (Springer, Berlin), pp. 258–277.

Ghosal, A. and Conti, M. (2020). Security issues and challenges in V2X: A survey, *Computer Networks* **169**, p. 107093.

Giraldo, J., Urbina, D., Cardenas, A., Valente, J., Faisal, M., Ruths, J., Tippenhauer, N. O., Sandberg, H., and Candell, R. (2018). A survey of physics-based attack detection in cyber-physical systems, *ACM Computing Surveys (CSUR)* **51**, 4, pp. 1–36.

Gottumukkala, R., Merchant, R., Tauzin, A., Leon, K., Roche, A., and Darby, P. (2019). Cyber-physical system security of vehicle charging stations, in *2019 IEEE Green Technologies Conference (GreenTech)* (IEEE, Lafayette, LA, USA), pp. 1–5.

Greene, K., Rodgers, D., Dykhuizen, H., McNeil, K., Niyaz, Q., and Al Shamaileh, K. (2020). Timestamp-based defense mechanism against replay attack in remote keyless entry systems, in *2020 IEEE International Conference on Consumer Electronics (ICCE)* (IEEE, Las Vegas, NV, USA), pp. 1–4.

Grover, K., Lim, A., and Yang, Q. (2014). Jamming and anti-jamming techniques in wireless networks: A survey, *International Journal of Ad Hoc and Ubiquitous Computing* **17**, 4, p. 197–215, doi:10.1504/IJAHUC.2014.066419.

Groza, B. and Murvay, P.-S. (2018). Security solutions for the controller area network: Bringing authentication to in-vehicle networks, *IEEE Vehicular Technology Magazine* **13**, 1, pp. 40–47.

Guo, L., Ye, J., and Yang, B. (2021). Cyberattack detection for electric vehicles using physics-guided machine learning, *IEEE Transactions on Transportation Electrification* **7**, 3, pp. 2010–2022, doi:10.1109/TTE.2020.3044524.

Gupta, K., Panigrahi, B. K., Joshi, A., and Paul, K. (2024). Demonstration of denial of charging attack on electric vehicle charging infrastructure and its consequences, *International Journal of Critical Infrastructure Protection* **46**, p. 100693.

Halder, S., Ghosal, A., and Conti, M. (2020). Secure over-the-air software updates in connected vehicles: A survey, *Computer Networks* **178**, p. 107343, https://api.semanticscholar.org/CorpusID:219928990.

He, X., Huang, W., and Lv, C. (2024). Toward trustworthy decision-making for autonomous vehicles: A robust reinforcement learning approach with safety guarantees, *Engineering* **33**, pp. 77–89.

Huang, K., Kockelman, K., and Gurumurthy, K. M. (2023). Innovations impacting the future of transportation: An overview of connected, automated, shared, and electric technologies, *Transportation Letters* **15**, 6, pp. 490–509.

Huang, X., Xu, C., Wang, P., and Liu, H. (2018). LNSC: A security model for electric vehicle and charging pile management based on blockchain ecosystem, *IEEE Access* **6**, pp. 13565–13574.

Hwang, H., Jung, G., Sohn, K., and Park, S. (2008). A study on MITM (man in the middle) vulnerability in wireless network using 802.1X and EAP, in *2008 International Conference on Information Science and Security (ICISS 2008)* (IEEE, USA), pp. 164–170.

Imteaj, A., Khan, I., Khazaei, J., and Amini, M. H. (2021a). Fedresilience: A federated learning application to improve resilience of resource-constrained critical infrastructures, *Electronics* **10**, 16, p. 1917.

Imteaj, A., Thakker, U., Wang, S., Li, J., and Amini, M. H. (2021b). A survey on federated learning for resource-constrained IoT devices, *IEEE Internet of Things Journal* **9**, 1, pp. 1–24.

Intelligent Transportation Society of America (2025). Could Neuro-Symbolic AI Change the Landscape of Autonomous Driving? https://itsa.org/could-neuro-symbolic-ai-change-the-landscape-of-autonomous-driving/, accessed: May 07, 2025.

International Organization for Standardization (2011). 26262: Road vehicles-functional safety, *International Standard ISO/FDIS* **26262**, pp. 53–54.

Iqbal, S., Ball, P., Kamarudin, M. H., and Bradley, A. (2022). Simulating malicious attacks on VANETs for connected and autonomous vehicle cybersecurity: A machine learning dataset, in *2022 13th International Symposium on Communication Systems, Networks and Digital Signal Processing (CSNDSP)* (IEEE, Porto, Portugal), pp. 332–337.

Jahangir, H., Lakshminarayana, S., and Poor, H. V. (2024). Charge manipulation attacks against smart electric vehicle charging stations and deep learning-based detection mechanisms, *IEEE Transactions on Smart Grid* **15**, 5, pp. 5182–5194.

Jiang, L., Xie, S., Maharjan, S., and Zhang, Y. (2019). Blockchain empowered wireless power transfer for green and secure internet of things, *IEEE Network* **33**, 6, pp. 164–171.

Jin, X., Haddad, W. M., Jiang, Z.-P., Kanellopoulos, A., and Vamvoudakis, K. G. (2019). An adaptive learning and control architecture for mitigating sensor and actuator attacks in connected autonomous vehicle platoons, *International Journal of Adaptive Control and Signal Processing* **33**, 12, pp. 1788–1802.

Johnson, J., Berg, T., Anderson, B., and Wright, B. (2022). Review of electric vehicle charger cybersecurity vulnerabilities, potential impacts, and defenses, *Energies* **15**, 11, p. 3931.

Khemissa, H. and Urien, P. (2022). Centralized architecture for ECU security management in connected and autonomous vehicles, in *2022 13th International Conference on Information and Communication Technology Convergence (ICTC)* (IEEE, Jeju Island, Republic of Korea), pp. 1409–1414.

Kim, K., Kim, J. S., Jeong, S., Park, J.-H., and Kim, H. K. (2021). Cybersecurity for autonomous vehicles: Review of attacks and defense, *Computers & Security* **103**, p. 102150.

Kitchenham, B., Pearl Brereton, O., Budgen, D., Turner, M., Bailey, J., and Linkman, S. (2009). Systematic literature reviews in software engineering – A systematic literature review, *Information and Software Technology* **51**, 1, pp. 7–15, doi:https://doi.org/10.1016/j.infsof.2008.09.009, Special Section - Most Cited Articles in 2002 and Regular Research Papers.

Koopman, P. and Wagner, M. (2016). Challenges in autonomous vehicle testing and validation, *SAE International Journal of Transportation Safety* **4**, 1, pp. 15–24.

Koscher, K., Czeskis, A., Roesner, F., Patel, S., Kohno, T., Checkoway, S., McCoy, D., Kantor, B., Anderson, D., Shacham, H., and Savage, S. (2010). Experimental security analysis of a modern automobile, in *2010 IEEE Symposium on Security and Privacy* (IEEE, Oakland, CA, USA), pp. 447–462, doi:10.1109/SP.2010.34.

Kumar, A., Varadarajan, V., Kumar, A., Dadheech, P., Choudhary, S. S., Kumar, V. A., Panigrahi, B. K., and Veluvolu, K. C. (2021). Black hole attack detection in vehicular ad-hoc network using secure AODV routing algorithm, *Microprocessors and Microsystems* **80**, p. 103352.

Kuznietsov, A., Gyevnar, B., Wang, C., Peters, S., and Albrecht, S. V. (2024). Explainable ai for safe and trustworthy autonomous driving: A systematic review, *IEEE Transactions on Intelligent Transportation Systems* **25**, 12, pp. 19342–19364, doi:10.1109/TITS.2024.3474469.

La Cour, A. S., Afridi, K. K., and Suh, G. E. (2021). Wireless charging power side-channel attacks, in *Proceedings of the 2021 ACM SIGSAC Conference on Computer and Communications Security*, CCS'21 (Association for Computing Machinery, New York, NY, USA), pp. 651–665, doi:10.1145/3460120.3484733, https://doi.org/10.1145/3460120.3484733.

La Fontaine, S., Muralidhar, N., Clifford, M., Eliassi-Rad, T., and Nita-Rotaru, C. (2022). Alternative route-based attacks in metropolitan traffic systems, in *2022 52nd Annual IEEE/IFIP International Conference on Dependable Systems and Networks Workshops (DSN-W)* (IEEE, Baltimore, MD, USA), pp. 20–27.

Lamssaggad, A., Benamar, N., Hafid, A. S., and Msahli, M. (2021). A survey on the current security landscape of intelligent transportation systems, *IEEE Access* **9**, pp. 9180–9208.

Laszka, A., Potteiger, B., Vorobeychik, Y., Amin, S., and Koutsoukos, X. (2016). Vulnerability of transportation networks to traffic-signal tampering, in *2016 ACM/IEEE 7th International Conference on Cyber-Physical Systems (ICCPS)* (IEEE), pp. 1–10.

Li, Y., Tu, Y., Fan, Q., Dong, C., and Wang, W. (2018). Influence of cyber-attacks on longitudinal safety of connected and automated vehicles, *Accident Analysis & Prevention* **121**, pp. 148–156.

Lokman, S.-F., Othman, A. T., and Abu-Bakar, M.-H. (2019). Intrusion detection system for automotive controller area network (can) bus system: A review, *EURASIP Journal on Wireless Communications and Networking* **2019**, 1, pp. 1–17.

Mia, J. and Amini, M. H. (2024). A secure object detection technique for intelligent transportation systems, *IEEE Open Journal of Intelligent Transportation Systems* **5**, pp. 495–508, doi:10.1109/OJITS.2024.3440876.

Miller, C. and Valasek, C. (2015). Remote exploitation of an unaltered passenger vehicle, *Black Hat USA* **2015**, S 91, pp. 1–91.

Mohd, T. K., Majumdar, S., Mathur, A., and Javaid, A. Y. (2018). Simulation and analysis of DDoS attack on connected autonomous vehicular network using OMNET++, in *2018 9th IEEE Annual Ubiquitous Computing, Electronics & Mobile Communication Conference (UEMCON)* (IEEE, New York, NY, USA), pp. 502–508.

Muhammad, Z., Anwar, Z., Saleem, B., and Shahid, J. (2023). Emerging cybersecurity and privacy threats to electric vehicles and their impact on human and environmental sustainability, *Energies* **16**, 3, p. 1113.

Murvay, P.-S. and Groza, B. (2018). Security shortcomings and countermeasures for the SAE J1939 commercial vehicle bus protocol, *IEEE Transactions on Vehicular Technology* **67**, 5, pp. 4325–4339.

Mwanje, M. D., Kaiwartya, O., Aljaidi, M., Cao, Y., Kumar, S., Jha, D. N., Naser, A., and Lloret, J. (2024). Cyber security analysis of connected vehicles, *IET Intelligent Transport Systems* **18**, 7, pp. 1175–1195.

National Highway Traffic Safety Administration (2022). Cybersecurity best practices for the safety of modern vehicles, Technical Report, U.S. Department of Transportation, https://www.nhtsa.gov/research/vehicle-cybersecurity, Updated 2022.

Nie, S., Liu, L., and Du, Y. (2017). Free-fall: Hacking Tesla from wireless to can bus, *Briefing, Black Hat USA* **25**, 1, p. 16.

Ozkan-Okay, M., Akin, E., Aslan, Ö., Kosunalp, S., Iliev, T., Stoyanov, I., and Beloev, I. (2024). A comprehensive survey: Evaluating the efficiency of artificial intelligence and machine learning techniques on cyber security solutions, *IEEE Access* **12**, pp. 12229–12256.

Pavlitska, S., Lambing, N., and Zöllner, J. M. (2023). Adversarial attacks on traffic sign recognition: A survey, in *2023 3rd International Conference on Electrical, Computer, Communications and Mechatronics Engineering (ICECCME)* (IEEE, Tenerife, Canary Islands, Spain), pp. 1–6.

Perrine, K. A., Levin, M. W., Yahia, C. N., Duell, M., and Boyles, S. D. (2019). Implications of traffic signal cybersecurity on potential deliberate traffic disruptions, *Transportation Research Part A: Policy and Practice* **120**, pp. 58–70.

Petersen, K., Vakkalanka, S., and Kuzniarz, L. (2015). Guidelines for conducting systematic mapping studies in software engineering: An update, *Information and Software Technology* **64**, pp. 1–18, doi:https://doi.org/10.1016/j.infsof.2015.03.007.

Poudel, S., Abouyoussef, M., Baugh, J. E., and Ismail, M. (2024). Attack design for maximum malware spread through EVs commute and charge in power-transportation systems, *IEEE Systems Journal* **18**, 3, pp. 1809–1820.

Punal, O., Pereira, C., Aguiar, A., and Gross, J. (2014). Experimental characterization and modeling of rf jamming attacks on VANETs, *IEEE Transactions on Vehicular Technology* **64**, 2, pp. 524–540.

Purohit, K. C., Dimri, S. C., and Jasola, S. (2017). Mitigation and performance analysis of routing protocols under black-hole attack in vehicular ad-hoc network (VANET), *Wireless Personal Communications* **97**, 4, pp. 5099–5114.

Purohit, S. and Govindarasu, M. (2022). Ml-based anomaly detection for intra-vehicular can-bus networks, in *2022 IEEE International Conference on Cyber Security and Resilience (CSR)*, IEEE (IEEE, Rhodes, Greece), pp. 233–238.

Razzaque, M. A., Khadem, S. K., Patra, S., Okwata, G., and Noor-A-Rahim, M. (2025). Cybersecurity in vehicle-to-grid (V2G) systems: A systematic review, *Applied Energy* **398**, p. 126364.

Rouf, I., Miller, R., Mustafa, H., Taylor, T., Oh, S., Xu, W., Gruteser, M., Trappe, W., and Seskar, I. (2010). Security and privacy vulnerabilities of in-car wireless networks: A tire pressure monitoring system case study, in *Proceedings of the 19th USENIX Conference on Security*, USENIX Security'10 (USENIX Association, USA), p. 21.

Saltaformaggio, B., Choi, H., Johnson, K., Kwon, Y., Zhang, Q., Zhang, X., Xu, D., and Qian, J. (2016). Eavesdropping on fine-grained user activities within smartphone apps over encrypted network traffic, in *Proceedings of the 10th USENIX Conference on Offensive Technologies*, WOOT'16 (USENIX Association, USA), pp. 69–78.

Sarker, A., Shen, H., Sen, T., and Uehara, H. (2020). An advanced black-box adversarial attack for deep driving maneuver classification models, in *2020 IEEE 17th International Conference on Mobile Ad Hoc and Sensor Systems (MASS)* (IEEE), pp. 184–192.

Schlenoff, C. I., Kootbally, Z., Rachakonda, P., Lightman, S., Vassilev, A., Wollman, D. A., and Griffor, E. (2024). Standards and performance metrics for on-road automated vehicles, doi:https://doi.org/10.6028/NIST.IR.8527, https://tsapps.nist.gov/publication/get_pdf.cfm?pub_id=957550.

Shalev-Shwartz, S., Shammah, S., and Shashua, A. (2017). On a formal model of safe and scalable self-driving cars, *arXiv preprint* arXiv:1708.06374.

Sharifi, I., Yildirim, M., and Fallah, S. (2023). Towards safe autonomous driving policies using a neuro-symbolic deep reinforcement learning approach, arXiv:2307.01316 [cs.RO], https://arxiv.org/abs/2307.01316.

Shen, J., Wang, N., Wan, Z., Luo, Y., Sato, T., Hu, Z., Gao, Y., Zeng, Q.-A., Liu, Y., and Zhang, Y. (2022). SoK: On the semantic AI security in autonomous driving, *CoRR* **abs/2203.05314**, arXiv:2203.05314 [cs.CR], https://arxiv.org/abs/2203.05314.

Söderhäll, A., Alimadadi, Z., and Papadimitratos, P. (2025). On the impact of sybil-based attacks on mobile crowdsensing for transportation, *arXiv preprint* arXiv:2506.20585.

Song, R., Ozmen, M. O., Kim, H., Muller, R., Celik, Z. B., and Bianchi, A. (2023). Discovering adversarial driving maneuvers against autonomous vehicles, in *32nd USENIX Security Symposium (USENIX Security 23)* (USENIX Association, Anaheim, CA), pp. 2957–2974, https://www.usenix.org/conference/usenixsecurity23/presentation/song.

Soykan, E. U., Bagriyanik, M., and Soykan, G. (2021). Disrupting the power grid via EV charging: The impact of the SMS phishing attacks, *Sustainable Energy, Grids and Networks* **26**, p. 100477.

Sumalee, A. and Ho, H. W. (2018). Smarter and more connected: Future intelligent transportation system, *IATSS Research* **42**, 2, pp. 67–71, doi:https://doi.org/10.1016/j.iatssr.2018.05.005.

Sun, X., Yu, F. R., and Zhang, P. (2021). A survey on cyber-security of connected and autonomous vehicles (CAVs), *IEEE Transactions on Intelligent Transportation Systems* **23**, 7, pp. 6240–6259.

Tappeta, V. S. R., Appasani, B., Patnaik, S., and Ustun, T. S. (2022). A review on emerging communication and computational technologies for increased use of plug-in electric vehicles, *Energies* **15**, 18, p. 6580.

Tian, Y., Tran, V., and Kuerban, M. (2019). DOS attack mitigation strategies on SDN controller, in *2019 IEEE 9th Annual Computing and Communication Workshop and Conference (CCWC)* (IEEE, Las Vegas, NV, USA), pp. 0701–0707.

Ukkusuri, S., Hamim, O. F., Lei, Z., Ka, E., Salek, M. S., Chowdhury, M., Amini, M. H., Cardenas, A., and Thuraisingham, B. (2025). Cybersecurity for next-generation road transportation: A review, *ACM Journal on Autonomous Transportation Systems* **3**, 2, doi:10.1145/3744352, https://doi.org/10.1145/3744352.

Underwood, S. (2014). Automated, connected, and electric vehicle systems: Expert forecast and roadmap for sustainable transportation, Technical Report, Graham Institute for Sustainability, University of Michigan, Ann Arbor.

Vitale, C., Piperigkos, N., Laoudias, C., Ellinas, G., Casademont, J., Escrig, J., Kloukiniotis, A., Lalos, A. S., Moustakas, K., Diaz Rodriguez, R., *et al.* (2021). Caramel: Results on a secure architecture for connected and autonomous vehicles detecting GPS spoofing attacks, *EURASIP Journal on Wireless Communications and Networking* **2021**, 1, pp. 1–28.

Wang, X., Yang, L. T., Xie, X., Jin, J., and Deen, M. J. (2017). A cloud-edge computing framework for cyber-physical-social services, *IEEE Communications Magazine* **55**, 11, pp. 80–85, doi:10.1109/MCOM.2017.1700360.

Ward, D. and Wooderson, P. (2021). *Automotive Cybersecurity: An Introduction to ISO/SAE 21434* (SAE International, USA).

Woo, S., Jo, H. J., and Lee, D. H. (2014). A practical wireless attack on the connected car and security protocol for in-vehicle can, *IEEE Transactions on Intelligent Transportation Systems* **16**, 2, pp. 993–1006.

Xuefeng, L. and Wei, Z. (2022). Risks of cyber threats and developing robust security protocols within electric vehicle charging infrastructure, *Journal of Sustainable Urban Futures* **12**, 12, pp. 16–31.

Yan, C., Xu, Z., Yin, Z., Mangard, S., Ji, X., Xu, W., Zhao, K., Zhou, Y., Wang, T., Gu, G., *et al.* (2022). Rolling colors: Adversarial laser exploits against traffic light recognition, in *31st USENIX Security Symposium (USENIX Security 22)* (USENIX Association, Boston, MA), pp. 1957–1974.

Yang, Z., Ying, J., Shen, J., Feng, Y., Chen, Q. A., Mao, Z. M., and Liu, H. X. (2023). Anomaly detection against GPS spoofing attacks on connected and autonomous vehicles using learning from demonstration, *IEEE Transactions on Intelligent Transportation Systems* **24**, 9, pp. 9462–9475.

Yeasmin, S. and Haque, A. (2023). Collaborative DDoS attack defense for OTA updates in cavs using hyperledger fabric blockchain, in *2023 International Conference on Smart Applications, Communications and Networking (SmartNets)* (IEEE, Istanbul, Turkiye), pp. 1–6.

Yilma, B. A., Panetto, H., and Naudet, Y. (2021). Systemic formalisation of cyber-physical-social system (CPSS): A systematic literature review, *Computers in Industry* **129**, p. 103458, doi:https://doi.org/10.1016/j.compind.2021.103458.

Zeng, J., Yang, L. T., Lin, M., Ning, H., and Ma, J. (2020). A survey: Cyber-physical-social systems and their system-level design methodology, *Future Generation Computer Systems* **105**, pp. 1028–1042, doi:https://doi.org/10.1016/j.future.2016.06.034.

Zeng, K. C., Shu, Y., Liu, S., Dou, Y., and Yang, Y. (2017). A practical GPS location spoofing attack in road navigation scenario, in *Proceedings of the 18th international workshop on mobile computing systems and applications* (Association for Computing Machinery, New York, NY, USA), pp. 85–90.

Zhang, Q., Jin, S., Zhu, R., Sun, J., Zhang, X., Chen, Q. A., and Mao, Z. M. (2024). On data fabrication in collaborative vehicular perception: Attacks and countermeasures, in *33rd USENIX Security Symposium (USENIX Security 24)*, pp. 6309–6326.

Zhou, C., Yan, Q., Yu, Z., Dixit, E., Zhang, N., Zeng, H., and Ghanhdari, A. S. (2023). ChargeX: Exploring state switching attack on electric vehicle charging systems, *arXiv preprint* arXiv:2305.08037.

https://doi.org/10.1142/9789819831548_0002

Chapter 2

Intelligent Transportation Systems: Emerging Cyberthreats and Innovative Defense Strategies

Chidozie Anadozie, Kushal Pokhrel, and Amjad Ali
Morgan State University, Baltimore, USA
amjad.ali@morgan.edu

Abstract
Intelligent transportation systems (ITS) are recently gaining attention in the transportation infrastructure, connecting vehicles, traffic and communication systems, thereby ensuring safer, more efficient and sustainable transportation. The increasing complexity and interconnectedness of ITS, however, create significant vulnerabilities, making them prime targets for sophisticated cyberattacks. The critical problem at stake is ensuring that as ITS continue to evolve, the pace of security measures must not lag behind, otherwise it could seriously compromise system integrity and user safety. This chapter presents a systematic review of recent ITS cybersecurity literature, supplemented with relevant standards and industry reports, to synthesize the emerging cyberthreats to ITS. It offers innovative defense strategies designed to secure ITS against prominent cyberattacks, including denial-of-service (DoS), sybil, and wormhole attacks. In addition, to make ITS secure, safer, and more resilient, the applications of innovative technologies, such as artificial intelligence (AI), blockchain, post-quantum cryptography (PQC), and game theory, are discussed. The detection of threats and their countermeasures can be realized in real-time through AI-based frameworks. Blockchain provides end-to-end data integrity and decentralized control, preventing the manipulation of critical vehicle data. Post-quantum cryptography addresses the protection of ITS infrastructure from future threats of quantum computers using quantum-resistant solutions. Game-theory formulations make efficient schemes of proactive

defense available, providing optimal response to cyberattacks. This chapter examines innovative defense measures and strategies to ensure ITS security in a rapidly evolving technology landscape of transportation systems.

Keywords: Artificial intelligence, Blockchain, Cybersecurity, Denial of service, Game theory, Post-quantum cryptography, Sybil attacks, Vehicle-to-everything.

2.1 Introduction

The intelligent transportation system (ITS) is a revolutionary movement in transportation management and urban transport optimization. It involves the integration of transport engineering and information technologies and electronics to enhance the transport system and minimize accidents (Elassy *et al.*, 2024). According to the European Directive 2010/40/EU, ITS covers a range of applications that provide advanced services in areas such as transport telematics, intelligent transportation systems, and intelligent vehicle projects related to various modes such as roads and railways (Lv and Shang, 2023; Pavão *et al.*, 2024). The objective of ITS is to provide people with more efficient, congestion free, and environmentally friendly transport services (Naeem *et al.*, 2024).

With smart cities becoming a reality, a significant evolution in ITS development is taking place, therefore relying on real-time communications and automated connectivity. The use of ITS in autonomous systems in vehicles, shared transport systems, and on-demand services has transformed urban transport by offering alternative transport options that are highly flexible (Ceder, 2021; Nikitas *et al.*, 2020). Vehicle tracking via GPS based on ITS data provides route guidance and traffic management services in terms of minimizing travel time and optimizing the overall commuter experience (Rocha *et al.*, 2021).

Modern transportation systems are highly interconnected, with various interdependencies on various technologies, including sensors, electronics, control, and communication (Oladimeji *et al.*, 2023). These technologies are noted to possess unmatched capabilities, including real-time information acquisition and predictive analytics. Nonetheless, there exist unchallenged weaknesses, including legacy and heterogeneous components, unsecured wireless protocols, and patching gaps, that amplify an attack surface favored by cybercriminals, leading to a corresponding increase in cyberattacks on the transportation system. Usama *et al.* (2024) reported a 48% significant increase in cyberattacks on transportation systems, coupled with an escalation in ransomware deployment. The transportation system ranked fifth among the international business machines (IBM) most attacked sectors in 2024, with a rapid increase in ransomware incidents, pointing to an escalation of ungauged vulnerabilities within transportation systems that potentially affect sustainability and business activities and put the lives of citizens in danger. The complexity and ubiquity of these cyberattacks have impelled investments in transportation cybersecurity, making it an absolute need.

Grand View Research estimates the global ITS market at USD 58.28 billion in 2025 and projects it to reach USD 128.17 billion by 2033, representing a CAGR of 10.8% from 2026 to 2033. North America accounted for 34.9% of global revenue in 2025. By mode of transportation, roadway represented the largest share in 2025 (39.0%). These market trends are illustrated in Fig. 2.1. An escalating growth in these markets necessitates a secure economy, prompting continuous investment in ITS global markets.

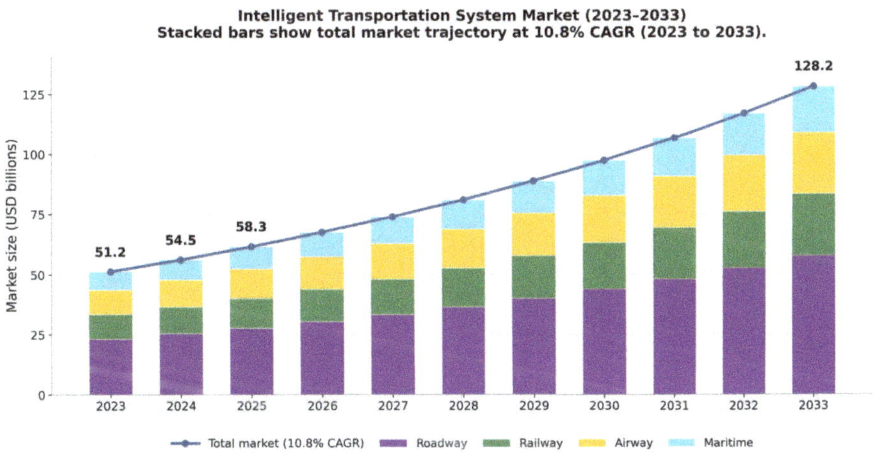

Fig. 2.1. Intelligent transportation system market (2023–2033).

Source: Author-generated plot; underlying market estimates adapted from Grand View Research (2025).

With the growing trend of cyberthreats, prevention techniques alone cannot sustain cybersecurity. Experts therefore advocate for a movement towards greater resilience within transport systems, enabling these systems to continue functioning when under attack and then quickly recover with minimal disruption to services. Current research has focused on the potential future trends between the requirements within the ITS environment that should ensure the systems involved are able to absorb and recover from the attacks while performing other basic functions (Alhidaifi *et al.*, 2024; Verma *et al.*, 2025). Achieving cyber resilience within the ITS environment has been challenging, considering the transport systems are highly interconnected, stretching from sensor devices to infrastructure, and leaving each device exposed as a vulnerable network endpoint.

Moreover, because of the dynamic nature of traffic and transit operations, static measures on ensuring system security may sometimes find it challenging to keep pace. As traffic flows, changes in user behavior as well as other surrounding environments are dynamic, with systems required to adapt through methods of exploitation. This calls for more progressive measures in securing a system capable of defending itself. It is in this context that this chapter will explore the emerging threats of cybersecurity to ITS systems, the vulnerabilities inherent in the architecture of the ITS system and explore innovative measures that can be put in place to ensure enhanced security of the system.

2.1.1 *Methodology*

This chapter's scope was defined through a systematic literature review following PRISMA guidelines (Preferred Reporting Items for Systematic Reviews and Meta-Analyses). A comprehensive search of relevant literature was conducted across key databases, including IEEE Xplore, ACM, SpringerLink, and Google Scholar, for publications published between 2020 and 2025. Search strings were developed through combinations of keywords and Boolean logic pertinent to both intelligent transportation systems and cybersecurity. Such as ITS, intelligent transportation system, security and cyber threats, along with specific phrases such as autonomous vehicle cybersecurity, blockchain and transportation, post-quantum cryptography (PQC) transportation, machine learning, deep learning,

V2X security, ITS security and game theory ITS security. These phrases were considered to capture both general concerns for ITS security, as well as the novel defense technologies that will be considered in this chapter (AI, blockchain, PQC, game theory). We also manually scanned the reference lists of recent survey papers for additional relevant studies.

The initial database search yielded approximately 780 records. After removing duplicate results, about 590 unique records remained. The titles and abstracts for these 590 records were then analyzed for relevance to ITS cybersecurity and defense technologies, leaving about 140 for full-text analysis. Each full-text document was then analyzed using inclusion criteria that included relevance to cyber threats in ITS, discussion of new defense methodologies, and relevance to the main themes for this chapter. Based on full-text analysis, a total of 85 references for this chapter were selected.

In summary, Intelligent Transportation Systems are highly efficient in terms of improving the efficiency of the transport sector with better connectivity capabilities, but at the same time, the degree of interconnectivity for such systems has also expanded the risks of possible cyber attacks. This represents a trade-off between embracing cutting-edge technologies for smart mobility and managing the resulting increase in cybersecurity vulnerabilities. Looking ahead, one of the biggest concerns for the efficiency of ITS systems is the ability to implement strong security systems for such systems to withstand the growing complexities of the systems with rapid advancements in the form of possible cyber attacks.

2.2 Intelligent Transportation System Architecture

Within this section, the basic architecture of intelligent transportation systems is introduced, breaking it down into its main layers and discussing how each layer contributes to ITS functionality. Intelligent Transportation Systems are an advance in transport management as they optimize traffic flow, enhance road safety, and drive sustainability. As an integrated system that leverages cutting-edge technology, ITS provides a symphony system that aims to maximize mobility, eliminate human errors, and mitigate risks to the environment (Elassy *et al.*, 2024). The primary objective of ITS architecture

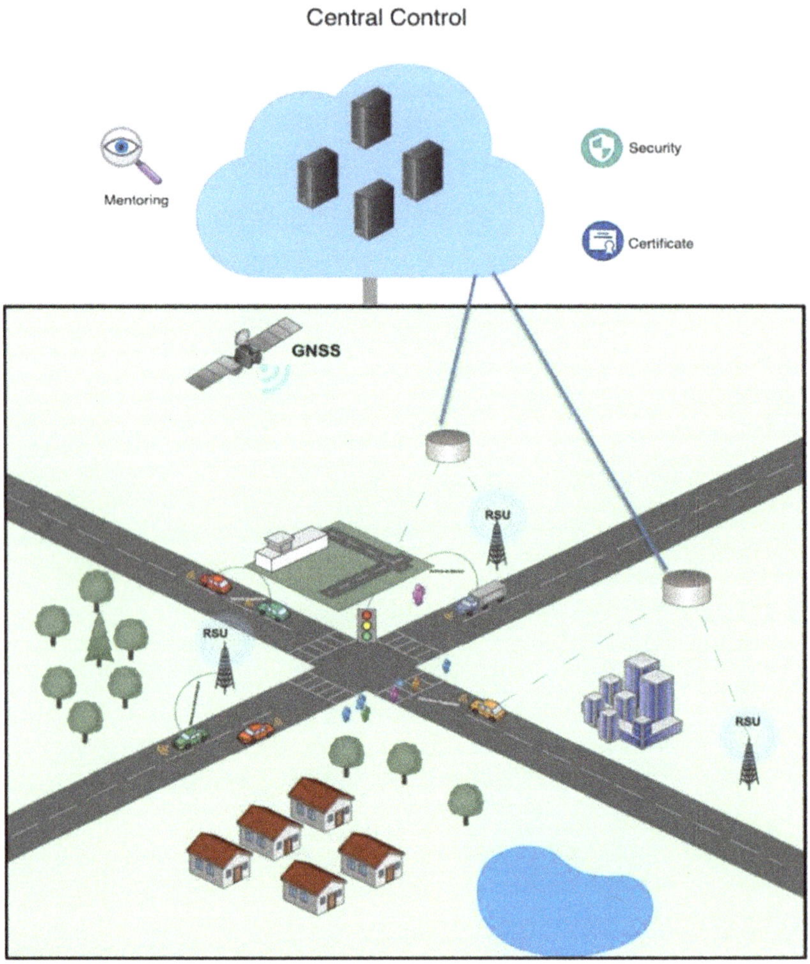

Fig. 2.2. Overview of intelligent transportation system architecture and communication pathways (vehicles, roadside units, traffic management/cloud, and V2X links).

is that it must be dynamic, seamless and interact with heterogeneous elements, thereby allowing specific functionality in real-time decision-making (Avcı and Koca, 2024), as shown in Fig. 2.2.

The ITS architecture consists of four primary layers: the perceptive layer, network transmission layer, database layer, and application layer, as shown in Fig. 2.3. These layers enable the overall processing and response of information in the transport network.

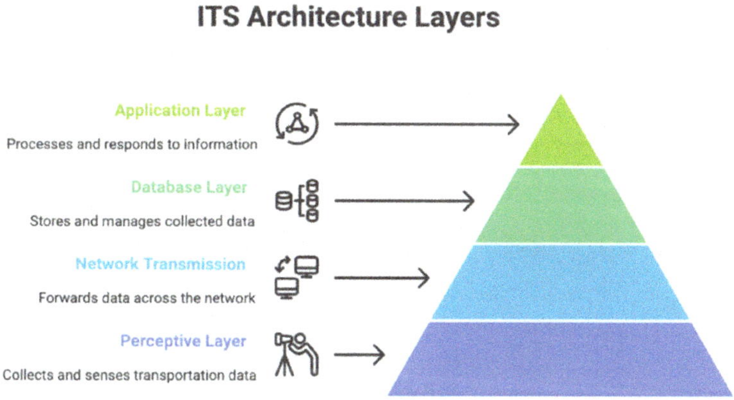

Fig. 2.3. ITS physical architecture illustrates a layered approach (perception, network transmission, database and application/control layers).

Source: Figure created by the authors using napkin.ai.

(a) **Perception layer:** This is the lowest level of the ITS and is made up of sensors, cameras, GPS, radar, light detection and ranging (LIDAR), and sensors on the roadside units (RSU), which are fitted on vehicles and along the roadsides. They provide the information on inputting data relating to the environment, such as traffic flow, vehicle speed, and road conditions (Lin *et al.*, 2023). The perception level is the starting point and is the first real-world interface where data is collected.

(b) **Network transmission layer:** This layer acts as the relay for the data transmitted from the perception layer to the concerned parties liable for dealing with the exchanged data. It makes use of dedicated short-range communication (DSRC), 5th generation (5G) cellular technology, Wi-Fi, and other V2X communication for vehicle to vehicle (V2V), vehicle to infrastructure (V2I), and infrastructure to infrastructure communication (I2I). The use of real-time communication facilitates convenient traffic management in addition to first response during emergencies or traffic obstruction (Vamsi *et al.*, 2024).

(c) **Database layer:** This layer provides the functionality to store and replicate the data input originating from the components of the ITS. The network transmission layer communicates the data input, and the database processes the historical and real-time data for storage and retrieval. These data form a base for decision operations and optimize traffic flow and predictive mobility requirements in the long run (Zeddini *et al.*, 2022).

(d) **Application control layer:** At the top of the ITS architecture sits the application control layer, where information is processed to provide decisions that can be used for actionable guidance. The application control layer includes traffic control centers, onboard vehicle systems, as well as cloud computing platforms that receive information on various mobility aspects such as signal adaptive signal timing, congestion-based diversion, and cooperative vehicle automation (Borcoci *et al.*, 2018). This is where the ITS is made to be responsive to traffic conditions while ensuring a well-managed transportation system.

The complex interdependency relation among various layers of ITS can lead to one layer acting as a trigger for other layers. For instance, if there are spoofing and jamming attacks on the network transmission layer, they can spread across the network and influence decision-making at the application layer. There are many vendors and jurisdictions that are involved in the application and use of ITS. These interdependencies add to complexity and increase the potential for attacks (Yang and Zhu, 2024). For example, weaknesses within the network communications that support data exchange between automobiles and between road-side facilities and traffic control centers can be attacked by intruders. If attackers breach network security at the network transmission layer, they can modify real-time traffic information, and as such, making improper decisions based on false data can lead to inefficient traffic management, or even accidents. Moreover, as most systems within ITS have become networked and geographically distributed, ensuring security across the entire system becomes even more challenging.

In short, the ITS architectural structure with its multi-layer structure from perception to application makes real-time operations possible with a greater degree of robustness but also creates complex interdependencies. The trade-off is that while multiple layers and diverse technologies improve transportation efficiency and capabilities, they simultaneously broaden the attack surface and complicate security management across the system. An open challenge moving forward is to ensure that each architectural layer, along with the interfaces between them, is secured against cyber threats without compromising the seamless operation and interoperability that ITS demands.

2.3 Cybersecurity Vulnerabilities and Threats to Intelligent Transportation Systems

This section introduces the cybersecurity vulnerabilities of modern ITS and the main cyberattacks that exploit these weaknesses. We examine the key vulnerabilities within the various levels of the ITS and then outline the emerging cyberattacks categorized by their impact on the availability, authenticity, confidentiality, and integrity of the ITS.

A myriad of cyberthreats has thus become feasible with the complexity of ITS and increased integration of real-time communications and data. These cyber threats can range from attacking various tiers within the ITS structure, including in-vehicle networks and the cloud-based data analytics platform. As the ITS continues to develop, offering real-time and data-driven solutions to urban mobility challenges, it becomes increasingly susceptible to a wide range of cyber threats. With greater connectivity among components of the ITS ecosystem, from vehicles to roadside devices, and from sensors to clouds and communication technologies, various cyber exploits become possible. Although these technologies improve transportation efficiency, they have fundamental operational flaws that cyber attackers seek to exploit (Rathore *et al.*, 2022; Thomas *et al.*, 2025). Today's cyberattacks more than ever rely on automated penetration, social engineering, and exploits of vulnerabilities in legacy Supervisory Control & Data Acquisition (SCADA) systems and other Internet of Things (IoT) devices, thereby creating the risk of cascading effects within critical infrastructure. Disruption of any single infrastructure sector, such as the energy sector, could lead to paralysis of other downstream infrastructure sectors such as Healthcare, Transportation, Supply Chain, and Public Services (Okur and Dener, 2025).

2.3.1 *In-Vehicle Vulnerabilities and Threat*

Increased reliance on real-time data transfer and communications for ITS has exposed the system to a world of cyber threats targeting various elements of the ITS infrastructure. The idea began with car-to-car networks based on the Controller Area Network (CAN) bus to facilitate communication among the myriad electronic control units (ECUs) of a car (Oladimeji *et al.*, 2023). Car-to-car networks based on the CAN bus are easily susceptible to cyberattacks because the system was designed without modern security mechanisms such as confidentiality, authentication, or privacy protection, leaving buses vulnerable to eavesdropping, spoofing, and message alteration attacks (Ouaissa *et al.*, 2024). The nature of the potential attack involves the installation of a malicious ECU by the attacker on the bus designed to broadcast malicious signals across the network to create chaos. The absence of confidentiality measures leaves the buses susceptible to eavesdropping by malicious sources intending to intercept vital information transmitted among the ECU units.

On the other hand, the CAN bus uses a priority message arbitration mechanism, with the highest priority accorded to messages with smaller identifiers. This implies that the CAN bus is susceptible to denial-of-service (DoS) attacks, in which an attacker can exploit the bus's arbitration mechanism by flooding it with forged high-priority messages, thereby delaying or dropping legitimate messages (Rogers and Rasmussen, 2022). This type of resource shortage can potentially disrupt a vehicle's critical systems. Furthermore, because the in-vehicle bus network is interconnected, a successful attack on an ECU can potentially jeopardize an entire vehicle, thereby threatening road safety. For instance, Iehira and Inoue (2024) conducted a combined bus-off attack as well as a spoofing attack that successfully disrupted the communication of legitimate messages on a vehicular CAN bus.

Besides the CAN, there are other in-vehicle bus systems, namely FlexRay and Local Interconnect Network (LIN), which are also vulnerable owing to the absence of built-in security measures. FlexRay, which operates in more advanced vehicles in time-sensitive systems, has the same vulnerable security points as the CAN (Rathore *et al.*, 2022). Just like the CAN, the FlexRay lacks encryption, authentication, or privacy mechanisms. There has been a proposal of a lightweight authentication scheme on the FlexRay using a modified version called the lightweight CAN authentication protocol (LCAP) as a mitigation against possible attacks. The Local

Interconnect Network, which operates in less critical areas like window and seat adjustment, can also be breached. Katsikeas *et al.* (2022) identified the possible injection of fake messages within the system through the error recovery procedure of the LIN Network.

2.3.2 *Vulnerabilities in Wireless Communication Technologies*

Wireless communication interfaces are spreading rapidly into vehicles, such as dedicated short-range communication systems (DSRC/WAVE), Cellular V2X solutions, Bluetooth, and Global Navigation Satellite Systems (GNSS). These interfaces provide several benefits but also pose a substantial security risk. Wireless communication systems create a novel attack surface that enables an external aggressor to engage with a closed internal vehicle network without any physical entry into the vehicle.

The Vehicular Ad Hoc Networks (VANETs) rely on IEEE 802.11p to enable a short-distance communication network among vehicles and between vehicles and the infrastructure. Although 802.11p is widely used in VANETs, it can be compromised by attackers. The use of omnidirectional antennas in IEEE 802.11p communication, for example, can compromise the network and make it susceptible to a jamming attack where the attacker sends interfering messages within the communication range to hinder communication (Saad *et al.*, 2017). This problem still persists despite the adoption of IEEE 802.11-OCB.

Similarly, the adoption of cellular networks (LTE/LTE-A and emerging 5G-based C-V2X) not only expands connectivity in the automotive industry but also poses risks such as the ability to read and interpret messages (eavesdropping), forge messages (spoofing), and a distributed Denial of Service (DDoS) attack where the attacker can hamper the intended communication in the vehicles (Wei *et al.*, 2024). Bluetooth technology, a frequently used short-distance communication method in vehicles, can be accessed, modified, and even halted by DoS and message-tampering attacks (El-Rewini *et al.*, 2020). Likewise, the GNSS system in vehicles can be easily jammed and spoofed by an attacker transmitting a slightly stronger signal. Therefore, the attacker can further compromise the weak signal and hinder the positioning and timing information (Radoš *et al.*, 2024).

2.3.3 *Software Vulnerabilities*

In modern vehicles, there are often more than 100 million lines of software code that control a wide range of networked systems, from infotainment to safety systems. Vulnerabilities in software, such as buffer overflows, code injection bugs, and inadequate access controls, are not uncommon in modern vehicles, allowing attackers to gain control over vehicle functions (Aslan *et al.*, 2023). For example, a buffer overflow vulnerability can allow a hacker to inject malicious code into a vehicle's systems, for which there is no easy solution for the manufacturer, as they cannot release security updates like software companies often do. In view of this, a hacker could potentially disable safety systems or even steal personal information stored in the vehicle through those software flaws. Also, inadequate input validation can result in attacks on the vehicle's infotainment system using SQL injection or XSS attacks. Interestingly, there was an XSS vulnerability found in Tesla Model 3 software that allowed a hacker to gain access to the users' private information (Xie *et al.*, 2023).

The addition of external communication interfaces and sophisticated software in contemporary vehicles has resulted in a broader attack surface. If these vulnerabilities are successfully exploited, they pose critical safety and privacy risks to vehicle occupants. It is essential that in-vehicle networks, including CAN, FlexRay, and LIN, as well as wireless communication systems such as IEEE 802.11p, Bluetooth, and GNSS, be secured to maintain vehicle safety in the increasingly interconnected world. In addition, mitigating software weaknesses through secure design and providing continuous updates is vital in preventing malicious actors from exploiting bugs in automotive software.

2.4 Emerging Cyberattacks Against Intelligent Transportation Systems

While technological innovation in ITS has greatly improved the management of traffic flow and road safety, it has nevertheless introduced some level of risk. Cyberattacks in the context of ITS have the capacity to threaten the safety as well as the privacy of road commuters as

Fig. 2.4. Classification of emerging cyber threats to ITS by impacted security objective (availability, authenticity, confidentiality, integrity) with representative attacks.

Source: Figure created by authors using napkin.ai.

shown in Fig. 2.4. As vehicles and infrastructure become more interconnected, the dangers of cyber-attacks against ITS infrastructure grow, making the system vulnerable to attacks targeting communications infrastructure and critical vehicle operations (El-Rewini *et al.*, 2020; Vellinga, 2022). Such cyberattacks can be classified according to the threat to the system's availability, authenticity, confidentiality, and integrity.

2.4.1 *Attacks on Availability*

Warding off attacks against the availability of resources in ITS services leads to disastrous results in terms of network downtime. This is due to the real-time services running in most ITS systems, which may fail due to such attacks, thus affecting related services such as traffic flow and emergency services in general. One common threat here is the DoS attack, in which an attacker obstructs the normal running of a network by flooding it with traffic. Such an attack could result in disastrous consequences in vehicular communication networks, such as delaying the dissemination of messages or accidents occurring in life-critical applications (Tariq *et al.*, 2023). The methods that have been adopted to prevent DoS attacks include developing enhanced network protocols. A proposed solution to mitigate DoS attacks in vehicular communication networks applies a puzzle-based co-authentication (PCA) scheme that delays attackers' ability to forge certificates, thereby lessening the impact of DoS attacks. Another innovation to prevent DoS attacks in vehicular communication networks involves applying a reCAPTCHA-like solution in filter actionable DDoS traffic proposed by Basuli and Padhya (2024).

Regarding notable trends in 2024, there was an evident rise in DoS and DDoS attacks, most of which targeted critical infrastructure such as sectors within the energy, transportation, and public administration sectors. In an extreme attack of this kind, the public transportation in Poland was affected by a DDoS attack, leading to delays across the country as operations were brought to a standstill (Swathika *et al.*, 2024). Figure 2.5 presents a schematic overview of denial-of-service (DoS) and distributed denial-of-service (DDoS) attacks targeting Intelligent Transportation Systems (ITS), illustrating the associated attack vectors, operational impacts, and mitigation categories documented in the literature.

Another type of availability attack is a black hole attack. Here, the attacker disguises itself as a legitimate node and uses the routing mechanism to redirect data traffic to the malicious node, thereby forming a black hole. In this attack, malicious nodes receive data packets intended for other nodes but discard them intentionally rather than routing them to their destinations, thereby reducing network performance (Reshi *et al.*, 2024). Black hole attacks are challenging to detect, as the attacker may periodically drop packets,

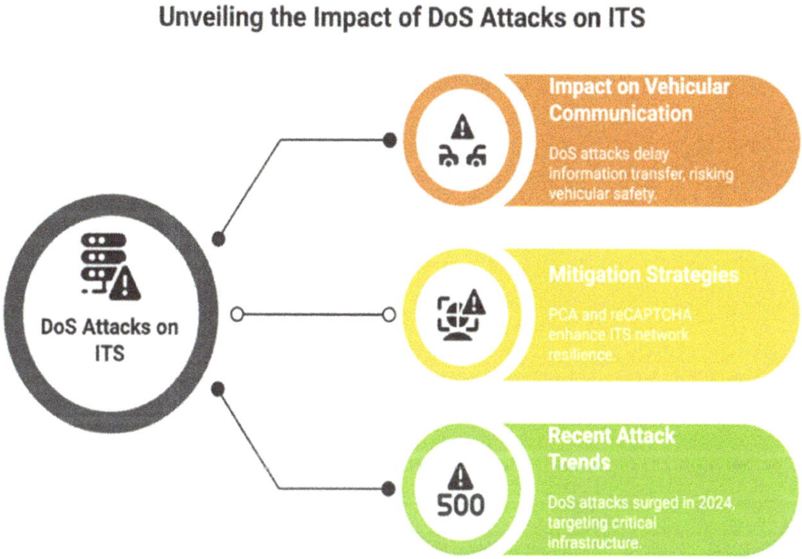

Fig. 2.5. DoS/DDoS attacks on ITS: typical attack vectors, operational impacts, and mitigation categories reported in the literature.

Source: Figure created by authors using napkin.ai.

which do not trigger standard monitoring mechanisms. Moumen *et al.* (2023) put forward a multi-stage defense mechanism against black hole attacks using route backtracking and blacklisting malicious nodes.

Malware attacks also pose a very real threat to availability. Viruses, worms, and Trojan horses can infect vehicle systems because of the high level of software used within a vehicle. When there is a high level of software involved, such as millions of lines of code for communication and vehicle control, a vehicle is vulnerable to a malware attack. The entry points for a vehicle may be telematics systems, entertainment/infotainment systems, or on-board diagnostic (OBD) ports.

These systems can be used to gain access to the vehicle system if they are compromised. Once inside, the malware may cause system malfunctions or take direct control of a vehicle system. In 2024, hacktivists sponsored by a Russian state actor were behind one of the most sophisticated and destructive attacks on Ukraine's energy and transportation sectors, where they used tools like Kapeka to make critical transportation services unavailable (Pollard, 2024). This attack had a noticeable effect on the resilience of the targeted infrastructure, leaving systems non-functional and their capacity in ruins. During roughly the same timeframe, there were ransomware attacks on the power and water utilities sectors, following a pattern seen on US infrastructure such as public transportation and water utilities. Incidents were observed to cluster in short succession, a serious concern for critical infrastructure (Skomra and Wojtasik, 2025).

2.4.2 *Attacks on Authenticity*

The goal for attacks on authenticity is for unauthorized entities to be able to masquerade as legitimate members within a network. A well-known attack is a Sybil attack, where an attacker creates multiple forged identities, commonly referred to as Sybil nodes, in an attempt to pose as multiple vehicles (Al Ani, 2024), as illustrated in Fig. 2.6. It becomes difficult for these attacks to be discovered, especially in a VANET, owing to their highly distributed and mobile environment (Ali *et al.*, 2025). Using multiple forged identities, an attacker could use these for a variety of purposes, for example, creating phantom traffic jams for directional purposes, while also using a DDoS attack for spam messages from many nodes.

Various detection schemes have also been proposed for combating Sybil attacks, but some of these detection schemes are not fully capable of combating the issue because of technical limitations. The identity registration schemes can be flawed if the attackers succeed in registering several identities for malicious purposes. The position verification techniques (using GPS or radio signals) are also inefficient because vehicles are quite agile on the roads, and the attackers can easily manipulate the signals. To overcome the above limitations, Baza *et al.* (2020) introduced a new efficient approach by amalgamating game theories, Proof of Work (PoW), and allocation verification techniques to resist Sybil attacks with low latency. In their approach, vehicles essentially play a proof of identity by solving hard computational puzzles (requirement of PoW); these are then checked for authenticity by analyzing the validity of signed timestamped tags associated with valid solutions. This prevents attackers from flooding the network with useless identities because only vehicles willing

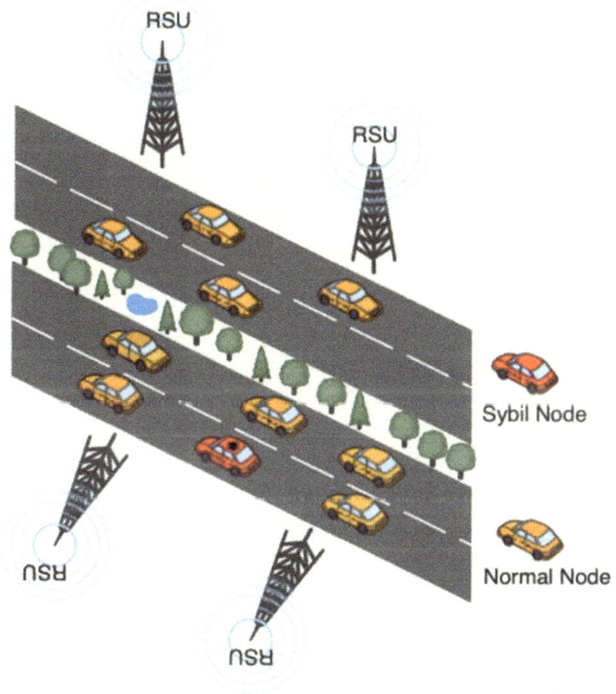

Fig. 2.6. Sybil attack in vehicular networks: a malicious participant presents multiple forged identities to influence network trust, routing, or traffic decisions.

to dedicate computational power (assumed to be trustworthy members) gain access to the network at the end of the day. The potential weakness of the approach still involves the computationally expensive nature of the PoW concept for vehicles at the end of the day because of high power consumption. To overcome the power consumption issue associated with the previous approach, Li (2024) recommended the application of the Proof of Stake or hybrid PoW/PoS protocols for the same purpose to minimize power consumption without compromising network security at the same time.

Also emerging as a challenge affecting authenticity within the network is the wormhole attack, which happens when the attacker forms a false connection between two remote regions within the network, essentially routing packets through an out-of-band channel.

The wormhole creates the impression that two nodes that are quite far are actually close and affects the routing protocol within the network (Bhatti *et al.*, 2024). The wormhole allows attackers to intercept packets over a large area. For dealing with wormhole attacks, a new protocol has recently been proposed by Hameed and Saini (2024) to defend against such attacks in VANETs. According to their defense mechanism, it becomes possible to defend against both colluding and cooperative wormholes. These colluding attacks include cases where more than two nodes collude with each other to increase the range of the malicious link.

2.4.3 *Attacks on Confidentiality*

The purpose of a confidentiality attack is to covertly obtain any sensitive and privileged information from the ITS. Within the ITS network, there is valuable information that, if disclosed, would result in breaches of privacy and secure communication channels between vehicles and infrastructure. The main attack on confidentiality is eavesdropping, which involves secretly listening to communications either among vehicles or among vehicles and roadside units (RSUs). Because many communications on VANETs are broadcast publicly, attackers only need to listen to obtain information. Eavesdropping is constructive for mounting sophisticated attacks like masquerade attacks and location tracking attacks. The effects on vehicle users would indirectly result in loss of privacy and privacy of travel habits. Confidentiality attacks in ITS would be addressed through the use of encryption and secure communication channels on V2V and V2I communication that would secure messages beyond interception and comprehension by any unauthorized party (Tariq *et al.*, 2023).

2.4.4 *Attacks on Integrity*

In integrity attacks, the emphasis is on altering or injecting wrong information into the ITS communication streams with the purpose of misguiding systems or users. Such attacks on data integrity can lead to dangerous situations. A typical example of an integrity attack is the replay attack, in which an attacker records valid communication packets and retransmits them later to mislead the system. In this attack, systems can end up accepting outdated or wrong information as if it were up-to-date information, an example of which is impersonating vehicles or RSUs (replaying their authentication messages). To mitigate this attack, using timestamping or sequence numbers and implementing timely constraints (e.g., checking whether the message timestamp is recent) can effectively mitigate this attack (Chandrasekaran *et al.*, 2020). Operational details on how the attacks can be performed based on the use of synchronized clocks in the ITS are as follows: Timing attacks rely on the use of synchronized clocks in the ITS. For example, an attack may be mounted by introducing delays or simulating the arrival of timing messages. This can impede various systems that rely on time-critical synchronization, such as platooning or signal control. Liu *et al.* (2025) in 2025 proposed methods to defend against these timing attacks. Their strategies include the careful design of system modules, including array accesses that provide known delays in the microarchitecture, and monitoring for unusual timing delays or jitter to detect malicious interference.

In light of growing connectivity and advanced communication systems in vehicles, it is important to counter attacks on integrity to ensure they do not happen. A particularly novel integrity focused threat is data poisoning, where malicious or false data is injected into a system to corrupt its decisions. As a result of the growing incorporation of AI and ML in ITS, such as traffic prediction, signal control, or autonomous vehicles, such systems are highly dependent on accurate input data. A poisoning attack can be done during the training stage or during real-time execution for AI models (Wang *et al.*, 2024). For instance, an attacker can launch such an attack during the training stage of an ML-powered crash detection or traffic prediction system, where they could manipulate the input data, such as adding erroneous sensor readings or image labels to create errors in such a system when deployed. Similarly, overloading a real-time traffic management system with erroneous vehicle traffic counts could result in such a traffic management system failing to make appropriate decisions, leading to traffic inefficiency or gridlocks (Fernández Llorca *et al.*, 2025). This attack is more sinister in that the system, despite its failure, will continue to operate normally in the background without immediately alerting users to the fact that it has been attacked until a serious failure has occurred.

Defending against data poisoning attacks in ITS requires securing many layers of the system. This includes strict data validation and filtering for any anomalies, securing data points for gathering data from potential sources such as sensors against any kind of tampering, and also the use of robust algorithms that are less sensitive to anomalous or outlier data. Crowdsourced traffic data and even maintenance data for smart transport analytics are also prone to data poisoning attacks if they are not properly secured, as described by Girdhar *et al.* (2023). Since data for ITS directly affects the safety and security of various citizens, their security and integrity are most essential. Data poisoning attacks are identified as part of other information attacks called deceitful information attacks and require much attention for future development at least as mentioned by Diro *et al.* (2025).

While more dynamic and damaging types of threats such as DDoS, Sybil, and wormhole attacks are being observed, less attention and, in fact, growing threats in the form of passive eavesdropping and data poisoning have emerged. An ITS that involves the use of real-time data communication can be easily affected and compromised by the above-mentioned new and emerging threats. To counter these new threats, a diverse and dynamic approach to cybersecurity strategy is required. Such a strategy should involve conventional measures and the latest technologies and innovations in the form of AI-powered intrusion detection and systems resilience testing. The evolving threat landscape underscores the need for continued research, innovation, and cross-sector collaboration to ensure ITS remains secure against both current and future cyberthreats.

2.5 Defense Strategies Against Cyber Threats to Intelligent Transportation Systems

This section discusses the more advanced defense strategies to secure the infrastructure as well as the vehicles against the cyber threats mentioned briefly above. With the evolution of ITS systems, their defense mechanisms against various major threats such as Sybil attacks, Wormhole attacks, and DoS attacks should also be more evolved to make way for novel defense techniques. The limitations of various defense techniques against such dynamically increasing threats present in the natural complexity of ITS systems make this section focus on four novel defense techniques: AI/Machine learning (ML) defense systems, data integrity with Blockchain technology,

PQC systems to secure against potential threats of the quantum world, and game-theory-based adaptive defense systems.

2.5.1 *AI-Based Defense Strategies in ITS*

Artificial intelligence, especially ML, is an integral part of the enhancement of ITS cyber defense systems. By enabling real-time reactiveness, uninterrupted malicious activity observation, and autonomous network defense fortification, AI-based systems were found to be very effective in recognizing unusual behaviors, pinpointing likely attacks, and triggering autonomous responses. These systems are very crucial in ensuring uninterrupted ITS functionality and data communication without human mediated delays, thereby increasing the level of automotive safety.

For instance, DoS attacks target network services, especially communication channels, through massive data flooding, generating network bottlenecks that can trigger network failures. This is particularly risky in an ITS network, which is extremely dependent on real-time communication between the vehicular environments. In this regard, a reinforcement learning algorithm was proposed to detect DoS attacks. This technique uses a Q-learning agent that continuously observes traffic in real time and identifies patterns indicative of a DoS attack. The agent examines network data traffic, scrutinizing data validity, thereby eliminating malicious packets before they actually cause network disruptions. This AI-enabled technique minimizes false alarms, ensuring high accuracy in recognizing attacks, making this technique a very effective measure in preventing DoS failures in an ITS network.

Basnet proposed an AI-based intrusion detection system for Electric Vehicle Charging Station (EVCS) under DoS attacks in 2022. The author argues that despite AI's effectiveness in detecting these types of attacks, it remains susceptible to innovative or stealthy forms of attacks. Basnet's work underlines the use of diverse and updated training sets to match AI's adaptability to innovative methods of attacks, which requires complex deep architectures in many cases.

Another solution proposed by Wang *et al.* (2025) targets vehicular networks' DoS mitigation via a vehicular communication trust model. This model is a Quality-of-Service (QoS) scheme that focuses on vehicular network nodes' trustworthiness regarding DoS mitigation.

The proposed AI model includes an automatic trust downgrading mechanism that automatically downgrades nodes showing any form of suspicious activity related to network DoS attacks, such as attempting to overload the network with data packets, and thus ignores them in network communication sessions without significantly affecting network performance due to malicious data transmissions. This solution seems quite valuable in DoS attacks in EVCS, though it has a limitation in that it remains unproven in highly connected vehicular networks and should therefore have improved AI-based trust calculation methods that do not hamper network performance due to complex trust evaluations. Pasupathi *et al.* (2025) utilize a network-based EVCS DoS defense technique that assigns special markers to packets transmitted in vehicular communications to detect DoS attacks by filtering any DoS-related packets in EVCS networks. This solution also appears highly effective at detecting DoS attacks in EVCS, based on projected network performance, yet it remains only resistant to AI-based methods that attempt to evade markers of DoS attacks.

Another major threat in ITS is the blackhole attack, in which an unauthorized node receives data and refuses to forward it, thereby inhibiting communication between the source and destination. This might cause loss of essential information with respect to location or road conditions, posing grave concerns regarding safety. Olanrewaju *et al.* (2023) designed an improved Ad-hoc On-demand Distance Vector (AODV) protocol to overcome blackhole attacks. In this optimized protocol, cryptographic functions are employed to authenticate packets, hence allowing only legitimate nodes to transfer information. It also anticipates verifying both source and destination nodes to participate in routing protocols. Thus, blackhole attackers are denied the opportunity to intervene in communication, thereby securing routing operations and ensuring safe data transfer in ITS.

Researchers have also observed that Automatic Intelligent Systems (AIS), which comprise AI components within vehicles, have the ability to apply preventive strategies based on the processing of sensor information to remove threats in real time. An AIS has the ability to apply personalized strategies to the vehicle in response to external dangers such as an unexpected barrier, roadblocks, or erratic motions in other vehicles, thus securing both the vehicle and its passengers. Even so, due to the high volumes of data produced from the sensors within the vehicle, high latency and computations could result in increased processing time from AI components within the vehicle, despite their capabilities in improving safety within ITS systems (Katiyar *et al.*, 2024). Edge computation and federated learning have been presented to eliminate the high latency associated with processing data from sensors within the vehicle.

Although AI-defenses are likely to significantly improve the security of ITS, there are also some issues related to these technologies. These include the need to provide instant responses to threats without introducing undue latency, dealing with the large volume of data being generated from the various sensors used in the ITS system, and keeping the models upgraded to respond to these threats. There is also the danger that more sophisticated attacks will be used against the models, such as adversarial attacks. Combining AI, such as in the use of blockchain for data, PQC communications, or processing in edge computing, may provide a more secure ITS.

From a practical standpoint, there is a need for a balance in implementing AI-based defense systems in ITS systems based on their detection efficiency. Realistic ITS deployments necessitate low latency detection, where AI algorithms should detect intrusion or anomalies almost instantly, preventing potential hazards, though this may be difficult owing to computational requirements for some ML algorithms. Furthermore, AI systems should be resistant to false positives; a highly sensitive intrusion detection system that tends to raise frequent false alerts about normal traffic activity may cause disruptions in transport services. Alternatively, non-cloud deep learning inference infrastructure for edge computing may be needed for implementing AI-based systems owing to their computational requirements, though this infrastructure can also be used for specialized hardware, for example, graphics processing units (GPUs) or tensor processing units (TPUs), for AI implementation. Finally, AI-based defenses consume data and must be updated frequently as new attack methodologies emerge, which poses logistical challenges when deploying updates across a vast number of vehicles and infrastructure nodes.

We, the authors of this chapter, built and tested the effectiveness of a lightweight AI-driven intrusion detection system (IDS) within a simulated ITS setting. The experiment combined the network simulator NS-3 and the traffic simulator SUMO to simulate attack scenarios against connected vehicle networks. The AI-driven IDS, utilizing multiple ML algorithms to detect intrusion attacks, including spoofing and DoS attacks, was built and included an automatic response module capable of instantly isolating malicious nodes within the network. The detection speed and accuracy of the AI-driven IDS proved highly beneficial in an ITS setting, accurately detecting attacks such as coordinated DDoS flooding and spoofing of traffic sensor readings within milliseconds, with minimal false positives, while automatically containing threats via communication routing or engaging vehicle safety systems.

Yet, the study did not escape the notice of the fact that while the use of AI-driven technologies would greatly improve the resilience of ITS networks, there would still exist concrete implementation realities, including the fact that processing requirements would increase substantially for each vehicle node within the simulated network, thus typically requiring upgrades and the ability to perform edge processing. Moreover, the detection of novel attack patterns, termed zero-day attacks, would not yet be possible, and the best way for the AI-driven IDS would still be within attack patterns for which the ML was trained, thus underlining the need for constant updates to the ML-driven model.

2.5.2 *Data Integrity and Secure ITS Communication Using Blockchain Technology*

Blockchain is a tamper-resistant, decentralized ledger solution that ensures end-to-end secure communication, integrity, and transparency in intelligent transportation systems (ITS). The decentralized nature of blockchain eliminates a central control point, thereby resisting single-point failures, data tampering, and privacy attacks, which are common in centralized systems (Essaid and Ju, 2025). This technology can be suitably deployed in security mechanisms in V2X communication systems, in which data is transferred between vehicles and infrastructure. Blackhole attacks, in which malicious nodes discard packets, have been a serious concern in vehicular ad hoc network systems. It was proposed by Inedjaren *et al.* (2021) that using blockchain security mechanisms in VANET routers can counter such attacks. In a blockchain-secured system of V2V communication, all transfers of data are authenticated and logged in the form of irreversible entries in a blockchain. The unalterable nature of a blockchain protects data integrity, as a malicious entity cannot manipulate previous transmissions or outrun the network's consensus. Thus, if any entity acts as a black hole, this can also be traced because the absence of data in the blockchain-secured communication is authentic, as data forwarding is obviously absent in the blockchain, allowing trust withdrawal from the malicious entity (Inedjaren *et al.*, 2021).

Zhang *et al.* (2024) analyzed the application of blockchain smart contracts for improving the security of ITS. Smart contracts, for instance, are programs running on blockchain technology, and they have the ability to enforce a reputation-driven access control scheme for vehicles and infrastructure nodes. For example, usage policies can be embedded, and any node not following the rules of the protocol, established based on on-chain information, will have its access rights revoked instantly. In this way, this scheme, through the distributed ledger, establishes a decentralized trust mechanism, where based on the distributed ledger, no node or entity will have the ability to change the rules, and all nodes will have the ability to verify the history of the actions. Zhang *et al.* specified that this strategy resists attacks of (data) tampering and fake data attacks since all the important traffic, location, and diagnostic information will be recorded into an immutable ledger and attempts for spoofing and tampering with this information will clearly fail (Zhang *et al.*, 2024).

A system called DrivMan was proposed by Sheikh *et al.* (2025), who constructed an in-vehicle management system based on blockchain that combines physically unclonable functions and public key infrastructure (PKI) for automotive identity verification. Here, each car has a physically unclonable function (PUF) derived key that is extremely hard to forge, and these keys are verified through blockchain transaction verification that is secured through public key infrastructure. The system resists imitation attacks or Sybil attacks much more thoroughly because the imitation key cannot fulfill the verification based on PUF on the blockchain. Sahih *et al.* (2024) also proposed a peer-to-peer trading system of resources for cars based on blockchain in 2024, where, although not a security system, they were able to verify that it is possible to record the transaction securely on the decentralized system.

Blockchain technology has also been considered an area with potential in securing newly proposed ITS services like peer-to-peer parking or tolling solutions, where personal data (location and identity) has to be securely protected. New cryptographic solutions, such as zero-knowledge proofs and attribute-based encryption in blockchains, can also assure users' privacy and provide verification (Wang *et al.*, 2025). There are several drawbacks to the implementation and operation of blockchains in ITS applications. Current blockchains like Proof of Works in general blockchains take too much time and energy for processing as an extra requirement for real-time communication in an ITS system. Even when faster approaches such as practical byzantine fault tolerance (PBFT), proof-of-stake (PoS), and others are used, there would be additional delays that could impact actual timing-critical vehicle communication as well. To overcome this, solutions by Reshi and Sholla (2024) include using Layer-2 solutions (off-chain transaction channels), parallel processing with "sharding" along with the main chain structure, and new algorithms like "proof of authority" (PoA) or delegated "proof of stake" (DPOS), for enhanced performance in an ITS setup. Even with this optimization, the existing high latency that takes place in milliseconds would not be fully eliminated. Hence, an additional post-processing background task or high-speed and zero-latency system architectures would be required, as in autonomous vehicle control or dynamic traffic flow.

Figure 2.7 shows a secure routing mechanism based on blockchain technology in a VANET (vehicle ad-hoc network), which protects from "black hole attacks" through a tamper-proof ledger to record routing information to guarantee that no malicious node can alter the routing of a single packet of data.

The secure transfer of vehicular data, combined with reputation management through blockchain, can complement the present security systems of ITS. Once authenticity and trust are established through the blockchain network, any attempt to distort the data stored within it is deemed impossible due to decentralization. In fact, Zhang *et al.* (2024) argue that this kind of decentralization can make a revolutionary change to the management of data within the context of ITS by providing greater security to data management while

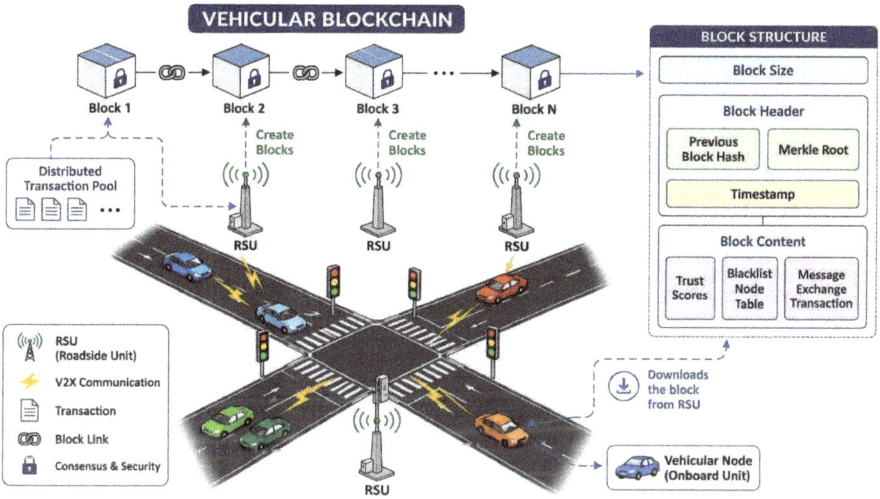

Fig. 2.7. Blockchain-based secure routing mechanism in VANETs: ledger-assisted route verification to deter or detect black hole/packet-dropping behavior.

lowering the risk of any cyberattacks associated with data silos managed through centralized systems. However, the scalability of the technology has remained a concern from the perspective of large-scale management systems like ITS. The cost of operations as well as the computational costs of a blockchain can create barriers for its usability. Current research is ongoing to make such technology more efficient through the application of layer-2 scaling or sharding (Artenie *et al.*, 2025; Quan *et al.*, 2024).

Table 2.1 presents a detailed study of the applications of blockchain technology in the IoT domain, which concerns ITS, focusing on how they enhance security, privacy, and information integrity. It highlights selected recent works (2022–2025) that demonstrate the importance of blockchain technology for enhancing the safety and decentralization features of ITS applications in the IoT domain. Key elements, such as secure information sharing, device authentication, and location privacy, are identified as significant characteristics of blockchain-based solutions for reliable V2V and V2I communication in ITS.

Table 2.1. Blockchain integration in IoT and ITS security applications (selected studies, 2022–2025).

Feature	Blockchain Security	Blockchain Application	IoT Application	IoT-Blockchain Integration in ITS	References
Blockchain-based IoT Device Security	Trust and authentication across IoT elements	Secure IoT device management, prevention of data breaches	ITS devices (vehicles, traffic sensors, infrastructure)	Ensures secure communication and data integrity for ITS devices	Bao and Lu (2022)
Location Privacy in IoT using LDP	Decentralized data storage, access control	Secure and private location data management	ITS (smart cars, traffic management)	Protects vehicle location privacy in ITS systems	He *et al.* (2022)
Blockchain & IoT for Food Supply Chain	Enhanced traceability & tamper-proof records	Food Supply Chain	Farm-to-Fork Tracking, Smart Contracts for Transactions	Ensures tamper-proof tracking of vehicle data in ITS	Kaur *et al.* (2022)
Device Shadow-Blockchain for Secure IoT Data	Decentralized & tamper-proof shadow data storage	Enhanced data security & privacy	Focus on Shadow Data	Secures ITS data transmissions through decentralized storage	Cp *et al.* (2022)
Integration of IoT with Blockchain	Secure IoT device management	Secure control using blockchain and smart contracts	ITS (vehicle management, traffic flow control)	Enhances control and authentication for ITS networks	Razaque *et al.* (2023)

Blockchain with IoT for Secure Manufacturing Supply Chain	Secure and tamper-proof transaction signing	Enhanced supply chain transparency & traceability	Manufacturing	Ensures transparency and security for the vehicle supply chain in ITS	Madhwal *et al.* (2023)
Blockchain for IoT security and privacy	Secure decentralization	✗	Smart objects	Secures communication and data in autonomous ITS	Sinha (2024)
Integration of IoT and blockchain with enhanced security	Decentralized and tamper-proof system	✗	Healthcare, smart cities, industrial automation	Enhances security and privacy in smart city ITS applications	Velamakanni and Patwal (2024)
Blockchain's Role in IoT Security	Immutable data records, auditable transactions	Secure data exchange, access control	Supply chain management, logistics tracking	Provides data integrity and transparency for ITS operations	Essaid and Ju (2025) and Rai *et al.* (2024)
Edge Computing & Blockchain for Secure IoT	Secure data storage & processing at the edge	Improved security & privacy across the network	Healthcare, Agriculture, Manufacturing, Smart Cities	Ensures secure data processing and privacy in real-time ITS environments	Zhukabayeva *et al.* (2025)

2.5.3 *Post-Quantum Security Strategy in ITS*

The advent of quantum computing challenges many of the existing cryptographic methods that are being used to protect ITS communications. Quantum computing algorithms, such as Shor's algorithm, would be able to break many of the existing public-key encryption schemes (e.g., Rivest-Shamir-Adleman, a.k.a. RSA, and elliptic-curve cryptography) in a reasonable time. However, current ITS communications across all vehicles to infrastructure and vehicle to vehicle communications depend heavily on many of the existing classical algorithmic methods for purposes of encryption and authentication. As a consequence, advances in quantum computing power are increasingly exposing them to threats. For example, as stated by Cyriac *et al.* (2024), a general-purpose quantum computer would be able to factor large numbers in RSA moduli or compute discrete logarithms of elliptic curves using Shor's algorithm, breaking the respective encryption schemes in minutes and seconds, respectively. Moreover, as stated by Faruk *et al.* (2022), a quantum-capable attack would be a threat to the security of ITS communications within a decade.

However, PQC has been identified as the main solution to this emerging threat. Post-quantum cryptography generally refers to a group of cryptographic algorithms that are resistant to attacks by quantum computers. Examples of promising contenders, which have been standardized by NIST, include lattice-based cryptography and hash-based signature schemes. These methods are based on mathematical problems like lattice problems, hash functions, code-based problems, and multi-variable equations, which cannot be easily resolved by existing quantum computing models. For instance, lattice-based cryptographic methods, like CRYSTALS-Kyber encryption and CRYSTALS-Dilithium encryption, are based on problems like Learning with Errors, which are believed to be resistant to quantum attacks. Hash signature methods, such as the extended Merkle signature scheme (XMSS) and Leighton-Micali signatures (LMS), take advantage of cryptographic hash functions, which are secure based on their ability to withstand hash collisions.

Nonetheless, several issues arise in the application of PQC schemes, such as computational efficiency and the use of larger keys, which can be considered restrictive in terms of resource constrained environments. Most of the PQC schemes have shown the production of considerably larger public keys/signatures in comparison to RSA/ECC, which can be challenging to the limited bandwidth of the vehicular environment. Code-based cryptosystems such as the use of the McEliece cryptosystem based on error-correcting codes typically involve extremely large public keys (with sizes of hundreds of kilobytes or more) that can be challenging to implement or transmit. Other schemes involving multivariate polynomials or isogeny-based cryptosystems have certain disadvantages, where some have been broken or made less secure as a result of recent developments.

Despite present difficulties, the transition for PQC is considered a necessity. Zeddini *et al.* (2022) indicate that traditional methods such as PKI and digital signatures available today are sufficient in securing car authentication processes (prevention of malicious nodes in VANETs), but fast developments in quantum computing make them inadequate in the near future. Baseri *et al.* (2025) emphasized moving to quantum-resistant settings as a requirement for maintaining security; their study provided a framework for quantum security risk assessment and strategies for quantum safe transition. A strategy that has been suggested to ensure the security of ITS network communications involves the implementation of quantum key distribution (QKD) protocols such as KeyShield (Al-Darwbi *et al.*, 2022), where the utilization of the phenomena of quantum physics ensures that any eavesdropping attempt is detectable. However, there are challenges to the implementation of QKD protocols in mobile communications, including ITS, that have recently been articulated by Aquina *et al.* (2025), suggesting that QKD protocols require optical devices that are incompatible with ITS devices that have restricted capabilities. Thus, it may complement PQC algorithms rather than replace them.

Another aspect is energy efficiency, as research by Irshad *et al.* (2023) suggests that the concept of blockchain hybridization in the field of ITS should revolve around a combination of classical and post-quantum algorithms. As stated, a blockchain with V2X security involves the combination of classical algorithms and PQC on the same network for the purpose of wallet addresses as well as digital signatures, hence providing security from both traditional as well as quantum opponents. A major advantage with hybrids is that they have the capability to remain safe even when one of the two is under attack. For instance, the impact of quantum attacks on the safety of RSA due to the intact status of the latter, the PQC layer still protects the transaction or message.

A complete transition to PQC in ITS would take time. In the immediate future, hybrid cryptographic schemes could also be used. For instance, a vehicle can broadcast a signature consisting of both ECDSA (elliptic curve) and PQC signatures; updated receivers would accept the message only if both checks pass, while backward compatibility is maintained for legacy vehicles, which simply disregard the PQC signature and verify only the ECDSA component. Later on, with the development of standards, specifically the primary PQC standards finalized by NIST in August 2024, a full transition can be made (Dang *et al.*, 2024). One of the essential areas to focus on here would be update strategies for vehicles, many of which support over-the-air updates, making it possible to update the cryptographic software on the vehicle if needed. In short, the goal of PQC in the ITS is to make the security of the vehicular networks as future-proof as possible. Although the current near-term scenario where quantum computers break existing public-key cryptosystems may still be several years away, the long-term lifetimes of transportation infrastructure and vehicles make the architectural decisions made in 2026 regarding PQC integration likely to affect the security landscape of the 2030s and beyond. Thus, the current need for the transition to PQC is indeed timely.

The conceptual inclusion of the use of PQC in the ITS networks based on the need for protecting V2V as well as V2I communications highlights the role of the use of quantum computing algorithms in the future as an alternative for classical computing for the authentication as well as the encryption components of the communication stacks for the ITS. Actually, the implementation of such quantum computing algorithms as scalable and compatible schemes poses a future area of concern for continued academic as well as technological efforts for their implementation in the current transportation community for the prevention of security threats without affecting the safety as well as the efficiency of the transportation processes and systems being developed or utilized.

2.5.4 *Game Theory and Adaptive Strategies*

Game theory is an effective methodology toolkit to represent the attacker-defender interactions in intelligent transportation systems using attack-defender models of game theory. Game theory enables the defenders to think ahead of the attacks to occur rather than simply reacting to the launched attacks. Game theory models assist in implementing optimal or resilient defenses against the likely actions of the attacker, enabling a response. According to Khalid *et al.* (2023), game-theoretic models should be used in ITS to enhance defenders' capacity against all attacker classes. These models allow defenders to respond to attackers' changing tactics over time, unlike traditional models. For example, for Sybil attacks in particular, the creation of several identities for malicious purposes, a game-theoretic solution might formulate a trust dynamics system where node trusts are updated over time based on behavior, making it harder for malicious nodes to maintain several honest-looking identities (Hassan *et al.*, 2024). In cases of prolonged deception attacks, or "sleeper" attacks, reputation systems could be built to allow for node trust escalation over time, so that any sudden change in behavior is observable (Patel and Jinwala, 2022). Such systems also need to counteract issues of false positives; game-theoretic solutions often model penalties for mistakenly labeling a genuine node malicious and work to optimize for them.

Game theory has also been used to address Sybil attack prevention by resource commitment. Baza *et al.* (2020) combined the game theory paradigm with the concept of Proof of Work to force the attacker to invest in each of their identities. In the resulting game, honest vehicles and Sybils play as rivals, where sincere nodes perform the least amount of work to verify their identities, and an attacker with multiple Sybils will incur increasing costs to verify each of their identities. With appropriate values for the cost settings, the game's result will limit the scale of the attack for the attacker and the number of Sybils. A major drawback of using proof of work in the context of vehicular networks has been the associated energy and computation overhead, which has also been studied in the context of game theory. Li (2024) has proposed the use of proof of stake and combined models to impact the attack's incentives, so as to provide decreased energy overhead for the sincere nodes, which will be costly for the attackers.

The existence of wormhole attacks has also been considered from the perspective of game theory. Stępień and Poniszewska-Marańda (2021) used game theory with geographic data for the detection and prevention of a wormhole attack. The system assumes communication timing and routing matching geographic locations (for example, directly connected nodes should be close in geographic locations). The game scenario involves the network (defender) trying to authenticate routes versus an attacker trying to mount a wormhole attack without being detected. The attacker's reliance on GPS information allows for the development of a game where any route path deviating from distance or timing information would receive a penalty (marked for possible wormhole attack). The attacker could attempt to deceive about location or timing information in order to go undetected, and then a cryptographic test of location authenticity (for example, signature of geographic locations checked for inconsistencies) would be employed. The net effect would be two-tier protection where the attacker succeeds at one level (for example, quick packet transfer) but fails at another (valid evidence of false geographic proofs), making it possible for detection.

Game theoretic approaches can also help in designing multi-layered defense networks. In this scenario, researchers like Sreevallabh Chivukula *et al.* (2022) defined a unified mathematical framework including the interplay between automobile authentication systems, networks, and data authentication. In this situation, the game is multi-level: if the attacker's focus is on violating authentication, efforts would be put into fortifying this tier; once their focus shifts to a data attack, authentication would become paramount, and so on. Based on the attacker's optimal path at each tier, designing a relevant defense strategy helps in collectively lowering risks. Morić *et al.* (2025) proposed the concept of deception in these game-theoretic paradigms, where the defending party could selectively "deceive" or set up traps, like honeypot nodes or fake vulnerabilities, that would mislead the attacker, which would again fall in this framework. In this scenario, if the attacker is fooled, the defending party could gain the upper hand in analyzing them while defending actual assets. This deception would further add to the complexity in the attacker's strategy, hopefully in the defending party's favor.

Game theory is also known to facilitate cooperative defense among trusted nodes. While a lone reaction mechanism involving vehicles and roadside units would seem inefficient, game theory would allow trusted nodes to pool their efforts and work together. A combined system involving reputation systems and GPS verification, for instance, would be a setting where trusted nodes, working together, would penalize and check suspicious parties. These systems, if tuned correctly, have proved credible for countering Sybil and wormhole attacks. However, they would add computational complexity (solving puzzles and building reputations) and are themselves exposed to underlying assumptions concerning GPS jamming and spoofing attacks and reputation systems that are slow to adapt to situation-specific malicious activities.

The implementation of a game theory defense mechanism within ITS systems raises a set of issues. There is a challenge involved in solving a complex game theory within a real-time environment. Theoretical methods may be computationally expensive when a large number of states and/or players are involved. Another challenge is the fact that attackers often do not strictly follow a purely logical strategy postulated by traditional game theory. Cyber attackers' actions may include unpredictable aspects or goals that are not contemplated within a given game theory framework. For instance, state-run attackers may choose the maximal level of disruption regardless of the associated costs. Moreover, the feasibility of integration will necessarily be a dimension worth considering. The implementation strategy for a given theory may include a coordination mechanism that varies strategies at a centralized or vehicle-distributed level.

Despite such challenges, game theory remains a sound base on which a secure network in ITS can be built through proactive detection and prevention. By modeling interactions between both attackers and defenders, a defensive strategy can be built to proactively look forward to predictions as opposed to just reacting to attacks. For instance, a network in the ITS can use a game theory algorithm to distribute security resources (processing power and bandwidth devoted to security messages) to various defense lines according to their level of threats; based on external intelligence information on a possible threat by trends, reinforcement of that sector can be undertaken proactively. While these defensive lines have been touted to have security in place, several issues stand out, which include high processing power, possible scalability problems in very large networks, and constant updates to improve on evolving approaches

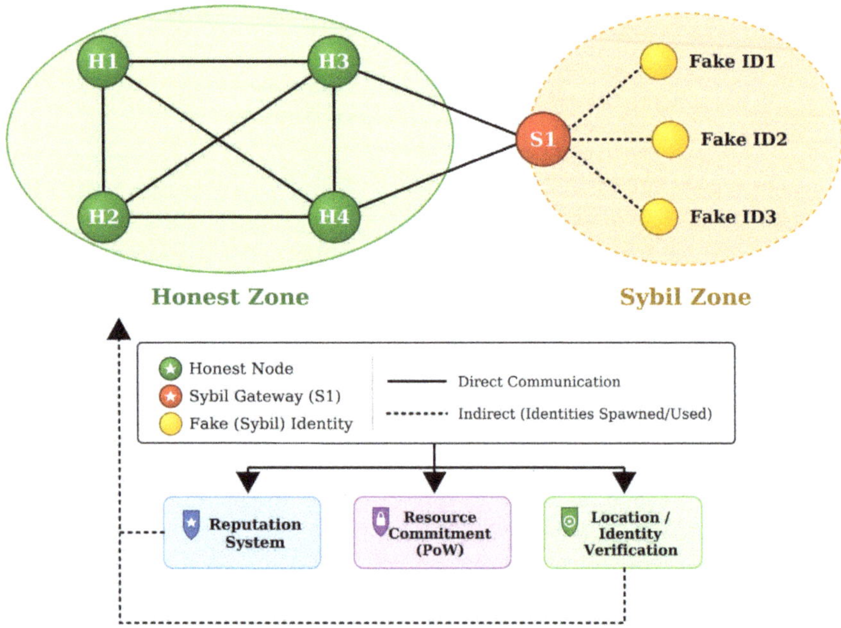

Fig. 2.8. Game-theoretic attacker vs. defender model for ITS security (Sybil mitigation using reputation, resource commitment, and location/identity verification concepts).

by attackers. Research continues to advance game theory models, which incorporate machine learning to forecast attacks by players or dynamic payoff models from real-time data to counter ever-changing attacks.

Figure 2.8 illustrates the game-theoretic presentation of Sybil attacks and how reputation systems, PoW, and GPS verification can collaborate to enhance ITS security. It describes a multi-layered defense system that simulates communication between genuine nodes and Sybil attacks, aiming to maximize ITS security.

From this figure, honest participants collaborate with the goal of unmasking the Sybils while maximizing the expenditure imposed on attackers to ensure the continued use of services for honest users. The key to using game theory effectively in the context of ITS security lies in the concept of dynamic, strategic defensive measures. Rather than relying on set in stone policies, a dynamic defensive mechanism responds to the question of what the next move of the attacker would

Table 2.2. Comparative analysis of four innovative defense strategies in ITS.

Defense Strategy	Primary Benefit in ITS	Key Limitations/Trade-offs	Ideal Application Context
Artificial Intelligence (AI) & Machine Learning	Enables real-time threat identification as well as adaptive response mechanisms in ITS networks.	Inherently requires large amounts of computation & data; susceptible to situations of false positives and drift due to evolution of threat profiles	Real-time intrusion prevention and V2X communications.
Blockchain	Keeps data integrity and decentralizes trust through the elimination of points of failure in ITS data transactions.	Scalability and latency: high overhead with respect to computation and storage for large-scale ITS implementations.	Secure messaging for vehicle to vehicle and vehicle to infrastructure communications; secure logging for significant ITS events and transactions.
Post-Quantum Cryptography	Provides quantum-resistant encryption to protect ITS communications against possible attacks using quantum computing.	Larger computational complexity and increased key size; still in the early stages of implementation.	Protection of V2X communications and ITS infrastructure communications against possible attacks using quantum computing.
Game Theory (Adaptive Defense)	Allows for proactive defending by predicting attacker behavior and cooperating on defending mechanisms.	Hard to achieve in real time, dependent on the attacker following rational game theory models, which is unlikely.	Adaptive intrusion response systems and collaboration among various ITS nodes for defending against threats like Sybil or wormholes.

be and how to effectively mitigate it. The convergence of AI-defenses, blockchain technology, PQC, and the concepts of game theory (as mentioned above) points towards a cohesive approach to achieving the best possible results in the field. Utilizing the mechanism of game theory to decide when to employ AI detection tools, when to log data into the blockchain, and so on. Through this mechanism, the defenses for ITS can become immune to a wide variety of threats.

The synergies that exist between AI-driven defense strategies, blockchain platforms, PQC, and game-theoretic models provide a complete approach to safeguarding the future of ITS. Real-time protection against threats is made easier through AI, which offers faster processing of big data for detecting any anomalies. Additionally, blockchain technology ensures that all deals involving vehicle data are recorded in an immutable record book, ensuring data authenticity and complete audibility. Post-quantum cryptography helps to safeguard data communications of ITS for the possibility of future advances in quantum computing, providing long-lasting security and authenticity as computing power increases. Game-theoretic models provide an approach that helps to predict all moves of attackers and provide proactive resources for defense accordingly. Although a composite approach to defending ITS offers huge potential, there are drawbacks to this method as well. As it has high computational requirements for implementation, it may encounter scalability issues and would require constant updating. Hence, there is a need for ongoing research and interdisciplinary collaboration between industry and academia to optimize and apply these tactics to real-world ITS platforms. Table 2.2 summarizes the comparative trade-offs analysis of these approaches, highlighting their benefits, key limitations, and ideal applications in intelligent transportation systems.

2.6 Conclusions

The evolution of Intelligent Transportation Systems (ITS) into highly networked, real-time environments has ushered in both remarkable opportunities and unprecedented cyber risks. As highlighted throughout this chapter, the increasing sophistication and interconnectivity of ITS make them attractive targets for a wide array of cyberattacks, including DoS, Sybil, and wormhole attacks.

Addressing these threats requires not only robust technical solutions but also a holistic, adaptive approach that ensures operational continuity.

The integration of artificial intelligence, blockchain, post-quantum cryptography (PQC), and game theory represents a new era in ITS defense strategies. AI and machine learning empower ITS to detect and respond to threats in real time, adapting to evolving attack patterns with minimal human intervention. Blockchain technology adds a critical layer of security, enabling tamper-proof data sharing and decentralized trust across vehicles and infrastructure. PQC is a proactive measure that prepares ITS for future challenges posed by quantum computing and ensures long-term cryptographic resilience. Meanwhile, game theory offers a strategic framework for anticipating attacker behavior and optimizing defensive measures.

Ultimately, the synergy among these advanced technologies forms a comprehensive, dynamic shield for ITS. However, the journey toward fully secure ITS demands ongoing innovation, interdisciplinary collaboration, and the continual refinement of both policies and technical solutions. As ITS becomes increasingly integral to modern mobility and urban life, a forward-thinking and resilient cybersecurity approach will be essential for safeguarding both the technology and the people who depend on it.

2.7 Recommendations for Future Research

Building on the comprehensive defense strategies discussed in this chapter, future research should focus on deepening the synergy, scalability, and real-world applicability of AI, blockchain, PQC, and game theory within ITS security. As cyber threats continue to evolve alongside advances in transportation technology, research must prioritize adaptive, proactive, and resilient security mechanisms that can keep pace with both current and emerging risks. The following recommendations outline critical directions for the next generation of ITS cybersecurity research:

Despite the promise that these novel defense schemes have, several challenges exist that need to be mitigated to make them practicable to be used. Computational complexity is a key issue, where many AI and cryptographic algorithms consume high amounts of processing power, which could be a limitation on edge platforms like vehicles. Scalability is another issue where many of these novel security solutions have to be scaled from a small-scale setting to a large-scale setting like a city-scale ITS network that scales to thousands of vehicles and sensors. Efficiency, especially in terms of energy, is a key challenge where electric vehicles cannot devote high amounts of energy to security where they have limited range. Future research work needs to focus on optimizing these solutions to fit a resource-limited and high-speed environment of ITS.

Another aspect of ITS security is the implementation of new security measures without disruptions. Improvements to ITS security should not require system-wide changes to existing infrastructure that might result in downtime or backward compatibility problems. Part of this process will involve work in partnership between industry and academia, as trials of various concepts will have to take place to test theories in real-world conditions. This might involve testing such projects as using blockchain security for V2V communications in test cities or using PQC-secured hardware in fleets of vehicles. This kind of work can also help to establish standards for acceptable security practices.

Moving forward, research and innovation are needed to further develop these cyber defense technologies. Some of the recommendations in this regard include: design of explainable AI for cybersecurity, which will allow human decision-makers in ITS to be assured of AI decision-making; lightweight designs for blockchain consensus algorithms for vehicular networks; quantum-resistant communication algorithms for V2X communication; and design of adaptive algorithms with the ability to be trained on the actions of attackers through game theory and machine learning techniques. The development of these technologies will ensure that the ITS ecosystem is not only intelligent but also secure.

Bibliography

Al-Darwbi, M. Y., Ghorbani, A. A., and Lashkari, A. H. (2022). QKeyShield: A practical receiver-device-independent entanglement-swapping-based quantum key distribution, *IEEE Access* **10**, pp. 107685–107702.

Al Ani, S. (2024). *A Predictive Approach to Detect and Mitigate Sybil Attacks in a Waiting Time-Based Adaptive Traffic Signal Controller* (The University of Alabama, Tuscaloosa, Alabama, USA).

Alhidaifi, S. M., Asghar, M. R., and Ansari, I. S. (2024). A survey on cyber resilience: Key strategies, research challenges, and future directions, *ACM Computing Surveys* **56**, 8, pp. 1–48.

Ali, S., Sharma, M., and Sinha, S. (2025). Implementation and strategy for Sybil attack and DDoS attack in blockchain-based secure routing protocol, in *Integrated Technologies in Electrical, Electronics and Biotechnology Engineering* (CRC Press), pp. 308–314.

Aquina, N., Cimoli, B., Das, S., Hövelmanns, K., Weber, F. J., Okonkwo, C., Rommel, S., Škorić, B., Tafur Monroy, I., and Verschoor, S. (2025). A critical analysis of deployed use cases for quantum key distribution and comparison with post-quantum cryptography, *EPJ Quantum Technology* **12**, 1, p. 51, https://doi.org/10.1140/epjqt/s40507-025-00350-5.

Artenie, A. C., Silaghi, D. L., and Popescu, D. E. (2025). Exploring the synergy between Ethereum layer 2 solutions and machine learning to improve blockchain scalability, *Computers* **14**, 9, p. 359, https://www.mdpi.com/2073-431X/14/9/359.

Aslan, Ö., Aktuğ, S. S., Ozkan-Okay, M., Yilmaz, A. A., and Akin, E. (2023). A comprehensive review of cyber security vulnerabilities, threats, attacks, and solutions, *Electronics* **12**, 6, p. 1333.

Avcı, İ. and Koca, M. (2024). Intelligent transportation system technologies, challenges and security, *Applied Sciences* **14**, 11, p. 4646.

Bao, L. and Lu, G. (2022). Blockchain for storing shadow data of IoT devices, in *Proceedings of the 2022 5th International Conference on Blockchain Technology and Applications*.

Baseri, Y., Chouhan, V., Ghorbani, A., and Chow, A. (2025). Evaluation framework for quantum security risk assessment: A comprehensive strategy for quantum-safe transition, *Computers & Security* **150**, p. 104272, https://doi.org/10.1016/j.cose.2024.104272.

Basnet, M. (2022). *Deep Learning-Powered Computational Intelligence for Cyber-Attacks Detection and Mitigation in 5G-Enabled Electric Vehicle Charging Station*, Doctoral dissertation, The University of Memphis (ProQuest Dissertations & Theses, No. 29994125).

Basuli, S. and Padhya, M. (2024). Botnet-based DDoS attack: Automatic detection, mitigation, and real-time traffic filtering in cloud environments, in *International Conference on Advanced Network Technologies and Intelligent Computing*.

Baza, M., Nabil, M., Mahmoud, M. M., Bewermeier, N., Fidan, K., Alasmary, W., and Abdallah, M. (2020). Detecting sybil attacks using proofs of work and location in vanets, *IEEE Transactions on Dependable and Secure Computing* **19**, 1, pp. 39–53.

Bhatti, D. S., Saleem, S., Imran, A., Kim, H. J., Kim, K.-I., and Lee, K.-C. (2024). Detection and isolation of wormhole nodes in wireless ad hoc networks based on post-wormhole actions, *Scientific Reports* **14**, 1, p. 3428.

Borcoci, E., Obreja, S. G., and Vochin, M. C. (2018). Functional layered architectures and control solutions in internet of vehicles-comparison, *International Journal on Advances in Internet Technology* **11**, pp. 31–43.

Ceder, A. (2021). Urban mobility and public transport: Future perspectives and review, *International Journal of Urban Sciences* **25**, 4, pp. 455–479.

Chandrasekaran, S., Ramachandran, K., Adarsh, S., and Puranik, A. K. (2020). Avoidance of replay attack in CAN protocol using authenticated encryption, in *2020 11th International Conference on Computing, Communication and Networking Technologies (ICCCNT)*.

Cp, V., Kalaivanan, S., Karthik, R., and Sanjana, A. (2022). Blockchain-based IoT device security, in *2022 2nd International Conference on Artificial Intelligence and Signal Processing (AISP)*.

Cyriac, R., Eswaran, S., Selvarajan, S., and Yuvaraj, D. (2024). Quantum computing in cryptographic systems, *International Journal of Advanced IT Research and Development*, **1**, 1, pp. 18–24.

Dang, T., Lichtinger, J., Liu, Y.-K., Miller, C., Moody, D., Peralta, R., Peralta, R., Perlner, R., and Robinson, A. (2024). Technology (NIST): Module-Lattice-Based Digital Signature Standard.

Diro, A., Kaisar, S., Saini, A., Fatima, S., Hiep, P. C., and Erba, F. (2025). Workplace security and privacy implications in the GenAI age: A survey, *Journal of Information Security and Applications* **89**, p. 103960, https://doi.org/10.1016/j.jisa.2024.103960.

El-Rewini, Z., Sadatsharan, K., Selvaraj, D. F., Plathottam, S. J., and Ranganathan, P. (2020). Cybersecurity challenges in vehicular communications, *Vehicular Communications* **23**, p. 100214.

Elassy, M., Al-Hattab, M., Takruri, M., and Badawi, S. (2024). Intelligent transportation systems for sustainable smart cities, *Transportation Engineering* **16**, p. 100252.

Essaid, M., and Ju, H. (2025). Blockchain solutions for enhancing security and privacy in industrial IoT, *Applied Sciences* **15**, 12, p. 6835, https://www.mdpi.com/2076-3417/15/12/6835.

Faruk, M. J. H., Tahora, S., Tasnim, M., Shahriar, H., and Sakib, N. (2022). A review of quantum cybersecurity: Threats, risks and opportunities, in *2022 1st International Conference on AI in Cybersecurity (ICAIC)*.

Fernández Llorca, D., Hamon, R., Junklewitz, H., Grosse, K., Kunze, L., Seiniger, P., Swaim, R., Reed, N., Alahi, A., Gomez, E., and Kriston, A. (2025). Testing autonomous vehicles and AI: Perspectives and challenges from cybersecurity, transparency, robustness and fairness, *European Transport Research Review* **17**, 1, p. 38, https://doi.org/10.1186/s12544-025-00732-x.

Girdhar, M., Hong, J., and Moore, J. (2023). Cybersecurity of autonomous vehicles: A systematic literature review of adversarial attacks and defense models, *IEEE Open Journal of Vehicular Technology* **4**, pp. 417–437.

Hameed, A. and Saini, I. (2024). Lightweight hybrid approach for DoS attack detection in VANET, in *2024 IEEE Middle East Conference on Communications and Networking (MECOM)*.

Hassan, J., Sohail, A., Awad, A. I., and Zaka, M. A. (2024). LETM-IoT: A lightweight and efficient trust mechanism for Sybil attacks in Internet of Things networks, *Ad Hoc Networks* **163**, p. 103576, https://doi.org/10.1016/j.adhoc.2024.103576.

He, M., Bai, F., Zhang, C., Shen, T., and Gong, B. (2022). A blockchain-enabled location privacy-preserving under local differential privacy for internet of vehicles, in *Proceedings of the 2022 4th Blockchain and Internet of Things Conference.*

Iehira, K. and Inoue, H. (2024). Spoofing attack detection method by estimating transmitting device on 10BASE-T1S, in *2024 IEEE 29th Asia Pacific Conference on Communications (APCC).*

Inedjaren, Y., Maachaoui, M., Zeddini, B., and Barbot, J.-P. (2021). Blockchain-based distributed management system for trust in VANET, *Vehicular Communications* **30**, p. 100350.

Irshad, R. R., Hussain, S., Hussain, I., Nasir, J. A., Zeb, A., Alalayah, K. M., Alattab, A.A., Yousif, A., and Alwayle, I. M. (2023). IoT-enabled secure and scalable cloud architecture for multi-user systems: A hybrid post-quantum cryptographic and blockchain-based approach toward a trustworthy cloud computing, *IEEE Access* **11**, pp. 105479–105498.

Katiyar, N., Shukla, A., Chawla, N., Singh, R., Singh, S. K., and Husain, M. F. (2024). AI in autonomous vehicles: Opportunities, challenges, and regulatory implications, *Educational Administration: Theory and Practice* **30**, pp. 6255–6264.

Katsikeas, S., Johnsson, P., Hacks, S., and Lagerström, R. (2022). VehicleLang: A probabilistic modeling and simulation language for modern vehicle IT infrastructures, *Computers & Security* **117**, p. 102705.

Kaur, A., Singh, G., Kukreja, V., Sharma, S., Singh, S., and Yoon, B. (2022). Adaptation of IoT with blockchain in food supply chain management: An analysis-based review in development, benefits and potential applications, *Sensors* **22**, 21, p. 8174.

Khalid, M. N. A., Al-Kadhimi, A. A., and Singh, M. M. (2023). Recent developments in game-theory approaches for the detection and defense against advanced persistent threats (APTs): A systematic review, *Mathematics* **11**, 6, p. 1353.

Li, L. (2024). Mitigating challenges in Ethereum's proof-of-stake consensus: Evaluating the impact of EigenLayer and Lido, *arXiv preprint* arXiv:2410.23422.

Lin, C., Sun, G., Wu, D., and Xie, C. (2023). Vehicle detection and tracking with roadside LiDAR using improved ResNet18 and the Hungarian algorithm, *Sensors* **23**, 19, p. 8143.

Liu, T., Wang, S., Dong, Z., Li, B., and He, T. (2025). From perception to computation: Revisiting delay optimization for connected autonomous vehicles, *ACM Computing Surveys* **57**, 8, pp. 1–45.

Lv, Z. and Shang, W. (2023). Impacts of intelligent transportation systems on energy conservation and emission reduction of transport systems: A comprehensive review, *Green Technologies and Sustainability* **1**, 1, p. 100002.

Madhwal, Y., Yanovich, Y., Balachander, S., Poojaa, K. H., Saranya, R., and Subashini, B. (2023). Enhancing supply chain efficiency and security: A proof of concept for IoT device integration with blockchain, *IEEE Access* **11**, pp. 121173–121189.

Morić, Z., Dakić, V., and Regvart, D. (2025). Advancing cybersecurity with honeypots and deception strategies, *Informatics*, **12**, 1. https://doi.org/10.3390/informatics12010014.

Moumen, I., Rafalia, N., Abouchabaka, J., and Chatoui, Y. (2023). AODV-based defense mechanism for mitigating blackhole attacks in MANET, in *E3S Web of Conferences.*

Naeem, M. A., Chaudhary, S., and Meng, Y. (2024). Road to efficiency: V2V enabled intelligent transportation system, *Electronics* **13**, 13, p. 2673.

Nikitas, A., Michalakopoulou, K., Njoya, E. T., and Karampatzakis, D. (2020). Artificial intelligence, transport and the smart city: Definitions and dimensions of a new mobility era, *Sustainability* **12**, 7, p. 2789.

Okur, C. and Dener, M. (2025). Symmetrical resilience: Detection of cyberattacks for SCADA systems used in IIoT in Big Data environments, *Symmetry* **17**, 4, p. 480, https://www.mdpi.com/2073-8994/17/4/480.

Oladimeji, D., Rasheed, A., Varol, C., Baza, M., Alshahrani, H., and Baz, A. (2023). Canattack: Assessing vulnerabilities within controller area network, *Sensors* **23**, 19, p. 8223.

Olanrewaju, O. M., Abdulwasiu, A. A. A., and Nuhu, A. (2023). Enhanced on-demand distance vector routing protocol to prevent blackhole attack in MANET, *International Journal of Software Engineering and Computer Systems* **9**, 1, pp. 68–75.

Ouaissa, M., Ouaissa, M., and Boulouard, Z. (2024). Secure vehicular networking: Architectures, applications, attacks, and challenges, in *Vehicular Networks-Principles, Enabling Technologies and Modern Applications* (IntechOpen, London, United Kingdom).

Pasupathi, S., Kumar, R., and Pavithra, L. (2025). Proactive DDoS detection: Integrating packet marking, traffic analysis, and machine learning for enhanced network security, *Cluster Computing* **28**, 3, p. 210.

Patel, A. and Jinwala, D. (2022). A reputation-based RPL protocol to detect selective forwarding attack in Internet of Things, *International Journal of Communication Systems* **35**, 1, p. e5007.

Pavão, J., Bastardo, R., and Rocha, N. P. (2024). Cyber resilience and intelligent transport systems, a scoping review, *Procedia Computer Science* **239**, pp. 139–148.

Pollard, M. (2024). A case study of Russian cyber-attacks on the Ukrainian power grid: Implications and best practices for the United States, *Pepperdine Policy Review* **16**, 1, p. 1.

Quan, B. L. Y., Wahab, N. H. A., Al-Dhaqm, A., Alshammari, A., Aqarni, A., Abd Razak, S., and Wei, K. T. (2024). Recent advances in sharding techniques for scalable blockchain networks: A review, *IEEE Access*, **13**, pp. 21335–21366.

Radoš, K., Brkić, M., and Begušić, D. (2024). Recent advances on jamming and spoofing detection in GNSS, *Sensors* **24**, 13, p. 4210.

Rai, H. M., Shukla, K. K., Tightiz, L., and Padmanaban, S. (2024). Enhancing data security and privacy in energy applications: Integrating IoT and blockchain technologies, *Heliyon* **10**, 19, p. e38917, https://doi.org/10.1016/j.heliyon.2024.e38917.

Rathore, R. S., Hewage, C., Kaiwartya, O., and Lloret, J. (2022). In-vehicle communication cyber security: Challenges and solutions, *Sensors* **22**, 17, p. 6679.

Razaque, A., Yoo, J., Bektemyssova, G., Alshammari, M., Chinibayeva, T. T., Amanzholova, S., Alotaibi, A., and Umutkulov, D. (2023). Efficient internet-of-things cyberattack depletion using blockchain-enabled software-defined networking and 6G network technology, *Sensors* **23**, 24, p. 9690.

Reshi, I. A. and Sholla, S. (2024). The blockchain conundrum: An in-depth examination of challenges, contributing technologies, and alternatives, *Concurrency and Computation: Practice and Experience* **36**, 8, p. e7987.

Reshi, I. A., Sholla, S., and Najar, Z. A. (2024). Safeguarding IoT networks: Mitigating black hole attacks with an innovative defense algorithm, *Journal of Engineering Research* **12**, 1, pp. 133–139.

Rocha, N. P., Bastardo, R., Pavao, J., Santinha, G., Rodrigues, M., Rodrigues, C., Queiros, A., and Dias, A. (2021). Smart Cities' applications to facilitate the mobility of older adults: a systematic review of the literature, *Applied Sciences* **11**, 14, p. 6395.

Rogers, M. and Rasmussen, K. (2022). Silently disabling ECUs and enabling blind attacks on the CAN bus, *arXiv preprint* arXiv:2201.06362.

Saad, A., Rahem, A. T., Hussein, A. J., and Mohammed, A. A. (2017). Vehicular ad hoc networks: Growth and survey for three layers. *International Journal of Electrical and Computer Engineering*, **7**(1), p. 271

Sahih, A. Z., Abbasi, A., and Ghasri, M. (2024). Blockchain-enabled solutions for fair and efficient peer-to-peer renewable energy trading: An experimental comparison, *Journal of Cleaner Production* **455**, p. 142301.

Sheikh, A. M., Islam, M. R., Habaebi, M. H., Zabidi, S. A., Bin Najeeb, A. R., and Kabbani, A. (2025). A survey on edge computing (EC) security challenges: Classification, threats, and mitigation strategies, *Future Internet* **17**, 4, p. 175.

Sinha, S. (2024). Blockchain for enhancing IoT privacy and security, *International Journal of Innovative Research in Computer Science and Technology* **12**, 2, pp. 106–110.

Skomra, W. and Wojtasik, K. (2025). Critical infrastructure as a target for hybrid operations. Case studies of attacks against the facilities and systems of CI, *Terroryzm-studia, analizy, prewencja,* Special Issue, pp. 13–34.

Sreevallabh Chivukula, A., Yang, X., Liu, B., Liu, W., and Zhou, W. (2022). Game theoretical adversarial deep learning, in *Adversarial Machine Learning: Attack Surfaces, Defence Mechanisms, Learning Theories in Artificial Intelligence* (Springer, Cham, Switzerland), pp. 73–149.

Stępień, K. and Poniszewska-Marańda, A. (2021). Security measures with enhanced behavior processing and footprint algorithm against sybil and bogus attacks in vehicular Ad Hoc network, *Sensors* **21**, 10, p. 3538.

Swathika, O. G., Karthikeyan, A., Rout, K., and Hatkar, S. (2024). Cybersecurity deployment in smart grids: Critical review, applications, protection and challenges, *IEEE Access*, **12**, pp. 113618–113641.

Tariq, U., Ahmed, I., Bashir, A. K., and Shaukat, K. (2023). A critical cybersecurity analysis and future research directions for the internet of things: A comprehensive review, *Sensors* **23**, 8, p. 4117.

Thomas, O., Salek, M. S., Tine, J.-M., Rahman, M., Hockstad, T., and Chowdhury, M. (2025). Cybersecurity in transportation systems: Policies and technology directions, *arXiv preprint* arXiv:2501.05356.

Usama, M., Ullah, U., and Sajid, A. (2024). Cyber attacks against intelligent transportation systems, in *Cyber Security for Next-Generation Computing Technologies* (CRC Press, Boca Raton, FL, USA), pp. 190–230.

Vamsi, B., Doppala, B. P., Mahanty, M., Veeraiah, D., Rao, J. N., and Rao, B. S. (2024). A detailed case study on various challenges in vehicular networks for smart traffic control system using machine learning algorithms, in *Artificial Intelligence and Machine Learning for Smart Community* (CRC Press, Boca Raton, FL, USA), pp. 51–87.

Velamakanni, R. and Patwal, D. (2024). Enhancing IoT security through experimental methods and blockchain integration, *Educational Administration: Theory and Practice* **30**, 5, pp. 8859–8870.

Vellinga, N. E. (2022). Connected and vulnerable: Cybersecurity in vehicles, *International Review of Law, Computers & Technology* **36**, 2, pp. 161–180, https://doi.org/10.1080/13600869.2022.2060472.

Verma, P., Newe, T., O'Mahony, G. D., Brennan, D., and O'Shea, D. (2025). Towards a unified understanding of cyber resilience: A comprehensive review of concepts, strategies, and future directions, *IEEE Access*, **13**, pp. 49945–49965.

Wang, F., Wang, X., and Ban, X. J. (2024). Data poisoning attacks in intelligent transportation systems: A survey, *Transportation Research Part C: Emerging Technologies* **165**, p. 104750.

Wang, X., Li, H., Yi, L., Ning, Z., Tao, X., Guo, S., and Zhang, Y. (2025). A survey on off-chain networks: Frameworks, technologies, solutions and challenges, *ACM Computing Surveys* **57**, 12, pp. 1–35.

Wei, X., Ma, J., and Sun, C. (2024). A survey on security of unmanned aerial vehicle systems: Attacks and countermeasures, *IEEE Internet of Things Journal*, **11**, 21, pp. 34826–34847.

Xie, X., Jiang, K., Dai, R., Lu, J., Wang, L., Li, Q., and Yu, J. (2023). Access your Tesla without your awareness: Compromising keyless entry system of Model 3, in *NDSS*.

Yang, Y.-T. and Zhu, Q. (2024). Game-theoretic foundations for cyber resilience against deceptive information attacks in intelligent transportation systems, *arXiv preprint* arXiv:2412.04627.

Zeddini, B., Maachaoui, M., and Inedjaren, Y. (2022). Security threats in intelligent transportation systems and their risk levels, *Risks* **10**, 5, p. 91.

Zhang, L., Hang, L., Zu, K., Wang, Y., and Yang, K. (2024). Dynamic vehicle reputation consensus: Enhancing iov communication with a blockchain algorithm, *IEEE Transactions on Vehicular Technology*, **74**, 3, pp. 4788–4806.

Zhukabayeva, T., Zholshiyeva, L., Karabayev, N., Khan, S., and Alnazzawi, N. (2025). Cybersecurity solutions for industrial Internet of Things–Edge computing integration: Challenges, threats, and future directions, *Sensors* **25**, 1, p. 213, https://www.mdpi.com/1424-8220/25/1/213.

https://doi.org/10.1142/9789819831548_0003

Chapter 3

Security Vulnerabilities in Software Supply Chain for Autonomous Vehicles

Md Waslul Haque*, Md Erfan†, Sagar Dasgupta*, Md Rayhanur Rahman†, and Mizanur Rahman*

**Department of Civil, Construction and Environmental Engineering, University of Alabama, Tuscaloosa, AL, USA*
†*Department of Computer Science, University of Alabama, Tuscaloosa, AL, USA*

Abstract
The interest in autonomous vehicles (AVs) for critical missions, including transportation, rescue, surveillance, reconnaissance, and mapping, is growing rapidly due to their significant safety and mobility benefits. AVs consist of complex software systems that leverage artificial intelligence (AI), sensor fusion algorithms, and real-time data processing. Additionally, AVs are becoming increasingly reliant on open-source software supply chains, such as open-source packages, third-party software components, AI models, and third-party datasets. Software security best practices in the automotive sector are often an afterthought for developers. Thus, significant cybersecurity risks exist in the software supply chain of AVs, particularly when secure software development practices are not rigorously implemented. For example, Upstream's 2024 Automotive Cybersecurity Report states that 49.5% of cyberattacks in the automotive sector are related to exploiting security vulnerabilities in software systems. In this chapter, we analyze security vulnerabilities in open-source software components in AVs. We utilize static analyzers on popular open-source AV software, such as Autoware, Apollo, and

openpilot. Specifically, this chapter covers: (1) prevalent software security vulnerabilities of AVs; and (2) a comparison of static analyzer outputs for different open-source AV repositories. The goal is to inform researchers, practitioners, and policymakers about the existing security flaws in the commonplace open-source software ecosystem in the AV domain. The findings would emphasize the necessity of security best practices earlier in the software development lifecycle to reduce cybersecurity risks, thereby ensuring system reliability, safeguarding user data, and maintaining public trust in an increasingly automated world.

Keywords: Autonomous vehicles (AVs), Autonomous vehicle software architecture, Common Weakness Enumeration (CWE), Continuous integration and deployment (CI/CD) security, Middleware security (ROS 2, Cyber RT), Open-source autonomous driving software, Safety-critical software systems, Security vulnerability, Software supply chain, Source code vulnerability, Static analysis, Vulnerability mitigation strategies.

3.1 Introduction

The rise of autonomous vehicles (AVs) represents one of the most profound technological shifts in modern transportation. Autonomous vehicles integrate perception sensors, precise localization and mapping, trajectory prediction, planning and control systems, real-time decision-making frameworks, and built-in safety redundancies to reduce human error, enhance mobility, and reshape urban ecosystems (Liu *et al.*, 2021; Shladover, 2018). Graphical representation of a generic autonomous vehicle along with the core hardware components is shown in Fig. 3.1. As vehicles are increasingly evolving into software-defined systems, much of their safety, performance, and functionality is dictated not only by mechanical design but also by the reliability and integrity of complex software stacks. This transformation, while enabling unprecedented flexibility and innovation, simultaneously exposes AVs to new classes of risks: software defects, insecure integration of third-party components, and vulnerabilities across the software supply chain (ENISA, 2021; National Institute of Standards and Technology, 2022). AVs represent a paradigm shift in transportation, offering the potential to fundamentally change how people and goods move within modern societies. Their popularity stems from the potential to mitigate human error, while improving traffic efficiency, reducing congestion, and enhancing mobility for individuals facing difficulty driving, such as the elderly or disabled

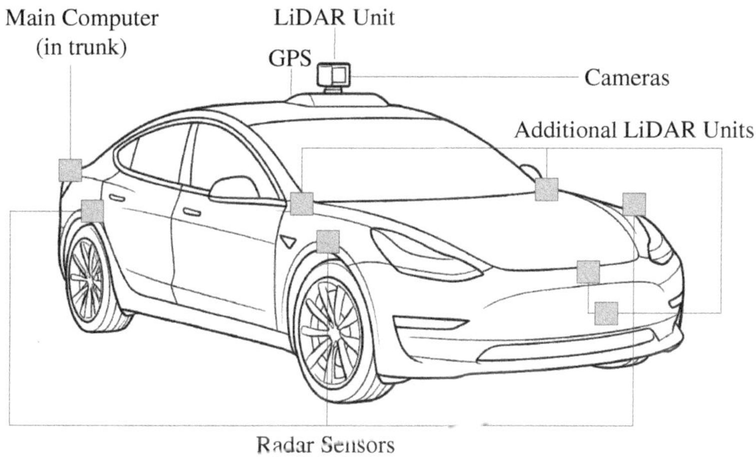

Fig. 3.1. Typical sensor and processing layout of an autonomous vehicle.

(Litman, 2017). Compared to conventional vehicles, AVs promise lower operational costs, increased fuel efficiency through optimized driving patterns, and environmental benefits from reduced emissions (Fagnant and Kockelman, 2015). In addition, the socio-economic potential of AVs is vast, ranging from rephrasing urban planning to enabling innovative mobility services such as autonomous taxis and freight delivery systems (Anderson *et al.*, 2014).

Unlike conventional vehicles, AVs are tightly coupled cyber-physical systems, where flaws in the software supply chain can manifest directly as a safety hazard on the road. For instance, compromised dependencies, misconfigured continuous integration (CI) pipelines, or malicious open-source packages may cascade from perception modules into planning and control, resulting in unsafe maneuvers and potential accidents (Guo *et al.*, 2023; Ladisa *et al.*, 2023). Reports already indicate that nearly half of all cyberattacks in the automotive sector exploit weaknesses in software systems, underscoring the urgency of securing AV supply chains (Durlik *et al.*, 2024). At the same time, the AV domain's reliance on both open-source ecosystems (e.g., Autoware, Apollo, openpilot) and proprietary platforms highlights the trade-offs between transparency, innovation, and the controlled assurance offered by closed industrial systems (Aliane, 2025; Kato *et al.*, 2018; Yurtsever *et al.*, 2020).

This chapter addresses these challenges by analyzing software supply chain vulnerabilities in AVs through a multi-layered approach.

We first examine the architecture of AVs as software-defined vehicles and identify how dependencies across hardware and software amplify supply chain risks. We then review existing vulnerability analysis methods, threat modeling tools, and standard risk assessment approaches (ISO/SAE, 2021), situating them in the context of safety-critical automotive systems. Building on this foundation, we apply static code analyzers to mainstream open-source AV repositories to uncover Common Weakness Enumerations (CWEs) within their code bases, and complement this with Software Bill of Materials (SBOM) tools to detect vulnerable third-party dependencies. Finally, we explore how these findings can be translated into meaningful risk scoring that reflects both technical severity and system-level safety impact (FIRST.Org, Inc., 2023).

By combining a structured review with empirical vulnerability analysis of widely used open-source AV software stacks, this work seeks to inform researchers, practitioners, and policymakers about the real and emerging risks in AV software supply chains. Ultimately, the goal is to emphasize the importance of embedding secure development practices early in the lifecycle of AV software, thereby enhancing resilience, maintaining public trust, and ensuring that the promise of autonomous mobility can be realized safely.

3.2 Autonomous Vehicles: Taxonomy, System Architecture, and Software Foundation

The development of AVs is often described using the Society of Automotive Engineers (SAE) taxonomy, which defines six levels of automation from Level 0 (no automation) to Level 5 (full automation) (ORAD Committee, 2021). These levels are briefly described at Fig. 3.2. At Level 1, vehicles support drivers with functions such as adaptive cruise control, while Level 2 combines features like lane-keeping with driver oversight. Level 3 introduces conditional automation, where the system manages most driving tasks but may require human intervention under certain conditions. Levels 4 and 5 represent high and full automation, respectively. Level 4 AVs can operate autonomously within geo-fenced areas or specific scenarios, while Level 5 AVs are envisioned to drive independently in all environments without human input (Shladover, 2018). This stratification highlights the gradual integration of automation

Level of Automation	Level 0	Level 1	Level 2	Level 3	Level 4	Level 5
Name	No Automation	Driver Assistance	Partial Automation	Conditional Automation	High Automation	Full Automation
Narrative	Full-time performance by the driver of all aspects of driving	Driving mode-specific control by an ADAS of either steering or speed	Driving mode-specific execution by one or more ADAS for both steering and speed	Driving mode-specific control by an ADAS of all aspects of driving		

Fig. 3.2. Level of autonomation of autonomous vehicles according to the SAE taxonomy.

<table>
<tr><th rowspan="2">Level</th><th colspan="3">Responsibility</th></tr>
<tr><th>Vehicle Direction and Speed</th><th>Monitoring Environment</th><th>Fallback</th></tr>
<tr><td>0</td><td>Driver</td><td rowspan="3">Driver</td><td rowspan="4">Driver</td></tr>
<tr><td>1</td><td>Driver and System</td></tr>
<tr><td>2</td><td rowspan="4">System</td></tr>
<tr><td>3</td><td rowspan="3">System</td></tr>
<tr><td>4</td><td rowspan="2">System</td></tr>
<tr><td>5</td></tr>
</table>

Fig. 3.3. Responsibilities of drivers vs. autonomous driving system in different levels of automation.

technologies and the increasing reliance on complex software systems as autonomous levels advance. Figure 3.3 depicts how the responsibilities change between the driver and the automation system across different levels.

Most recent vehicles of the current time are certified as *Partial* or *Conditional Automation* (Level 2 or 3). As of April 2024, only Honda and Mercedes have sold or leased Level 3 cars (Honda Motor Co., 2020; Mercedes-Benz Group, 2024; Wahba, 2024). Waymo offers Level 4 robotaxi services in parts of Arizona, California, Texas, Florida, Georgia, and Tennessee without safety drivers (Kusano *et al.*, 2024; Shepardson, 2020; Waymo LLC, 2024). Recent research and development are introducing near-future and experimental products. However, the ultimate goal for autonomous driving is to progress from the existing solutions of Levels 2 to 4, towards achieving Level 5 Full Automation.

The architecture of an AV is a tightly coupled ecosystem of hardware and software, designed for safety-critical decision-making in dynamic and uncertain environments. A high-level flow of the core hardware and software modules is illustrated in Fig. 3.4. Hardware components include exteroceptive sensors (cameras, LiDAR, radar, ultrasonic sensors) for perception, proprioceptive sensors (inertial measurement units, GPS, odometry) for localization and motion tracking (Ulbrich *et al.*, 2017), communication bus, and computational hardware. These components feed data into a layered software stack encompassing perception, localization, prediction, planning, and control modules (Yurtsever *et al.*, 2020). The decision-making framework integrates outputs from these modules to balance safety, efficiency, and legal/ethical considerations in real time, a safety-critical function even amid uncertain and dynamic environments (Liu *et al.*, 2021; Pendleton *et al.*, 2017). Figure 3.5 illustrates examples of common hardware and software components of an AV. Along with the sensors, the communication units traverse data through different components and the computational unit helps to process useful data to provide necessary decisions.

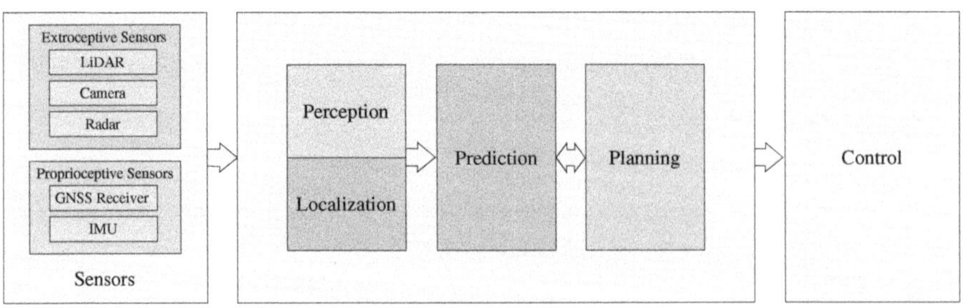

Fig. 3.4. Example general architecture of an AV automation components.

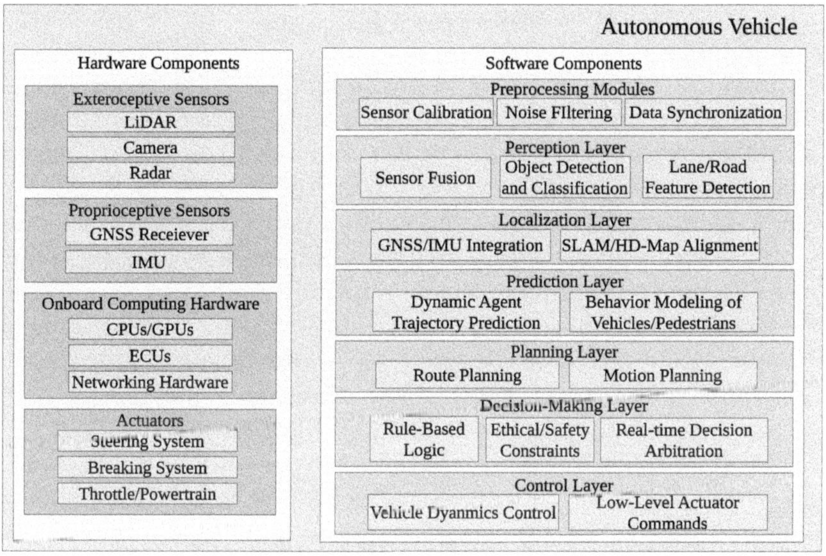

Fig. 3.5. Example hardware and software components of AVs.

AVs are inherently Software Defined Vehicles (SDVs). An SDV is a vehicle whose core functionality, performance, and user experience are primarily controlled and upgraded through software rather than fixed mechanical or hardware systems. Unlike conventional vehicles, where hardware changes dictate capability, SDVs decouple functionality from hardware by leveraging centralized computing platforms, over-the-air (OTA) updates, and modular software architecture. Perception, decision-making, path planning, and control functionalities of the AVs are entirely orchestrated through complex software stacks that interpret multi-sensor inputs and generate safety-critical actuation commands. This integration makes the software stack the operational backbone of the AVs (Teixeira *et al.*, 2024). The advantages of SDVs include significant flexibility, seamless distribution of new features (i.e., AI models, user interface, algorithms, etc.), improved resource utilization, and support for a wide ecosystem of vehicular services (Liu *et al.*, 2022). However, this reliance on software also exposes vehicles to heightened cybersecurity and privacy concerns, including over-the-air vulnerabilities, third-party supply-chain risks, and data inference threats, necessitating robust multi-layered defense strategies. System safety, quality assurance, and resilience to software and hardware anomalies are paramount to ensuring functional correctness and preventing catastrophic failure modes (De Vincenzi *et al.*, 2024).

Hardware components, including exteroceptive sensors such as LiDAR, radar, and cameras, along with proprioceptive units like inertial measurement units (IMUs), and GNSS receivers, generate high-volume, heterogeneous data streams that must be fused, interpreted, and acted upon by the vehicle's software stack (Grigorescu *et al.*, 2020). The data-driven nature of decision-making requires these raw sensor inputs to undergo multi-layered software processing: perception algorithms convert sensor data into scene representation; prediction modules estimate the trajectories of surrounding agents; and planning frameworks translate this contextual understanding into safe, executable motion strategies (Pendleton *et al.*, 2017).

This interdependency means that data originating from one hardware subsystem is often indispensable for the correct functioning of another. For instance, LiDAR point clouds may be fused with camera imagery through software pipelines to enhance object classification accuracy, which in turn informs motion planning and low-level control executed via actuator hardware (Wang *et al.*, 2019). A smooth flow of information between sensing hardware and computational modules is thus fundamental to safe and efficient AV behavior.

Critically, due to the software and data-driven nature of the AVs, it can be stated that sufficient hardware alone is inadequate to ensure safety. Even state-of-the-art sensors are limited by noise, occlusion, adverse weather, and bandwidth constraints. Without robust software to filter out errors, compensate for uncertainties, and enable redundancy, safety cannot be comprehensively achieved (Grigorescu *et al.*, 2020).

3.3 Software Supply Chain of Autonomous Vehicles

The software supply chain in AVs encompasses the full spectrum of activities involved in the development, integration, distribution, and maintenance of vehicle software systems. This supply chain can be broadly described as: (i) upstream open-source libraries (e.g., middleware such as ROS 2, open-source perception or control libraries, different packages, etc.), (ii) third-party commercial modules (e.g., mapping services, localization services, etc.), (iii) proprietary in-house codebases developed by manufacturers for planning, decision-making, and system integration, (iv) development and integration pipelines (e.g., CI/CD infrastructure, automated testing, containerization platforms), (v) distribution and update mechanisms (e.g., signed binarie distribution, OTA updates, etc.), and (vi) operational services that support runtime (e.g., cloud-based services and simulation, remote monitoring and diagnostics, etc.). In traditional automotive systems, software is confined to limited embedded functions. In contrast, AVs are software-centric systems that require continuous updates, third-party integrations, and cloud-based services for perception, mapping, connectivity, and safety validation. Weaknesses in any stage of this supply chain, whether from insecure code dependencies, compromised libraries, or misconfigured update mechanisms, pose systemic risks to the safety and reliability of the AVs and passenger safety. As a result, secure life-cycle practices, including software provenance, rigorous testing, and continuous monitoring, are critical (ISO/SAE, 2021). Organizations such as the National Institute of Standards and Technology (NIST) and the European Union Agency for Cybersecurity (ENISA) highlight the software supply chain assurance critically (ENISA, 2021; National Institute of Standards and Technology, 2022). Considering all these, the life-cycle of AV software can be summarized as: (i) requirements and hazard analysis, (ii) secure design and coding, (iii) integration of third-party components, (iv) testing and verification, (v) signed builds and distribution, and (vi) deployment with post-deployment monitoring and patching. Ensuring integrity across the life cycle requires provenance tracking (e.g., Software Bill of Materials (SBOM)), secure build pipelines, and rigorous update mechanisms.

AV software stacks exist in both open-source and closed-source forms. Open-source repositories like Autoware, Apollo, and openpilot make their codebases publicly accessible, which brings advantages such as transparency, peer review, faster debugging, reproducibility for academic studies, and flexibility for integration with diverse hardware platforms (Chen *et al.*, 2022; Kato *et al.*, 2018; Raju *et al.*, 2019). These advantages make these platforms valuable for academic research and early pilot deployments. However, open-source reliance comes with many challenges: security vulnerabilities are publicly visible, maintenance might depend on small developer communities, and performance tuning in real-time environments can be inconsistent. In contrast, closed platforms used by companies like Waymo or Tesla develop their stacks with large in-house engineering teams and controlled hardware environments. This model ensures tighter integration, dedicated security resources, and industrial-scale quality assurance, but at the expense of transparency and community evaluation. Even though they are closed source, they also have dependency over third-party modules and components. Thus, both open and closed source approaches highlight the trade-offs between innovation, transparency, and operational assurance in the AV software supply chain.

The most widely studied open-source platforms, Autoware, Apollo, and openpilot, highlight the diversity of software supply chain design. Autoware is the first widely adopted "all-in-one" open-source AV stack, implemented on ROS/ROS2. It contains modular components for perception, localization, planning, and control, and has been validated in real-vehicle demonstrations, including embedded implementations (Kato *et al.*, 2018). Apollo, released by Baidu, is oriented towards production-scale deployment and integrates perception with deep-learning models, high-definition mapping, and planning frameworks. A key distinction is its Cyber RT middleware, which replaces ROS and provides a static scheduling model to improve real-time performance. The Apollo Cyber RT framework is the world's first open-source, high-performance runtime framework designed specifically for the development of autonomous driving technologies. It was introduced for the first time with the release of the Apollo 3.5 open-source platform (Apollo Baidu Team, 2025; Belluardo *et al.*, 2021). openpilot has a much narrower scope, limited to Level 2 driver assistance (adaptive cruise control, lane keeping, driver monitoring), but demonstrates how lightweight open-source systems can support consumer use cases (Saez-Perez *et al.*, 2025). Comparative studies show that middleware choice can strongly affect performance: for example, ROS 2/DDS as used in Autoware suffers from variable latency under heavy load, while Apollo's Cyber RT achieves lower latency and more deterministic throughput (AndreiGavrilov *et al.*, 2021). Technical comparison between the three platforms is summarized in Table 3.1. These architectural differences demonstrate why supply chain robustness depends not only on code provenance but also on middleware performance and update infrastructure, both of which are integral to safe and scalable AV deployment.

Table 3.1. Technical comparison of Autoware, Apollo, and openpilot.

Feature	Autoware	Apollo	openpilot
Primary Language	C++ and Python	C++ and Python	C++ and Python
Target Autonomy Level	L3–L4	L3–L4	L2
Middleware	ROS / ROS 2 (DDS-based, pub-sub)	Cyber RT (custom, static scheduling, high-throughput)	Lightweight in-house messaging framework
Core Software Modules	Perception, Localization, Planning, Control	Perception (DL models), Localization, Prediction, Planning, Control, Simulation, HD Maps	Lane Keeping Assist, Adaptive Cruise Control, Driver Monitoring
Performance Considerations	ROS latency under load; DDS QoS tuning critical (Ye *et al.*, 2023)	Cyber RT deterministic scheduling, higher throughput than ROS (AndreiGavrilov *et al.*, 2021)	Lightweight but limited scalability beyond L2
Community and Ecosystem	Open-source, large academic and research community	Open-source, strong industry backing (Baidu), contributors from partners	Open-source but maintained primarily by comma.ai

3.4 Autonomous Vehicles Software Supply Chain Vulnerabilities

Across modern software stacks, supply-chain risks dominate because attackers can compromise systems before code even runs in production, i.e., through malicious packages, typosquatting, dependency confusion, compromised maintainers, poisoned build or CI pipelines, or tampered distribution artifacts. A comprehensive systemization by Ladisa *et al.* cataloged 107 attack vectors across the contribution, build, and distribution stages and linked them to 94 real incidents, portraying that these threats are not tied to any one programming language or ecosystem (Ladisa *et al.*, 2023). These "SoK" (Systematization of Knowledge) studies and follow-ups show that practical defenses must combine provenance, anomaly detection for packages, and build-time hardening, rather than only relying on a single safeguard (Ohm *et al.*, 2020).

Open-source AV stacks (e.g., Autoware, Apollo, openpilot) are primarily written in C/C++ (performance-critical perception, planning, and control) plus Python (tooling, glue code, ML workflows). Language-specific risks compound general supply-chain issues, which are discussed in Table 3.2. In Python, large-scale empirical analyses of the PyPI ecosystem identified 4,669 malicious package files, with 75% of the flagged packages reaching end-user projects (often through pip source installs), and frequent behaviors such as credential exfiltration and command execution can clear routes to compromise AV developer environments and CI (Guo *et al.*, 2023). In C/C++, peer-reviewed studies and surveys continue to find memory-safety flaws (buffer overflows, use-after-free (UAF), integer overflows, etc.) to be prevalent and exploitable despite mitigation efforts (Butt *et al.*, 2022); recent evaluations document how attackers still bypass defenses or trigger UAFs that enable control-flow hijack (Zhao *et al.*, 2024b). Together, these findings imply that UAFs that AV codebases inherit both ecosystem-level (package/mirror) and language-level (memory-unsafety) risks, even when upstream modules are popular or long-standing.

For AVs, the quintessential cyber-physical systems, software defects or compromised dependencies can propagate into unsafe vehicle behavior, such as corrupt perception or localization pipelines misclassify/lose obstacles, poisoned models or ROS/ROS 2 message paths desynchronize modules, and runtime memory errors in C/C++ planning/control code can induce denial-of-service or arbitrary actuation. Security studies document exploitable weakness in ROS/ROS 2 isolation and multi-tenant settings (Teixeira *et al.*, 2020; Xia *et al.*, 2025). Automotive cybersecurity reviews show how attacks on sensors

Table 3.2. Common software supply chain vulnerabilities in AVs: Impacts and mitigations [Ohm, Marc, *et al.* (2020)].

Vulnerability Type	**Example/ Source**	**Impact on AV Stack**	**Safety Implications**	**Mitigation Strategies**
Supply-chain attacks (agnostic)	Dependency confusion, typosquatting	Malicious code injected into OSS modules or CI/CD	Unsafe updates, backdoored perception/ control code	Provenance checks, SBOMs, reproducible builds
Python package threats	Malicious PyPI packages	Compromised perception/ ML pre-processing pipelines	Data poisoning, system hijack	Package vetting, signed package registries
C/C++ memory-safety flaws	Buffer overflows, UAFs	Exploits in planning/ control loops	DoS, arbitrary actuation commands	Fuzzing, sanitizers, memory-safe rewrites (Rust modules)
Middleware weaknesses	ROS 2 isolation flaws	Inter-process message tampering, desync	Loss of perception-control synchronization	DDS security, isolation, QoS tuning
Sensor/ perception attacks	Spoofing or jamming	Corrupted perception $\rightarrow$ wrong planning inputs	Pedestrian misdetection, unsafe maneuvers	Sensor fusion, anomaly detection, redundancy

and perception cascade into hazardous decisions, bridging software vulnerabilities and physical safety harms (Durlik *et al.*, 2024; Islam *et al.*, 2023). Consequently, mitigation is safety-relevant: AV teams must combine supply-chain controls with language-aware hardening and middleware-level protections to prevent defects from becoming hazardous vehicle actions. Table 3.2 illustrates the impact of these vulnerabilities and possible mitigation strategies.

Vulnerabilities are particularly critical in safety-critical systems such as AVs, whether they originate in hardware or software. Since this chapter emphasizes the software supply chain, it is essential to recognize that analyzing vulnerabilities at the system level provides a foundation for mitigating many of these risks. As AVs evolve into SDV, systematic risk and threat analysis becomes indispensable to ensure the integrity and resilience of their software supply chain. The following sections, therefore, examine existing analysis methods, modeling tools, and standard approaches that can be applied to vulnerability assessment in the AV context.

3.4.1 *Existing Vulnerability Analysis Methods*

Software risk or threat analysis systematically identifies plausible attack scenarios, estimates the likelihood or impact, and prioritizes mitigation across the software life cycle. Core steps include: (i) asset or dependency identification, (ii) threat elicitation, (iii) vulnerability enumeration, (iv) risk estimation, and (v) control selection, as formalized in NIST SP 800-30 (National Institute of Standards and Technology, 2022) and widely reused in safety-critical domains. In practice, organizations blend threat-modeling methods with risk analysis: STRIDE (property-centric enumeration of Spoofing, Tampering, Repudiation, Information disclosure, Denial of service, Elevation of privilege), PASTA (Process for Attack Simulation and Threat Analysis; seven-stage, risk-centric "attacker perspective"), LINDDUN (Linking, Identifying, Non-repudiation, Detecting, Data Disclosure, Unawareness, and Non-compliance; privacy-threat focus), attack trees/graphs, and SPTA-Sec-style (Systems-Theoretic Process Analysis for Security) system-hazard analyses for Cyber-Physical Systems. Systematic reviews document method families, steps, and comparative coverage of security properties (Lowe *et al.*, 2009; Wolf *et al.*, 2021; Xiong *et al.*, 2019). These methods complement supply-chain assurance by surfacing risks introduced by third-party libraries, build/CI infrastructure, middleware, and OTA channels (Ross, 2012).

In addition to model-based and process-oriented methods, static analysis tools are widely used to automatically detect code-level weaknesses by scanning source code against known patterns of vulnerabilities. For AV software stacks implemented largely in C/C++ and Python, static analyzers can be applied to the open-source repositories to identify Common Weakness Enumerations (CWEs). Empirical studies demonstrate that static analysis can effectively surface categories of flaws such as buffer overflows, use-after-free, or input-validation errors, which are particularly relevant in performance and safety-critical C/C++ modules (Ayewah and Pugh, 2010; Beller *et al.*, 2016; Livshits and Lam, 2005). While static analysis can suffer from false positives and may not capture runtime or configuration-level threats, it provides an essential first layer of vulnerability discovery by systematically narrowing the search space for deeper investigation.

Widely used analyzers in this context include Clang Static Analyzer, Cppcheck, Coverty, and CodeQL for C/C++, as well as Bandit and PyLint for Python (Beller *et al.*, 2016). Complementing these, dynamic analysis tools such as Valgrind, AddressSanitizer, and AFL (American Fuzzy Loop) provide runtime monitoring and fuzzing-based testing to uncover memory errors, resource leaks, and input-handling vulnerabilities that a static analyzer may be unable to identify (Lattner and Adve, 2004; Serebryany *et al.*, 2012; Zalewski, 2017). In practice, static analysis outputs are most effective when used as inputs to subsequent dynamic testing, manual triage, and system-level threat modeling, rather than as definitive indicators of exploitability or safety impact. Accordingly, for this chapter, we limit our empirical analysis to static analyzers and position the resulting findings as a structured baseline that informs prioritization and motivates deeper, AV-specific security assessment in later stages of the software lifecycle.

3.4.2 *Threat Modeling Tools*

Several tools operationalize the methods above with varying depth and automation. Model-based attack-graph simulators such as securiCAD generate attack paths over formalized meta-models to estimate probabilities, time-to-compromise, and control effects; peer-reviewed studies show their use for enterprise and vehicle domains (Katsikeas *et al.*, 2024; Mao *et al.*, 2019; Xiong *et al.*, 2019).

Strengths include explicit path analysis and "what-if" control evaluation; weaknesses include modeling cost, assumptions about attacker capabilities, and sensitivity to parameterization. Surveyed practice also highlights diagram-centric tools aligned to STRIDE, or PASTA, or LINDDUN that facilitate data-flow modeling and catalog-based threat enumeration. Literatures report strong onboarding and communication benefits, but limited quantification and coverage bias without domain-specific tailoring (Xiong *et al.*, 2019). On the other hand, static analyzers are not threat-modeling tools, but the vulnerability data they produce, e.g., CWE mappings, can be integrated into higher-level threat models or attack graphs. For AV stacks, tools that (i) capture supply-chain structure (dependencies, build provenance), (ii) reflect middleware/runtime interactions (ROS 2/DDS, Cyber RT), and (iii) support attack-graph reasoning are most suitable for analyzing cascading safety effects from software compromise.

3.4.3 *Standard Approaches*

Standardized approaches provide the foundation for assessing risks in any system by offering structured, repeatable processes. In the context of AVs, these approaches are essential for aligning general software assurance practices with domain-specific safety and cybersecurity requirements. This section outlines both general frameworks and automotive-specific practices, before considering their adaptation to the AV software stacks:

3.4.3.1 *General Software Analysis Standards*

The NIST Secure Software Development Framework (SSDF) prescribes organization and build-level practices (e.g., hardened builds, code provenance, review/testing, vulnerability response) and explicitly maps to federal acquisition expectations. It is framework-agnostic and intended to overlay any Software Development Life Cycle (SDLC) (Souppaya *et al.*, 2022). For risk assessment, NIST SP 800-30 remains the canonical method (risk framing, assessment, response, monitoring) and is often combined with control catalogs or domain guidance (Ross, 2012). SBOM standards underpin supply-chain transparency by tracking upstream components and transitive dependencies (Xia *et al.*, 2023).

3.4.3.2 *Autonomous Vehicle Specific Practices*

In road vehicles, ISO/SAE 21434 mandates TARA (Threat Analysis and Risk Assessment) integrated across concept, development, production, and post-production; UN R155 further requires a Cybersecurity Management System (CSMS) for type approval, operationalizing risk-based processes over the vehicle life cycle. Academic evaluations propose and compare automotive TARA methods, e.g., HEAVENS 2.0 aligns with ISO/SAE 21434 while addressing gaps (asset/threat cataloging, attack-feasibility, risk scoring) and has seen industrial uptake; case studies combine TARA with STRIDE for connected vehicle functions (Boi *et al.*, 2023; Lautenbach *et al.*, 2021). These standards/frameworks primarily target system-level cyber-risk and interface threats; they do not fully standardize software supply-chain assurance, necessitating adoption of SSDF or SBOM controls alongside ISO, SAE, and UN requirements for AV software.

3.4.3.3 *Adapting to AV Software Stacks*

For perception-localization-planning-control pipelines, combining automotive TARA (safety system context) with software supply-chain controls (SSDF tasks; SBOM for dependency provenance) produces traceability from third-party components to safety goals (National Institute of Standards and Technology, 2022; Xia *et al.*, 2023). The advantages are end-to-end coverage from upstream code through middleware/runtime, measurable build attestations, and repeatable patch/recall decisions, although it can cause modeling overhead.

While standards such as ISO/SAE 21434 (ISO/SAE, 2021), UN R155 (UNECE, 2021), and NIST SSDF (Souppaya *et al.*, 2022) provide structured processes for conducting vulnerability and risk assessments, they do not in themselves resolve the question of how to prioritize risks across the software supply chain. Static analyzers, SBOM-based provenance tracking, and model-driven assessments surface numerous weaknesses that are often mapped to CWEs; organizations require a systematic way to compare and rank these findings according to both technical severity and safety impact. This gap is addressed through risk scoring frameworks, which translate identified vulnerabilities into actionable metrics that guide mitigation priorities.

Risk can be scored qualitatively for speed and stakeholder alignment, or quantitatively to express loss exposure and support cost-benefit trade-offs. Common Vulnerability Scoring System (CVSS) v4.0 standardizes vulnerability severity at the component level with Base/Threat/Environment/Supplemental metrics. It is useful for triaging AV stack dependencies, but does not capture system-level safety impact by itself (Aggarwal, 2023; FIRST.Org, Inc., 2023). On the other hand, Factor Analysis of Information Risk (FAIR) style quantitative risk analysis models the frequency and magnitude of loss. Peer-reviewed work shows Bayesian/FAIR approaches can support monetary risk decisions and investment optimization, complementing qualitative TARA outputs (He *et al.*, 2024; Wang *et al.*, 2020).

3.5 Case Study: Vulnerability Analyses of Autoware, Apollo, and openpilot Platforms

In this section, we present an approach for vulnerability analysis for three different open-source AV software packages (i.e., Autoware, Apollo, and openpilot) and corresponding findings related to code and third-party library vulnerabilities.

3.5.1 *Approach for Vulnerability Analyses*

The approach for analyzing vulnerabilities in AV software stack consists of a set of steps, as illustrated in Fig. 3.6. The workflow follows a structured, stepwise approach from repository selection to actionable recommendations. The first step involves selecting the AV software repositories for analysis. In this case study, we select three open-source repositories: openpilot, Autoware, and Apollo. These repositories provide a diverse set of Python and C/C++ codebases representative of common AV software components. Once projects are selected, static analysis is conducted to detect potential vulnerabilities in the code and associated dependencies. This phase has two main components: (i) code vulnerability identification, and (ii) third-party library vulnerability identification. Tools, such as

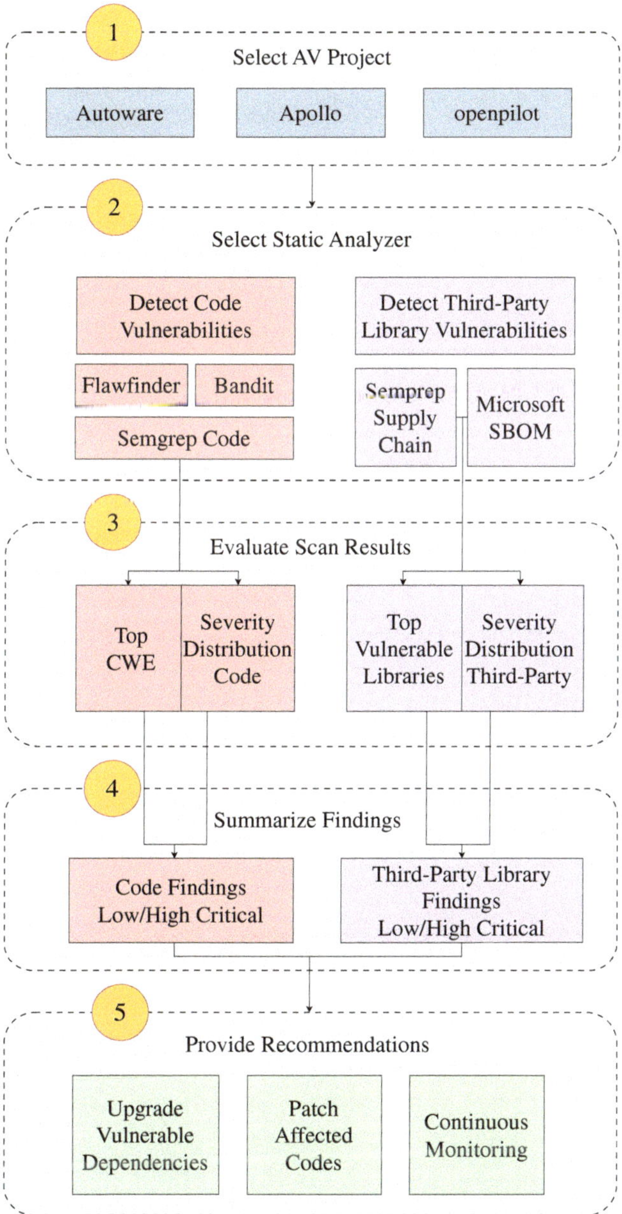

Fig. 3.6. Vulnerability analysis workflow.

Flawfinder (for C/C++), Bandit (for Python) and Semgrep, are employed to identify coding vulnerabilities, including buffer overflows, command injections, insecure input handling, and other critical weaknesses. Dependencies and third-party packages are analyzed using Semgrep's supply chain rules and Microsoft SBOM tools to detect outdated, insecure, or vulnerable components. After analyzing, the results are evaluated to understand the severity and distribution of vulnerabilities. The findings are categorized by CWEs, and severity distributions are analyzed to identify the most critical issues. Dependency vulnerabilities are prioritized based on severity and reachability, highlighting the most at-risk components. The critical and high-severity vulnerabilities from both code and third-party library analyses are summarized. This provides a comprehensive overview of the major risks present across the selected AV repositories and facilitates comparison between different types of vulnerabilities. Finally, actionable recommendations are provided to mitigate the identified risks, including upgrading vulnerable dependencies to patched versions, patching affected code to resolve critical and high-severity issues, and continuous monitoring of codebases and dependencies to proactively address future vulnerabilities.

The outputs of static analysis tools are treated as candidate vulnerabilities rather than confirmed exploits. Given the size and complexity of modern AV software repositories, the objective of this analysis is to surface recurring weakness patterns and high-risk components that demand deeper inspection.

3.5.2 *Attacker Model and Assumptions*

To contextualize the reported vulnerability findings and their potential impact on AV software systems, we define a minimal attacker model. The goal is not to exhaustively enumerate adversaries, but to make explicit the assumptions under which static analysis results can be interpreted and prioritized. We consider two primary attacker classes relevant to AV software supply chains:

(1) Runtime attacker: This attacker interacts with the AV system after deployment, leveraging exposed interfaces, such as inter-process communication channels, configuration files, or

externally supplied data. The majority of the source code vulnerabilities identified in this chapter are most relevant under this threat model.

(2) Supply-chain attacker: This attacker targets software development and the integration pipeline rather than deploying the attack in the vehicle directly. Capabilities include compromising third-party dependencies, injecting malicious code into upstream repositories, exploiting vulnerabilities in build scripts or CI/CD infrastructure, and distributing tampered artifacts through update mechanisms. Such an attacker may gain indirect execution on developer machines, build servers, or onboard compute platforms via vulnerable dependencies or tools. The majority of third-party library vulnerabilities identified in this chapter are most relevant under this threat model.

3.5.3 *Method and Experimental Setup*

This section illustrates the configurations, specific versions, and data processing logic to reproduce our analysis.

3.5.3.1 *Open-Source Repository Selection and Scope*

We select Autoware, Apollo, and openpilot as representative open-source autonomous driving software stacks based on both their community popularity and prominence in autonomous vehicle research. These repositories are among the most actively starred and widely referenced open-source projects in the self-driving domain. In addition to the GitHub activity, these platforms are recurrent subjects in the autonomous driving literature. Recent surveys and comparative studies of open-source autonomous driving systems specifically highlight Autoware and Apollo for their use in perception, planning, middleware, and system-level research, and note openpilot for its broad adoption in research on driver-assistance and real-world deployments (Aliane, 2025; Jung *et al.*, 2025; Zhao *et al.*, 2024a). Collectively, they provide a diverse cross-section of architectural designs and development practices, enabling a comprehensive assessment of software quality and security patterns in open-source AV development.

Autoware: Managed by the Autoware Foundation (Autoware Foundation, 2024), this repository contains the stable ROS 2 packages that form the essential backbone of the Autoware ecosystem. While experimental features live in the Universe branch, Core focuses on standardized, finalized architectures for reliable autonomous driving. We used the autoware_core[1] GitHub repository from the main branch, cloned on September 18, 2025, for our study.

Apollo: Developed by Baidu, Apollo (Baidu Apollo, 2024) is a high-performance, industrial-grade autonomous driving platform designed to accelerate the development and deployment of autonomous vehicles. The architecture is primarily built on C++ (74.5%) and utilizes the specialized CyberRT middleware to handle complex urban driving scenarios and real-time data processing. We analyse the Apollo[2] GitHub repository from the main branch, cloned on September 18, 2025.

Openpilot: Developed by comma.ai, the open-source ADAS (Advanced Driver Assistance System) are widely used globally in over 300 supported vehicles (comma.ai, 2024). It is developed primarily in Python and C++, and it emphasizes safety and reliability through compliance and rigorous hardware-in-the-loop testing. We used the openpilot[3] GitHub repository from the main branch, cloned on September 18, 2025, for static and SBOM analysis in our study.

3.5.3.2 *Tool Configuration and Rulesets*

In our study, the following tools were configured to prioritize actionable and high-severity vulnerabilities. This filtering process was essential to distinguish between risks and exploitable weaknesses within the AV codebases.

Flawfinder: This tool was executed with a minimum risk threshold. We utilized the default internal ruleset, which performs lexical scanning to identify high-risk C/C++ functions, such as `memcpy`, `strcpy`, and `scanf`. The findings were automatically mapped to the **Common Weakness Enumeration** (CWE) standards to facilitate systematic categorization and risk scoring.

---

[1]https://github.com/autowarefoundation/autoware_core.
[2]https://github.com/ApolloAuto/apollo.
[3]https://github.com/commaai/openpilot.

Bandit: We use Bandit for static analysis of Python source code, utilizing specialized plugins to identify vulnerabilities. The output was filtered to their severity and confidence levels to minimize false positives and prioritize critical logic flaws within the system's core layers.

Semgrep: Our analysis integrated security and supply-chain rulesets to audit both internal logic and external dependency risks into the selected open-source projects. We identified when vulnerable library functions were actually invoked, which tells the results focused exclusively on actionable security threats within the codebase.

Software Bill of Material (SBOM) Analysis (Syft and Grype): For the software supply chain audit, we utilized Syft to generate detailed SBOMs in the industry-standard CycloneDX format. The resulting SBOMs were processed by Grype, which was configured to query the National Vulnerability Database (NVD) and GitHub Security Advisories. Our analysis focused specifically on vulnerabilities with available patches to provide an assessment of the platforms' maintenance status and update cycles.

3.5.3.3 *Handling Overlapping Findings*

To prevent duplicate counting, we segregated tool execution by language, applying Flawfinder solely to C/C++ source files and Bandit/Semgrep to Python files. For the Python analysis, findings were cross-referenced by line number and CWE type to ensure that overlapping reports from different tools were treated as a single security flaw in the final count.

3.5.4 *Source Code Findings*

We applied multiple static analysis tools, i.e., Flawfinder, Bandit, and Semgrep, across Apollo, Autoware, and openpilot. Using a combination of language-specific and cross-language analyzers enabled us to capture a wide range of vulnerabilities, from low-level memory errors in C/C++ to insecure coding practices in Python. These tools allowed us to identify recurring weaknesses, map them to CWE categories, and assess their severity to better understand the security posture of each repository. This comparative analysis not only reveals repository-specific risks but also uncovers broader vulnerability patterns, offering valuable insights for AV researchers and practitioners to strengthen security during development and deployment.

3.5.4.1 *Flawfinder and Bandit Analysis Overview*

Static analysis tools like Bandit (PyCQA Bandit Developers, 2025), and Flawfinder (Wheeler, 2023) provide developers with automated ways to identify potential security vulnerabilities in their code before deployment. Bandit focuses on Python, scanning source files to identify insecure coding patterns, cryptographic weaknesses, and hard-coded secrets, while Flawfinder analyzes C/C++ code to detect common risks, such as buffer overflows, format string issues, race conditions, and unsafe system calls. Both tools generate prioritized reports that help developers focus on the most critical security issues, though they are not substitutes for secure coding knowledge or human review. In Table 3.3, we present a summary of the Flawfinder and Bandit analyses conducted on three open-source AV repositories: Apollo, Autoware_core, and Openpilot. Table 3.3 captures both C/C++ and Python files, showing the total number of analyzed files, statistical metrics (mean, median, maximum, and minimum) for the number of lines per vulnerable file, and the total number of identified vulnerabilities for each repository and tool. This summary provides a quick overview of the scale of analysis and the distribution of vulnerabilities across different languages and repositories.

3.5.4.2 *Vulnerability Density Analysis*

We calculated the vulnerability density by normalizing the total number of identified CWEs against the thousands of lines of code (KLOC) to provide a standardized security metric for comparing the C++ and Python components across all three platforms. The following table summarizes the findings from Flawfinder (C/C++) and Bandit (Python). Vulnerability Density (VD) is calculated as:

$$\text{VD} = \frac{\text{Total CWEs}}{(\text{Analyzed Files} \times \text{Mean LOC})/1000}$$

From Table 3.3, we can have several observations. Firstly, Openpilot C/C++ files have the highest mean and maximum number of

Table 3.3. Summary of Flawfinder and Bandit analysis with vulnerability density.

Project	Language	Files	Mean LOC	Median	Max	Min	Vuln. Files	CWEs	CWE/ KLOC
Apollo	C/C++	1957	1471.52	991.5	4523	11	424	604	**0.21**
	Python	314	182.50	92	956	5	121	121	**2.11**
Autoware	C/C++	605	154.44	155	593	19	25	26	**0.28**
	Python	25	78.44	111	128	18	9	9	**4.59**
Openpilot	C/C++	432	2906.45	329	16827	10	227	310	**0.25**
	Python	481	123.43	93	808	1	1179	1179	**19.86**

lines per vulnerable file, indicating that some files are very large and potentially complex. Apollo Python files show a moderate number of vulnerabilities with a relatively low mean and median number of lines, suggesting smaller but frequent issues. Autoware-core Python files have a very small number of analyzed files and vulnerabilities, but the median and maximum line counts indicate that a few files contain larger segments of vulnerable code. Overall, the total vulnerabilities are highest in openpilot Python files, reflecting the intensive analysis of Python code in that repository. The statistical spread between mean, median, min, and max provides insight into variability, showing that some repositories have large outliers in file size while others are more consistent.

The vulnerability density analysis shows a contrast in security posture across the three projects and their respective programming languages. While the C/C++ components consistently maintain low density levels (ranging from 0.21 to 0.28 CWE/KLOC), the Python codebases exhibit higher flaw concentrations, with Openpilot reaching a peak of 19.86 CWE/KLOC. These differences stem from a combination of different tool sensitivities between Flawfinder and Bandit, as well as the inherent architectural differences between the performance-critical C++ core layers and the more script-intensive Python orchestration layers.

3.5.4.3 *Analysis of CWEs*

Figure 3.7 presents a detailed view of the CWE issues identified across three AV open-source software packages: Apollo,

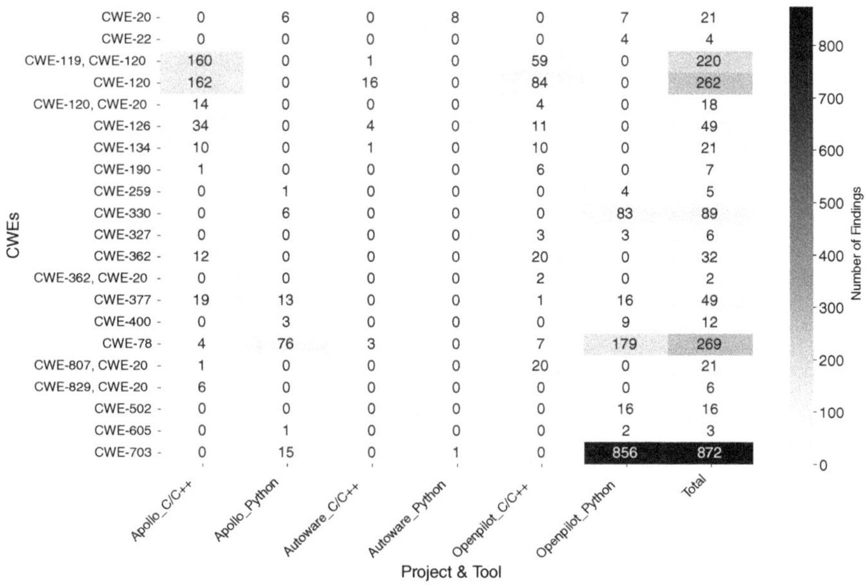

Fig. 3.7. Distribution of CWE issues detected across projects and tools.

Autoware_core, and openpilot. Both C/C++ and Python files were analyzed using Flawfinder and Bandit, respectively. These data capture the distribution of vulnerabilities across different CWE types for various AV repositories and programming languages. The heatmap provides a visual representation of the relative frequency of each CWE, with the maximum occurrences highlighted to facilitate quick insights into the most prevalent vulnerabilities.

Table 3.4 lists the selected CWEs that were most frequently observed across the analyzed AV repositories. Each CWE is accompanied by a brief description and notes on its potential impact, providing readers with a quick reference to understand the nature of common vulnerabilities in both C/C++ and Python code. Among the most frequent issues across all repositories are CWE-703, CWE-120, and CWE-119/CWE-120, indicating recurring patterns of vulnerabilities in both C/C++ and Python code. Examining repository-specific trends, Openpilot Python files exhibit the highest total number of vulnerabilities, largely driven by CWE-703,

Table 3.4. Selected CWEs.

CWE ID	Name	Impact/Notes
CWE-20 (MITRE, 2023f)	Improper Input Validation	Root cause of many attacks (injection, deserialization, etc.)
CWE-22 (MITRE, 2023g)	Path Traversal	Arbitrary file access possible
CWE-78 (MITRE, 2023q)	OS Command Injection	Remote code execution risk
CWE-119 (MITRE, 2023a)	Improper Restriction of Operations within Memory Buffer	Parent category of buffer overflows
CWE-120 (MITRE, 2023b)	Buffer Copy without Checking Size of Input	Classic buffer overflow
CWE-126 (MITRE, 2023c)	Buffer Over-read	Can cause leaks of sensitive info (e.g., Heartbleed bug)
CWE-134 (MITRE, 2023d)	Use of Externally-Controlled Format String	Can lead to code execution/memory corruption
CWE-190 (MITRE, 2023e)	Integer Overflow or Wraparound	May cause memory corruption or bypass checks
CWE-259 (MITRE, 2023h)	Use of Hard-coded Password	Credential disclosure risk
CWE-327 (MITRE, 2023i)	Use of Broken or Risky Cryptographic Algorithm	Weak crypto (e.g., MD5, DES)
CWE-330 (MITRE, 2023j)	Use of Insufficiently Random Values	Predictable keys/tokens
CWE-362 (MITRE, 2023k)	Race Condition	Can cause data corruption or privilege escalation
CWE-377 (MITRE, 2023l)	Insecure Temporary File	Exploitable in multi-user environments
CWE-400 (MITRE, 2023m)	Uncontrolled Resource Consumption (DoS)	Denial-of-service risk
CWE-502 (MITRE, 2023n)	Deserialization of Untrusted Data	Remote code execution
CWE-605 (MITRE, 2023o)	Multiple Binds to the Same Port	Service hijacking or denial of service
CWE-703 (MITRE, 2023p)	Improper Check or Handling of Exceptional Conditions	Error handling flaws leading to bypasses
CWE-807 (MITRE, 2023r)	Reliance on Untrusted Inputs in Security Decision	May allow privilege escalation
CWE-829 (MITRE, 2023s)	Inclusion of Functionality from Untrusted Control Sphere	Attacker-controlled code execution

which reflects the intensive use of Python code with security-relevant issues. In contrast, Apollo C/C++ files show a substantial number of CWE-119 and CWE-120 vulnerabilities, suggesting that certain legacy C/C++ modules may require closer review. Tool-specific patterns also become evident: some CWEs, such as CWE-330 and CWE-362, are primarily detected in Python files, highlighting the importance of using Bandit alongside Flawfinder to cover language-specific security weaknesses. Finally, distribution insights from the heatmap reveal both the concentration and spread of vulnerabilities, with red-highlighted maximum values indicating the CWE-repository combinations that are most critical, thereby helping prioritize remediation efforts.

3.5.5 *Top 10 CWE Issues Across AV Repositories*

To better visualize the distribution of the most prevalent vulnerabilities across the three AV open-source repositories, i.e., Apollo, Autoware_core, and Openpilot, a pie chart was generated for the top 10 CWE issues in Fig. 3.8. The chart summarizes the total occurrences of each CWE type, combining both C/C++ and Python analyses from Flawfinder and Bandit. This graphical representation helps quickly identify which vulnerabilities are most common and potentially critical, providing an at-a-glance understanding of the security landscape across these repositories.

The pie chart highlights several key insights from the top 10 CWEs across three AV repositories. CWE-703 dominates the distribution, largely driven by Openpilot Python files, indicating frequent use of insecure code patterns that could affect runtime security. Following this, CWE-120 and CWE-119/CWE-120 are the next most prevalent issues, primarily observed in C/C++ files, pointing to potential buffer overflow and memory management risks in legacy code. Other CWEs, such as CWE-330, CWE-78, and CWE-377, occur moderately across three AV repositories, reflecting recurring weaknesses that affect both Python and C/C++ components. Overall, the chart provides a visual ranking of vulnerabilities, enabling one to quickly identify the most critical weaknesses that require attention in these AV repositories.

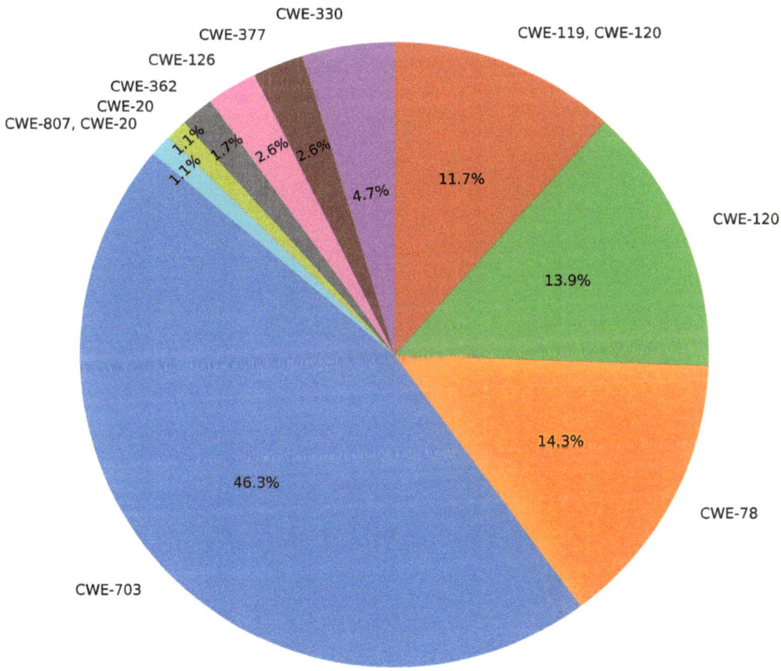

Fig. 3.8. Top-10 CWE issues across all projects.

3.5.5.1 *Semgrep Code Finding Analysis*

Semgrep (2025) is a lightweight, open-source static analysis tool that helps developers detect bugs, enforce coding standards, and uncover security vulnerabilities. Unlike simple text-based search tools, Semgrep understands code structure and semantics, making it effective for scanning across more than 30 programming languages. Its rule-based approach allows flexible customization, making it widely used for secure software development. In our analysis, we applied Semgrep to all three AV repositories. The tool reported both code findings (direct vulnerabilities in the source code) and third-party library findings (issues originating from external dependencies). The results reveal varying risk profiles across AV repositories, with Apollo showing the highest number of both code and third-party library issues, while Autoware Core had comparatively fewer findings. The Semgrep

Table 3.5. Semgrep static analysis findings across projects.

Project	Code Findings	Supply Chain Findings	Scanned Dependencies
Apollo	71	91	1,515
Openpilot	57	1	314

results indicate that Apollo has the largest security exposure, with 71 code findings and 91 third-party library issues across a wide dependency base of over 1,500 libraries. Openpilot shows a notable number of 57 code findings, but only one third-party issue, suggesting its risks are more localized within its own source code. In contrast, Autoware Core has very limited findings, with just seven code-related issues and no third-party library vulnerabilities, which may reflect a smaller or more contained codebase.

3.5.5.2 *Semgrep Findings*

To better understand the security weaknesses identified by Semgrep, the code findings were mapped to their corresponding CWE identifiers as shown in Table 3.5. Table 3.6 presents this mapping, showing the repository, CWE, severity, and the number of code findings for each issue. This layout highlights both the type and severity of vulnerabilities across the three AV repositories, as well as the frequency of their occurrence, allowing one to quickly identify the most prevalent and critical security issues.

3.5.5.3 *Semgrep's Top 10 CWEs*

In order to highlight the most frequent security weaknesses detected by Semgrep, we summarized the top 10 CWE identifiers based on coding findings across the three AV repositories. This table shows both the type of vulnerability and the number of occurrences, providing a clear overview of the most critical coding issues. From Table 3.7, several observations can be made. CWE-78 (OS Command Injection) is the most frequent issue, indicating that multiple modules are vulnerable to command execution risks. CWE-22 (Path Traversal) is

Table 3.6. Semgrep coding findings mapped to CWEs.

Project	CWE	Severity	No.
Autoware	CWE-78	High	4
Autoware	CWE-611	High	2
Autoware	CWE-913	Medium	1
Openpilot	CWE-476	High	3
Openpilot	CWE-78	High	12
Openpilot	CWE-78	High	3
Openpilot	CWE-427	High	2
Openpilot	CWE-22	Medium	2
Openpilot	CWE-798	High	1
Openpilot	CWE-611	High	1
Openpilot	CWE-125	Medium	1
Openpilot	CWE-502	Medium	9
Openpilot	CWE-353	Medium	6
Openpilot	CWE-706	Medium	5
Openpilot	CWE-939	Medium	3
Openpilot	CWE-327	Medium	1
Openpilot	CWE-276	Medium	1
Openpilot	CWE-95	Low	1
Openpilot	CWE-78	Low	4
Openpilot	CWE-319	Low	2
Apollo	CWE-416	High	2
Apollo	CWE-22	High	1
Apollo	CWE-78	High	7
Apollo	CWE-79	High	4
Apollo	CWE-319	High	4
Apollo	CWE-611	High	2
Apollo	CWE-120	Medium	3
Apollo	CWE-125	Medium	2
Apollo	CWE-22	Medium	13
Apollo	CWE-20	Medium	9
Apollo	CWE-676	Medium	4
Apollo	CWE-116	Medium	2
Apollo	CWE-353	Medium	2
Apollo	CWE-668	Medium	1
Apollo	CWE-96	Medium	1
Apollo	CWE-95	Medium	1
Apollo	CWE-706	Medium	1
Apollo	CWE-1333	Medium	1
Apollo	CWE-78	Low	4
Apollo	CWE-134	Low	7

Table 3.7. Top 10 coding findings mapped to CWEs.

S. No	CWE	Coding Findings
1	CWE-78	34
2	CWE-22	16
3	CWE-502	9
4	CWE-20	9
5	CWE-353	8
6	CWE-134	7
7	CWE-706	6
8	CWE-319	6
9	CWE-611	5
10	CWE-79	4

the second most common vulnerability, highlighting potential risks related to unauthorized file access. Other vulnerabilities, such as CWE-502, CWE-20, and CWE-353, are also notable, reflecting issues in deserialization, input validation, and improper data handling. Overall, these top 10 CWEs emphasize areas where developers should focus code review and remediation efforts to enhance the security of the AV repositories.

3.5.5.4 *CWE and Severity Distribution in Autonomous Vehicle Projects*

To better understand the security issues detected by Semgrep, we plotted pie charts at Fig. 3.9 showing the distribution of vulnerabilities by CWE (inner layer) and severity (outer layer) for each AV repository. From the pie charts, we observe that for Autoware, the majority of issues are medium to high severity, mostly related to CWE-78 and CWE-22. Openpilot shows a mix of high and medium severity issues, with CWE-78, CWE-502, and CWE-353 being more frequent. Apollo has several high-severity findings in CWE-78 and CWE-79, but also a considerable number of medium-severity issues. Overall, the charts help to quickly identify which repositories have more critical weaknesses and which CWE categories need more attention for fixes and code review. The severity distribution outer layer is especially useful to spot the critical areas quickly.

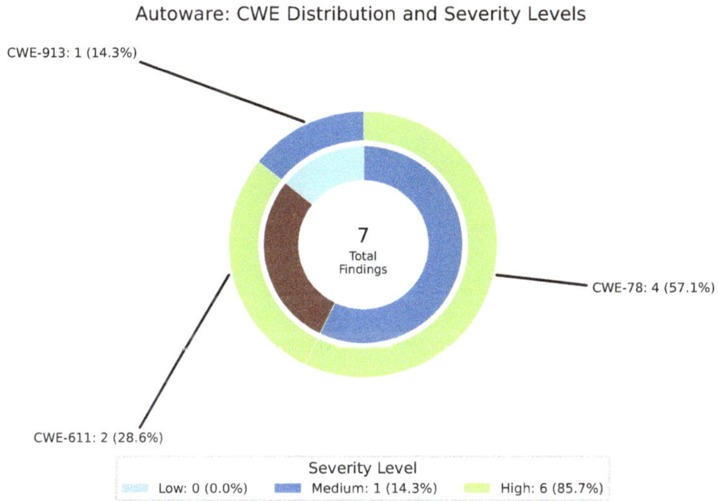

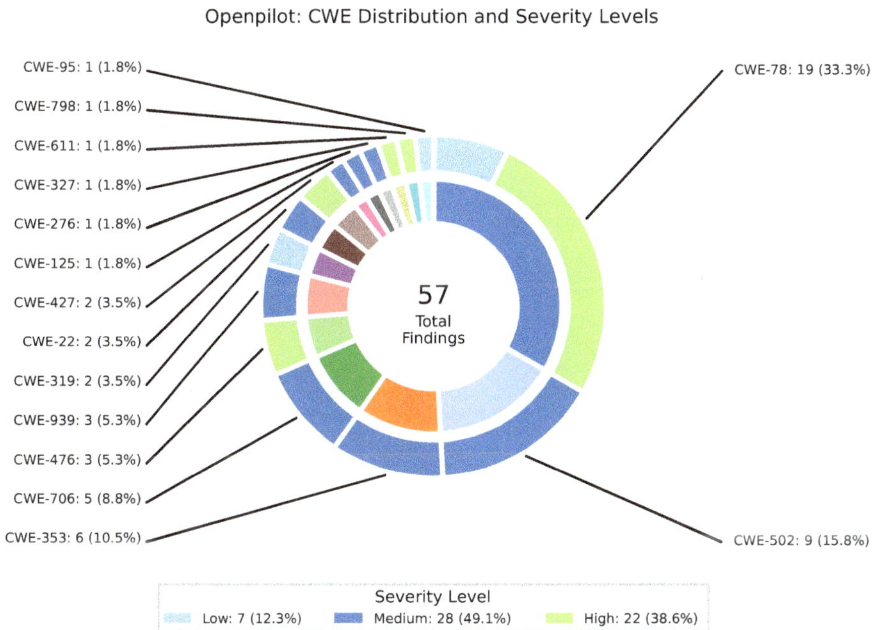

Fig. 3.9. CWE (inner) and severity (outer) distribution for all AV repositories. Top: Autoware, Middle: Openpilot, Bottom: Apollo.

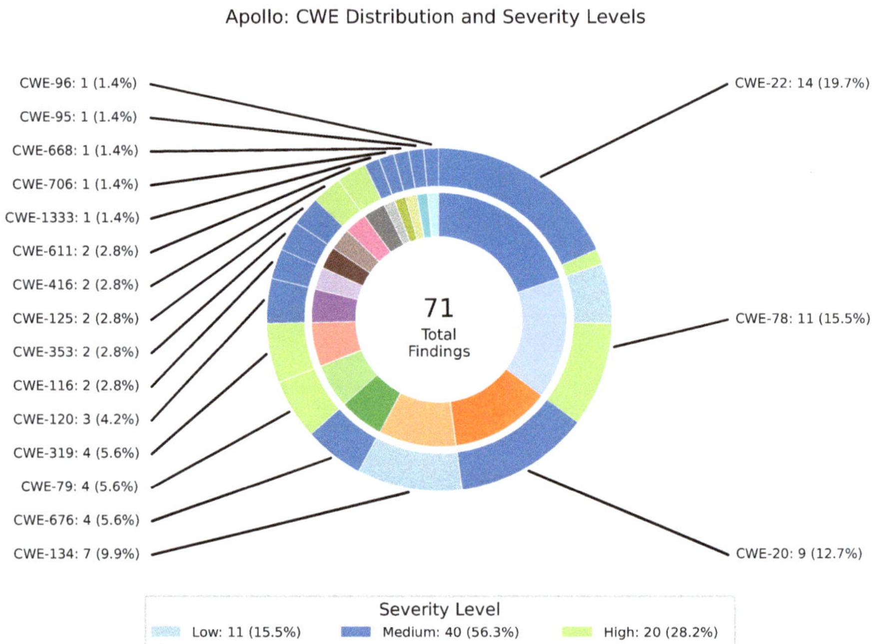

Fig. 3.9. (Continued)

3.5.5.5 *Key Findings of Code Vulnerability*

Across all three AV repositories, several CWEs appear repeatedly in both coding findings (Semgrep) and static analysis (Flawfinder/Bandit). CWE-78 (OS Command Injection) and CWE-22 (Path Traversal) are consistently among the top vulnerabilities, indicating risks in both Python and C/C++ code. Other recurring issues include CWE-502, CWE-20, and CWE-353, reflecting common input validation and deserialization weaknesses. Buffer and memory-related CWEs, such as CWE-119, CWE-120, and CWE-126, are notable in legacy C/C++ modules. Overall, these recurring CWEs suggest that certain classes of vulnerabilities are pervasive across repositories and languages, and they should be prioritized during code review and remediation.

It is necessary to note that this study does not perform component-level attribution of vulnerabilities. While the preceding analysis reports vulnerabilities by CWE category, their practical significance in AV software modules depends strongly on the functional role of the affected software modules. Vulnerabilities in the perception and planning modules are expected to be particularly impactful and safety-critical due to the size, complexity, and reliance on external libraries, data files, and scripting utilities of the components (Pendleton *et al.*, 2017). In contrast, issues in localization and control components are expected to be more sensitive to real-time execution constraints, where crashes or undefined behavior may directly affect vehicle stability or actuation (Pendleton *et al.*, 2017). These observations indicate that vulnerability risk in AV software is shaped less by raw frequency and more by where in the pipeline the weakness occurs.

3.5.6 *Third-Party Library Vulnerabilities*

Modern AV software stacks make extensive use of third-party libraries and dependencies. While this speeds up development and brings useful functionality, it also creates significant risks in the software supply chain. Even if the main application code is carefully written and reviewed, weaknesses in external components can expose the system to serious threats such as privilege escalation or remote code execution. To understand these risks, we examined the selected AV repositories with Semgrep SBOM scanning and the Microsoft SBOM tool combined with Grype, which allowed us to detect vulnerabilities in dependencies, group them by severity, and evaluate their potential impact on overall system security.

3.5.6.1 *Semgrap Supply Chain Vulnerability Analysis*

During the Semgrep SBOM scan of the selected repositories, we observed notable differences in dependency vulnerabilities. For the Apollo AV repositories, the scan identified a total of 91 dependencies that require further review. Among these, some dependencies were marked as reachable, indicating that they are actively used in the code and could potentially be exploited, while others were marked as need-review, requiring manual assessment to determine their actual impact. Additionally, several dependencies were unreachable, meaning they are present in the repository but not currently used, though they still pose a potential risk if brought into use in the future. In contrast, the openpilot AV repositories had only

Table 3.8. Vulnerabilities categorized by severity for apollo project.

Severity	Count
Critical	15
High	37
Medium	29
Low	10

1 reachable dependency, and Autoware_Core had none. The vulnerabilities in the Apollo AV repositories, categorized by severity, are shown in Table 3.8. For example, the tar-fs package contains a high-severity vulnerability related to path traversal. The output looks like,

```
>>> tar-fs - CVE-2024-12905
          Severity: HIGH
          Affected versions of tar-fs are vulnerable to Improper
          Limitation of a Pathname to a Restricted Directory ('Path
          Traversal'). tar-fs improperly resolves symlinks and fails
          to restrict pathnames during archive extraction. This
          vulnerability can allow an attacker to craft a tar archive
          that writes or overwrites files outside the intended
          directory, potentially compromising the host system.
            Fixed for tar-fs at versions: 1.16.4, 2.1.2, 3.0.7
          10516| tar-fs@^2.0.0, tar-fs@^2.1.1:
```

Affected versions of tar-fs fail to properly restrict pathnames during archive extraction, allowing attackers to overwrite files outside the intended directory. This issue has been fixed in versions 1.16.4, 2.1.2, and 3.0.7. Based on the Semgrep SBOM scan of the Apollo AV repositories, dependencies flagged as critical must be addressed immediately by upgrading to the latest secure versions, as they represent the highest potential risk if exploited. High-severity dependencies could be prioritized for short-term mitigation or upgrades, while medium- and low-severity dependencies could be monitored and updated regularly to prevent cumulative security exposure. Reachable dependencies require particular attention since they are actively used in the codebase, whereas dependencies marked as needing review should be assessed manually to determine their exploitability. Even unreachable dependencies could be periodically reviewed to ensure that future use does not introduce vulnerabilities. Overall, maintaining a routine of SBOM scanning and timely dependency updates is essential to minimize risk and ensure that both direct and transitive dependencies remain secure.

3.5.6.2 *Microsoft SBOM Analysis*

We performed an SBOM analysis on multiple open-source repositories using Microsoft's sbom-tool (Microsoft, 2025) and vulnerability scanning with Grype (Anchore, 2025). The goal was to identify known vulnerabilities in third-party components and dependencies within these repositories. For each repository, the workflow involved cloning the repository, generating the SBOM using sbom-tool, and scanning it with Grype. The sbom-tool parameters included specifying the build path (-bc), output folder (-b), package name (-pn), version (-pv), and namespace URL (-nsb). For example, the result generation process for the SBOM tool for the Apollo repository is,

```
sbom-tool generate \
  -b ./sbom-output \
  -bc . \
  -pn "apollo" \
  -pv "1.0.0" \
  -ps "ApolloAuto" \
  -nsb "https://github.com/ApolloAuto/apollo"
```

For Autoware_Core, the SBOM analysis did not detect any vulnerabilities, suggesting that the dependencies in this repository are either up-to-date or do not have known security issues listed in the vulnerability databases. In the case of openpilot, the scan detected 3 vulnerabilities, with severity distribution: 1 high and 2 medium. Notable components included `yajl-ruby` and `rake` as provided in Table 3.9. The SBOM scan for Apollo identified 1,532 components, of which 84 had known vulnerabilities. The severity distribution included 10 critical, 35 high, 29 medium, 10 low, and 0 negligible vulnerabilities. Selected critical and high-severity components are summarized in Table 3.10. The analysis suggests that immediate attention should be given to critical vulnerabilities, such as those found in `ejs` and `loader-utils`, by applying patches or upgrading

Table 3.9. Openpilot vulnerabilities detected from SBOM scan.

Component	Installed	Fixed In	Type	Vulnerability	Severity	EPSS	Risk
yajl-ruby	1.2.1	1.3.1	gem	GHSA-wwh7-4jw9-33x6	High	1.0%	0.7
yajl-ruby	1.2.1	1.4.3	gem	GHSA-jj47-x69x-mxrm	Medium	1.1%	0.6
rake	10.3.2	12.3.3	gem	GHSA-jppv-gw3r-w3q8	Medium	$< 0.1\%$	< 0.1

Table 3.10. Selected critical and high vulnerabilities in Apollo from SBOM scan.

Name	Installed	Fixed In	Type	Vulnerability	Severity	EPSS	Risk
ejs	2.7.4	3.1.7	npm	GHSA-phwq-j96m-2c2q	Critical	93.5% (99th)	87.9
json5	0.5.1	1.0.2	npm	GHSA-9c47-m6qq-7p4h	High	40.5% (97th)	29.6
json5	1.0.1	1.0.2	npm	GHSA-9c47-m6qq-7p4h	High	40.5% (97th)	29.6
json5	2.2.1	2.2.2	npm	GHSA-9c47-m6qq-7p4h	High	40.5% (97th)	29.6
loader-utils	0.2.17	1.4.1	npm	GHSA-76p3-8jx3-jpfq	Critical	15.8% (94th)	14.9
loader-utils	1.4.0	1.4.1	npm	GHSA-76p3-8jx3-jpfq	Critical	15.8% (94th)	14.9
loader-utils	2.0.2	2.0.3	npm	GHSA-76p3-8jx3-jpfq	Critical	15.8% (94th)	14.9

to secure versions. High-severity vulnerabilities, including multiple versions of `json5`, could be addressed in the short term to minimize potential exploitation. Furthermore, it is recommended to perform regular SBOM scans and maintain up-to-date dependencies to continuously reduce security risks and ensure the overall integrity of the software repositories.

3.5.6.3 *Key Findings of Third-Party Library Vulnerabilities*

The analysis of the selected repositories using Semgrep and Microsoft SBOM tools revealed notable third-party library vulnerabilities across multiple dependencies. Several dependencies were flagged as reachable, needing review, or unreachable, with issues spanning Critical, High, Medium, and Low severity levels. Critical and high-severity vulnerabilities were primarily associated with outdated or insecure package versions, such as tar-fs, node-forge, and postcss, potentially enabling attacks like path traversal, prototype pollution, or cryptographic signature forgery. These results emphasize the importance of proactive dependency management, timely patching, and continuous monitoring to reduce security risks in the software supply chain.

3.5.7 *From Static Vulnerabilities to AV-Specific Attack and Safety Scenarios*

Several of the frequently observed CWEs can be mapped to realistic AV-specific misuse and attack scenarios. For example, CWE-78 (OS Command Injection) (MITRE, 2023q), which appears repeatedly in both Python and C/C++ codebases, is particularly concerning when it occurs in development tooling, update mechanisms, or runtime management scripts. In an AV ecosystem, such components may handle model updates, configuration files, map ingestion, or logging pipelines. A command-injection vulnerability in these paths could enable an attacker to execute arbitrary commands on the onboard compute platform or developer environment, potentially leading to tampered perception models, corrupted configuration parameters, or unauthorized modification or planning logic. In a safety-critical setting, such manipulation could indirectly alter safety-critical and navigation tasks, such as obstacle detection, route planning, or vehicle behavior without triggering immediate fault alarms.

Similarly, CWE-22 (Path Traversal) (MITRE, 2023g) vulnerabilities are identified both in application code and third-party dependencies. This can be directly linked to AV-specific risk involving map files, calibration data, or recorded sensor logs. Autonomous driving stacks rely heavily on structured files for localization, high-definition (HD) maps, and system configuration. A path traversal flaw in a file-handling routine or third-party extraction library could allow an attacker to overwrite or replace trusted data artifacts, leading to incorrect localization or degraded perception accuracy. In downstream modules, these errors may propagate into unsafe trajectory planning or control decisions, illustrating how seemingly generic file-system vulnerabilities can translate into concrete safety hazards in AV deployments.

Low-level memory safety issues, such as CWE-119 (MITRE, 2023a) and CWE-120 (MITRE, 2023b) (buffer overflows), which are still prevalent in legacy C/C++ components, are particularly critical in performance-sensitive planning and control modules. In AV systems, crashes or undefined behavior in these components do not merely cause application failure; they may result in denial-of-service conditions, delayed control commands, or inconsistent actuation outputs. When such faults occur within real-time loops, they can compromise the vehicle's ability to respond to dynamic environments, increasing the chances of collisions or unsafe maneuvers. Overall, static analysis results have the potential to serve as a structured foundation for identifying where generic software weaknesses intersect with AV-specific safety and security concerns.

3.5.8 *Interpretation and Next Steps*

While the individual CWE categories surfaced by static analysis tools are common across large C/C++ and Python codebases, their value in the AV context lies in how they establish a structured vulnerability baseline for safety-critical, cyber-physical software stacks. Static analysis provides a scalable, language-aware mechanism to systematically enumerate candidate weaknesses across different modules, such as perception, localization, planning, and control, as well as their supporting tooling and middleware. In AV systems, these components are tightly coupled through real-time data flows and execution dependencies; therefore, even generic flaws become safety-relevant when they occur in modules that influence sensor fusion, trajectory generation, or actuation. By mapping static findings to specific software layers and execution roles within the AV architecture, the results serve as an initial triage step that narrows the search space for deeper investigation (i.e., using dynamic analysis), rather than a standalone security assessment.

The workflow presented in this chapter illustrates how code-level findings and third-party dependency issues can be collected early in the development lifecycle. In practice, CI/CD pipelines can leverage this information by escalating findings that affect runtime critical modules, reachable dependencies, or update-related tooling, while deprioritizing low-impact issues confined to offline utilities or test code. Such prioritization supports scalability in large AV codebases without conflating minor code smells and safety-relevant risks.

Moreover, this static vulnerability baseline enables targeted follow-on analyses that are uniquely meaningful in the AV domain. High-frequency CWEs identified statically can be prioritized for dynamic analysis (e.g., fuzzing, sanitizers, and runtime monitoring) in modules with real-time constraints or direct actuator influence, where exploitability and failure modes differ fundamentally from conventional software. Similarly, CWE distributions and SBOM-driven dependency risks can be injected into threat modeling and attack-graph analyses to reason about multi-stage attack paths, cascading failures, and safety hazards, such as how a command injection in a development toolchain or a memory error in localization could propagate into unsafe vehicular behavior. In this sense, static analysis is not the endpoint but the foundation upon which AV-specific exploitability assessment, attack modeling, and risk-to-safety mapping can be constructed, enabling precise prioritization and mitigation in subsequent phases of the security life-cycle.

3.6 Threats to Validity

While our study provides insights into code and third-party library vulnerabilities in AV open-source repositories, several factors could influence the validity and generalizability of our findings, such as,

(i) internal validity; (ii) external validity; (iii) construct validity; and (iv) reliability and reproducibility. Internal validity concerns the accuracy of our vulnerability detection. Although we used well-established tools, such as Flawfinder, Bandit, Semgrep, Microsoft SBOM tool, and Grype, no automated tool guarantees a comprehensive vulnerability identification. False positives and false negatives may exist, potentially skewing our quantitative findings. Additionally, mapping code findings to CWE identifiers may introduce minor misclassifications due to differing tool interpretations. Our study focuses on three open-source AV repositories: Apollo, Openpilot, and Autoware. While these AV repositories are representative of widely used autonomous software, the results may not generalize to other autonomous systems or repositories in different programming languages. Moreover, dependency and supply chain risks observed here might not apply to other environments with different libraries, versions, or configurations. Construct validity relates to whether our measures accurately capture the security properties of interest. CWEs, severity levels, and dependency classifications provide standardized measures, but they may not fully represent runtime exploitation risks. Our experiments depend on specific repository versions and tool versions. Reproducing the study exactly requires access to the same code snapshots and tool configurations. The datasets, analysis scripts, and supplementary materials used in this study are publicly available on GitHub: https://github.com/uaerfan/cav-security-vulnerabilities.git. Although we provide the GitHub link and Docker images, scripts to facilitate replication, differences in environments, future dependency updates, or changes in tool rules may lead to slightly different results.

This chapter does not attempt to manually validate or exploit individual vulnerabilities identified by static analysis tools. The reported findings should be interpreted as a screening-level assessment intended to inform prioritization rather than as a definitive measure of exploitability. False positives, low-impact code smells, and non-reachable findings may exist, and distinguishing these from genuinely exploitable flaws requires targeted manual triage, dynamic analysis, and coordinated disclosure processes that are beyond the scope of this chapter.

3.6.1 *Open-Source Bias and Generalizability*

This study focuses exclusively on open-source AV software stacks, including Autoware, Apollo, and openpilot. While these platforms are widely used in academic research, prototyping, and early deployments, most large-scale commercial AV systems rely on closed-source or hybrid software architectures. As a result, the findings reported in this chapter should be interpreted with awareness of potential open-source bias.

The use of open-source repositories enables transparent static analysis, reproducibility, and fine-grained inspection of source code and third-party dependencies. However, the visibility afforded by openness may lead to a high apparent number of reported vulnerabilities compared to proprietary systems, where source code and dependency structures are not publicly accessible. Consequently, the presence of vulnerabilities in open-source AV stacks should not be interpreted as evidence of higher intrinsic risk, but rather as an artifact of greater auditability.

3.7 Summary, Limitations and Future Directions

This study analyzed the software supply chain of AVs through static vulnerability assessment of three widely used open-source platforms, i.e., Autoware, Apollo, and openpilot, revealing recurring CWEs and dependency-related risks that highlight the safety-critical nature of securing AV software ecosystems. Our findings underscore that while static analyzers, such as Flawfinder and Bandit, provide meaningful insights into prevalent vulnerabilities, they also illustrate that open-source AV software stacks remain exposed to systematic supply-chain threats that must be addressed proactively.

A primary limitation of this study is that the vulnerability assessment is based exclusively on static analysis and SBOM-based dependency scanning. This chapter does not perform manual triage or validation of individual findings to distinguish false positives, low-impact code smells, and genuinely exploitable weaknesses. Static analysis tools may overapproximate risk by flagging defensive code paths, dead code, or benign usage patterns that do not lead to practical exploitation. Accordingly, the findings presented in this chapter should be interpreted as a structured baseline of code-level and supply-chain-related risks, rather than as a comprehensive assessment of operational security. Static analysis serves as an initial screening layer that highlights vulnerability classes and components warranting closer attention. Complementary techniques, such as dynamic analysis, fuzzing, runtime monitoring, and system-level attack modeling, are necessary to evaluate runtime exploitability and safety impact.

Future works should focus on validating a representative subset of high-severity and high-reachability findings through dynamic analysis techniques, such as fuzzing, runtime sanitization, and fault injection, with particular emphasis on modules that influence perception, planning, and control. In addition, integrating static vulnerability outputs into attack-graph models and automotive threat analysis frameworks (e.g., ISO/SAE 21434-aligned TARA ISO/SAE, 2021) will enable systematic reasoning about exploit chains and safety impact. These steps will help translate vulnerability discovery into actionable mitigation strategies and responsible disclosure workflows for AV software ecosystems. Despite these threats and limitations, the study provides meaningful insights into prevalent vulnerabilities and supply chain risks, highlighting areas for remediation and further research.

Bibliography

Aggarwal, M. (2023). A study of CVSS v4. 0: A CVE scoring system, in *2023 6th International Conference on Contemporary Computing and Informatics (IC3I)*, Vol. 6 (IEEE), pp. 1180–1186.

Aliane, N. (2025). A survey of open-source autonomous driving systems and their impact on research, *Information* **16**, 4, p. 317.

Anchore, I. (2025). Grype: A vulnerability scanner for container images and filesystems, https://github.com/anchore/grype, accessed: September 18, 2025.

Anderson, J. M., Nidhi, K., Stanley, K. D., Sorensen, P., Samaras, C., and Oluwatola, O. A. (2014). *Autonomous Vehicle Technology: A Guide for Policymakers* (Rand Corporation, Santa Monica, California, USA).

AndreiGavrilov, M. B., Tinyakov, S., and Krinkin, K. (2021). Analysis of robotic platforms: Data transfer performance evaluation, *architecture* Conference of Open Innovations Association, FRUCT. No. 29. FRUCT Oy, 2021. **17**, p. 18.

Apollo Baidu Team (2025). Apollo Cyber RT framework, https://developer.apollo.auto/cyber.html, accessed on September 9, 2025.

Autoware Foundation (2024). Autoware: The world's first "all-in-one" open-source software for self-driving vehicles, https://github.com/autowarefoundation, accessed: January 2026.

Ayewah, N. and Pugh, W. (2010). The Google FindBugs fixit, in *Proceedings of the 19th International Symposium on Software Testing and Analysis*, pp. 241–252.

Baidu Apollo (2024). Apollo: An open autonomous driving platform, https://github.com/ApolloAuto/apollo, accessed: January 2026.

Beller, M., Bholanath, R., McIntosh, S., and Zaidman, A. (2016). Analyzing the state of static analysis: A large-scale evaluation in open source software, in *2016 IEEE 23rd International Conference on Software Analysis, Evolution, and Reengineering (SANER)*, Vol. 1 (IEEE), pp. 470–481.

Belluardo, L., Stevanato, A., Casini, D., Cicero, G., Biondi, A., and Buttazzo, G. (2021). A multi-domain software architecture for safe and secure autonomous driving, in *2021 IEEE 27th International Conference on Embedded and Real-Time Computing Systems and Applications (RTCSA)* (IEEE), pp. 73–82.

Boi, B., Gupta, T., Rinhel, M., Jubea, I., Khondoker, R., Esposito, C., and Sousa, B. M. (2023). Strengthening automotive cybersecurity: A comparative analysis of ISO/SAE 21434-compliant automatic collision notification (ACN) systems, *Vehicles* **5**, 4, pp. 1760–1802.

Butt, M. A., Ajmal, Z., Khan, Z. I., Idrees, M., and Javed, Y. (2022). An in-depth survey of bypassing buffer overflow mitigation techniques, *Applied Sciences* **12**, 13, p. 6702.

Chen, L., Tang, T., Cai, Z., Li, Y., Wu, P., Li, H., Shi, J., Yan, J., and Qiao, Y. (2022). Level 2 autonomous driving on a single device: Diving into the devils of openpilot, *arXiv preprint* arXiv:2206.08176.

comma.ai (2024). openpilot: An open source driver assistance system, https://github.com/commaai/openpilot, accessed: January 2026.

De Vincenzi, M., Pesé, M. D., Bodei, C., Matteucci, I., Brooks, R. R., Hasan, M., Saracino, A., Hamad, M., and Steinhorst, S. (2024). Contextualizing security and privacy of software-defined vehicles: State of the art and industry perspectives, *arXiv preprint* arXiv:2411.10612.

Durlik, I., Miller, T., Kostecka, E., Zwierzewicz, Z., and Łobodzińska, A. (2024). Cybersecurity in autonomous vehicles—are we ready for the challenge? *Electronics* **13**, 13, p. 2654.

ENISA (2021). Threat landscape for supply chain attacks, Technical Report, European Union Agency for Cybersecurity (ENISA), Heraklion, Greece.

Fagnant, D. J. and Kockelman, K. (2015). Preparing a nation for autonomous vehicles: Opportunities, barriers and policy recommendations, *Transportation Research Part A: Policy and Practice* **77**, pp. 167–181.

FIRST.Org, Inc. (2023). *Common Vulnerability Scoring System v4.0: Specification Document*, https://www.first.org/cvss/v4.0/specification-document, accessed: September 17, 2025.

Grigorescu, S., Trasnea, B., Cocias, T., and Macesanu, G. (2020). A survey of deep learning techniques for autonomous driving, *Journal of Field Robotics* **37**, 3, pp. 362–386.

Guo, W., Xu, Z., Liu, C., Huang, C., Fang, Y., and Liu, Y. (2023). An empirical study of malicious code in PyPI ecosystem, in *2023 38th IEEE/ACM International Conference on Automated Software Engineering (ASE)* (IEEE), pp. 166–177.

He, Y., Xin, T., and Luo, C. (2024). Cybersecurity investments metrics using FAIR-ROSI.

Honda Motor Co. (2020). Honda sensing elite with traffic jam pilot: World's first approval of level 3 automated driving system, https://global.honda/en/newsroom/news/2020/4201111eng.html.

Islam, T., Sheakh, M. A., Jui, A. N., Sharif, O., and Hasan, M. Z. (2023). A review of cyber attacks on sensors and perception systems in autonomous vehicle, *Journal of Economy and Technology* **1**, pp. 242–258.

ISO/SAE (2021). ISO/SAE 21434:2021 Road Vehicles – Cybersecurity Engineering, Standard ISO/SAE 21434:2021, International Organization for Standardization (ISO) and SAE International, Geneva, Switzerland.

Jung, H.-Y., Paek, D.-H., and Kong, S.-H. (2025). Open-source autonomous driving software platforms: Comparison of Autoware and Apollo, *arXiv preprint* arXiv:2501.18942.

Kato, S., Tokunaga, S., Maruyama, Y., Maeda, S., Hirabayashi, M., Kitsukawa, Y., Monrroy, A., Ando, T., Fujii, Y., and Azumi, T. (2018). Autoware on board: Enabling autonomous vehicles with embedded systems, in *2018 ACM/IEEE 9th International Conference on Cyber-Physical Systems (ICCPS)* (IEEE), pp. 287–296.

Katsikeas, S., Ling, E. R., Johnsson, P., and Ekstedt, M. (2024). Empirical evaluation of a threat modeling language as a cybersecurity assessment tool, *Computers & Security* **140**, p. 103743.

Kusano, K. D., Scanlon, J. M., Chen, Y.-H., McMurry, T. L., Chen, R., Gode, T., and Victor, T. (2024). Comparison of Waymo rider-only crash data to human benchmarks at 7.1 million miles, *Traffic Injury Prevention* **25**, sup1, pp. S66–S77.

Ladisa, P., Plate, H., Martinez, M., and Barais, O. (2023). Sok: Taxonomy of attacks on open-source software supply chains, in *2023 IEEE Symposium on Security and Privacy (SP)* (IEEE), pp. 1509–1526.

Lattner, C. and Adve, V. (2004). Llvm: A compilation framework for lifelong program analysis & transformation, in *International Symposium on Code Generation and Optimization, 2004. CGO 2004* (IEEE), pp. 75–86.

Lautenbach, A., Almgren, M., and Olovsson, T. (2021). Proposing heavens 2.0–an automotive risk assessment model, in *Proceedings of the 5th ACM Computer Science in Cars Symposium*, pp. 1–12.

Litman, T. (2017). Autonomous vehicle implementation predictions.

Liu, Q., Li, X., Yuan, S., and Li, Z. (2021). Decision-making technology for autonomous vehicles: Learning-based methods, applications and future outlook, in *2021 IEEE International Intelligent Transportation Systems Conference (ITSC)* (IEEE), pp. 30–37.

Liu, Z., Zhang, W., and Zhao, F. (2022). Impact, challenges and prospect of software-defined vehicles, *Automotive Innovation* **5**, 2, pp. 180–194.

Livshits, V. B. and Lam, M. S. (2005). Finding security vulnerabilities in java applications with static analysis. in *USENIX Security Symposium*, Vol. 14, pp. 271–286.

Lowe, H. J., Ferris, T. A., Hernandez, P. M., and Weber, S. C. (2009). Stride–An integrated standards-based translational research informatics platform, in *AMIA Annual Symposium Proceedings*, Vol. 2009, p. 391.

Mao, X., Ekstedt, M., Ling, E., Ringdahl, E., and Lagerström, R. (2019). Conceptual abstraction of attack graphs-a use case of securiCAD, in *International Workshop on Graphical Models for Security* (Springer), pp. 186–202.

Mercedes-Benz Group (2024). Drive pilot approved for level 3 automated driving in California and Nevada, https://group.mercedes-benz.com/innovation/product-innovation/autonomous-driving/drive-pilot-california.html.

Microsoft (2025). Sbom tool, https://github.com/microsoft/sbom-tool, accessed: September 18, 2025.

MITRE (2023a). CWE-119: Improper restriction of operations within the bounds of a memory buffer, https://cwe.mitre.org/data/definitions/119.html.

MITRE (2023b). CWE-120: Buffer copy without checking size of input, https://cwe.mitre.org/data/definitions/120.html.

MITRE (2023c). CWE-126: Buffer over-read, https://cwe.mitre.org/data/definitions/126.html.

MITRE (2023d). CWE-134: Use of externally-controlled format string, https://cwe.mitre.org/data/definitions/134.html.

MITRE (2023e). CWE-190: Integer overflow or wraparound, https://cwe.mitre.org/data/definitions/190.html.

MITRE (2023f). CWE-20: Improper input validation, https://cwe.mitre.org/data/definitions/20.html.

MITRE (2023g). CWE-22: Path traversal, https://cwe.mitre.org/data/definitions/22.html.

MITRE (2023h). CWE-259: Use of hard-coded password, https://cwe.mitre.org/data/definitions/259.html.

MITRE (2023i). CWE-327: Use of a broken or risky cryptographic algorithm, https://cwe.mitre.org/data/definitions/327.html.

MITRE (2023j). CWE-330: Use of insufficiently random values, https://cwe.mitre.org/data/definitions/330.html.

MITRE (2023k). CWE-362: Concurrent execution using shared resource with improper synchronization ('race condition'), https://cwe.mitre.org/data/definitions/362.html.

MITRE (2023l). CWE-377: Insecure temporary file, https://cwe.mitre.org/data/definitions/377.html.

MITRE (2023m). CWE-400: Uncontrolled resource consumption, https://cwe.mitre.org/data/definitions/400.html.

MITRE (2023n). CWE-502: Deserialization of untrusted data, https://cwe.mitre.org/data/definitions/502.html.

MITRE (2023o). CWE-605: Multiple binds to the same port, https://cwe.mitre.org/data/definitions/605.html.

MITRE (2023p). CWE-703: Improper check or handling of exceptional conditions, https://cwe.mitre.org/data/definitions/703.html.

MITRE (2023q). CWE-78: OS command injection, https://cwe.mitre.org/data/definitions/78.html.

MITRE (2023r). CWE-807: Reliance on untrusted inputs in a security decision, https://cwe.mitre.org/data/definitions/807.html.

MITRE (2023s). CWE-829: Inclusion of functionality from untrusted control sphere, https://cwe.mitre.org/data/definitions/829.html.

National Institute of Standards and Technology (2022). Software supply chain security guidance under executive order (EO) 14028 Section 4(e), Technical Report, NIST, February 4, 2022.

Ohm, Marc, *et al.* (2020). "Backstabber's knife collection: A review of open source software supply chain attacks." *International Conference on Detection of Intrusions and Malware, and Vulnerability Assessment.* Cham: Springer International Publishing.

Ohm, M., Sykosch, A., and Meier, M. (2020). Towards detection of software supply chain attacks by forensic artifacts, in *Proceedings of the 15th International Conference on Availability, Reliability and Security*, pp. 1–6.

On-Road Automated Driving (ORAD) Committee (2021). *Taxonomy and Definitions for Terms Related to Driving Automation Systems for On-Road Motor Vehicles* (SAE International, Warrendale, PA, USA).

Pendleton, S. D., Andersen, H., Du, X., Shen, X., Meghjani, M., Eng, Y. H., Rus, D., and Ang, M. H. (2017). Perception, planning, control, and coordination for autonomous vehicles, *Machines* **5**, 1, p. 6.

PyCQA Bandit Developers (2025). Bandit: A security linter from PyCQA, https://bandit.readthedocs.io/en/latest/index.html, © Copyright 2025, Bandit Developers. Accessed: September 18, 2025.

Raju, V. M., Gupta, V., and Lomate, S. (2019). Performance of open autonomous vehicle platforms: Autoware and apollo, in *2019 IEEE 5th International Conference for Convergence in Technology (I2CT)* (IEEE), pp. 1–5.

Ross, R. S. (2012). Guide for conducting risk assessments.

Saez-Perez, J., Diez-Tomillo, J., Tena-Gago, D., Alcaraz-Calero, J. M., and Wang, Q. (2025). Design, implementation and validation of a level 2 automated driving vehicle reference architecture, *Expert Systems* **42**, 6, p. e70050.

Semgrep, Inc. (2025). Semgrep: Static analysis for code security, https://semgrep.dev/, accessed: September 18, 2025.

Serebryany, K., Bruening, D., Potapenko, A., and Vyukov, D. (2012). {AddressSanitizer}: A fast address sanity checker, in *2012 USENIX Annual Technical Conference (USENIX ATC 12)*, pp. 309–318.

Shepardson, D. (2020). Waymo opens driverless robo-taxi service to the public in phoenix, *Reuters* https://www.reuters.com/article/technology/waymo-opens-driverless-robo-taxi-service-to-the-public-in-phoenix-idUSKBN26T2Y3/.

Shladover, S. E. (2018). Connected and automated vehicle systems: Introduction and overview, *Journal of Intelligent Transportation Systems* **22**, 3, pp. 190–200.

Souppaya, M., Scarfone, K., and Dodson, D. (2022). Secure software development framework (SSDF) version 1.1, *NIST Special Publication* **800**, 218.

Teixeira, P. V., Raposo, D., Lopes, R., and Sargento, S. (2024). Deterministic and reliable software-defined vehicles: Key building blocks, challenges, and vision, *arXiv preprint* arXiv:2407.17287.

Teixeira, R. R., Maurell, I. P., and Drews, P. L. (2020). Security on ROS: Analyzing and exploiting vulnerabilities of ROS-based systems, in *2020 Latin American Robotics Symposium (LARS), 2020 Brazilian Symposium on Robotics (SBR) and 2020 Workshop on Robotics in Education (WRE)* (IEEE), pp. 1–6.

Ulbrich, S., Reschka, A., Rieken, J., Ernst, S., Bagschik, G., Dierkes, F., Nolte, M., and Maurer, M. (2017). Towards a functional system architecture for automated vehicles, *arXiv preprint* arXiv:1703.08557.

UNECE (2021). UN regulation No. 155-cyber security and cyber security management system, *The United Nations Economic Commission for Europe (UNECE).*

Wahba, P. (2024). Mercedes becomes the first automaker to sell autonomous cars in the U.S. *Fortune*, https://fortune.com/2024/04/18/mercedes-self-driving-autonomous-cars-california-nevada-level-3-drive-pilot/.

Wang, J., Neil, M., and Fenton, N. (2020). A bayesian network approach for cybersecurity risk assessment implementing and extending the fair model, *Computers & Security* **89**, p. 101659.

Wang, Z., Wu, Y., and Niu, Q. (2019). Multi-sensor fusion in automated driving: A survey, *IEEE Access* **8**, pp. 2847–2868.

Waymo LLC (2024). Waymo one: Autonomous rides in Phoenix, San Francisco, Los Angeles, https://waymo.com/rides/phoenix/.

Wheeler, D. A. (2023). Flawfinder: A static analysis tool for C/C++, https://dwheeler.com/flawfinder/, accessed: September 18, 2025.

Wolf, A., Simopoulos, D., D'Avino, L., and Schwaiger, P. (2021). The PASTA threat model implementation in the IoT development life cycle, in *INFORMATIK 2020* (Gesellschaft für Informatik, Bonn), pp. 1195–1204.

Xia, B., Bi, T., Xing, Z., Lu, Q., and Zhu, L. (2023). An empirical study on software bill of materials: Where we stand and the road ahead, in *2023 IEEE/ACM 45th International Conference on Software Engineering (ICSE)* (IEEE), pp. 2630–2642.

Xia, L., Gao, X., and Shi, W. (2025). Investigating security threats in multi-tenant ROS 2 systems, in *2025 IEEE International Conference on Robotics and Automation (ICRA)* (IEEE), pp. 16441–16448.

Xiong, W., Krantz, F., and Lagerström, R. (2019). Threat modeling and attack simulations of connected vehicles: A research outlook, in *ICISSP*, pp. 479–486.

Ye, Y., Nie, Z., Liu, X., Xie, F., Li, Z., and Li, P. (2023). ROS2 real-time performance optimization and evaluation, *Chinese Journal of Mechanical Engineering* **36**, 1, p. 144.

Yurtsever, E., Lambert, J., Carballo, A., and Takeda, K. (2020). A survey of autonomous driving: Common practices and emerging technologies, *IEEE Access* **8**, pp. 58443–58469.

Zalewski, M. (2017). American Fuzzy Lop.
Zhao, H., Meng, M., Li, X., Xu, J., Li, L., and Galland, S. (2024a). A survey of autonomous driving frameworks and simulators, *Advanced Engineering Informatics* **62**, p. 102850.
Zhao, X., Qu, H., Yi, J., Wang, J., Tian, M., and Zhao, F. (2024b). A fuzzer for detecting use-after-free vulnerabilities, *Mathematics* **12**, 21, p. 3431.

Part II

Risk Modeling, Analysis, and Platforms

https://doi.org/10.1142/9789819831548_0004

Chapter 4

Proactive Risk Assessment of Multi-Vehicle Transportation Systems through Reachability Analysis under Stealthy Attacks

Sounghwan Hwang, Minhyun Cho, Inseok Hwang
School of Aeronautics and Astronautics, Purdue University
West Lafayette, IN, USA

Abstract
This chapter addresses critical cybersecurity challenges in modern transportation systems, with a focus on urban air mobility (UAM), where dense and dynamic environments significantly increase the risk of stealthy attacks. Such cyber threats are particularly concerning as they can evade conventional monitoring while degrading vehicle dynamics and mission-level performance. To address this, we propose a Lyapunov-based risk assessment framework for multi-vehicle systems under sensor stealthy attacks. By leveraging linear matrix inequality (LMI) techniques and ellipsoidal reachable set computations, the method provides a conservative yet systematic characterization of adversarially induced state deviations. This transforms qualitative cyber-physical vulnerabilities into quantifiable security metrics, enabling risk evaluation at both vehicle and fleet levels. The resulting metrics support proactive defense by enabling early anticipation and mitigation of risks before they escalate into safety-critical failures. The framework captures the impact of stealthy attacks on key mission functions such as formation control,

collision avoidance, and trajectory tracking. A 3D UAM case study in an urban-like environment demonstrates the effectiveness of the proposed approach in identifying vulnerable operating conditions and enabling real-time adaptive responses. Overall, the proposed framework provides a principled and scalable foundation for proactive risk assessment in cyber-physical transportation systems, enabling security-aware control and planning for autonomous vehicles and improving the reliability and safety of future mobility infrastructures in adversarial environments.

Keywords: Urban Air Mobility (UAM), Cybersecurity, Stealthy attacks, Reachability analysis, Lyapunov-based risk assessment.

4.1 Introduction

Modern transportation systems are currently undergoing a profound transformation, driven by automation, electrification, and digital connectivity (Deka *et al.*, 2018; Islam *et al.*, 2023; Picone *et al.*, 2015). Fleets of connected cars on highways, truck platoons in freight corridors, and emerging formations of urban air mobility (UAM) service in city skies illustrate a shift from isolated vehicles to highly networked, cooperative transportation ecosystems. Technologies such as vehicle-to-vehicle (V2V) and vehicle-to-infrastructure (V2I) communication, often referred to collectively as vehicular ad hoc networks (VANETs), enable coordinated maneuvers like cooperative adaptive cruise control, synchronized merging, and collision avoidance. Similarly, in the airspace domain, aerial vehicles (AVs) rely on wireless communication to maintain separation, share intent, and navigate urban corridors safely. These innovations have been propelled by the rise of cyber-physical systems (CPSs), which tightly integrate sensing, computation, and control with physical vehicles and infrastructure (Xiong *et al.*, 2015). Broadly defined, CPSs are sophisticated systems in which cyber layers (e.g., digital controllers and communication modules) are tightly integrated with physical layers (e.g., actuators and sensors), enabling them to address a wide range of complex engineering challenges through coordinated computation, control, and interaction with the physical environment (Lee, 2015; Liu *et al.*, 2017; Rani *et al.*, 2025). By embedding advanced perception (e.g., camera, LiDAR, radar, and GPS),

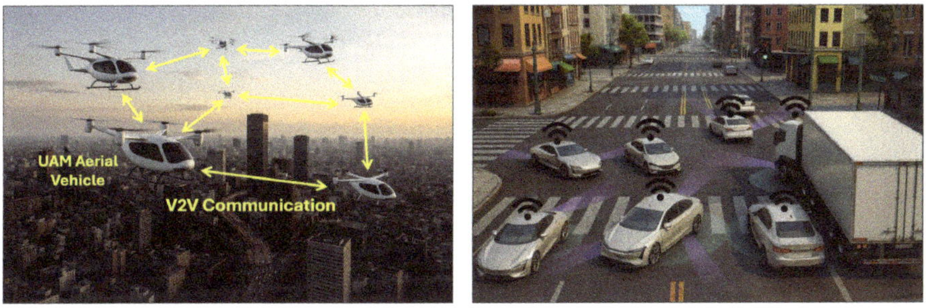

Fig. 4.1. Examples of modern transportation systems within a multi-vehicle system framework, such as urban air mobility (UAM) (Hill *et al.*, 2020) (left) and connected autonomous vehicles (CAVs) in urban environments (Islam *et al.*, 2023) (right). The operation of vehicles—such as UAM aerial vehicles (AVs) and autonomous ground vehicles—relies on inter-vehicle communication within a defined radius, enabling the exchange of information (e.g., position and velocity) necessary for collaborative missions. This figure was generated based on an AI-assisted drawing tool.

real-time decision-making, and wireless networking into transportation platforms, CPS technologies are redefining how mobility is delivered across ground, air, and water, as shown in Fig. 4.1. In particular, UAM has emerged as a CPS-enabled alternative to traditional ground transportation, addressing critical limitations such as traffic congestion, increasing travel delays, and urban air pollution. Its viability fundamentally relies on CPS capabilities, including real-time decision-making for efficient planning, wireless networking for air traffic management, and multi-sensor integration to ensure safe and reliable operation in complex urban environments (Freeman and Lewis, 2022; Hill *et al.*, 2020; Safwat *et al.*, 2024).

While CPS-enabled mobility promises safer and efficient operations, it simultaneously exposes fleets and platoons to new forms of security vulnerabilities (Duo *et al.*, 2022; Kayan *et al.*, 2022; Meng, 2023). The tight coupling of cyber components with physical processes makes these modern transportation systems susceptible to various cyber threats such as denial-of-service (DoS), false-data-injection (FDI), Byzantine faults, and stealthy attacks (Dibaji *et al.*, 2019). Autonomous vehicles increasingly depend on GPS, radar, LiDAR, and cooperative communication to maintain formations, synchronize velocities, and avoid collisions in dense environments

(Patil *et al.*, 2025). If these channels are compromised by attackers, the consequences can be catastrophic. For example, FDI attacks may distort sensor readings, DoS attacks may block coordination signals and/or data packets, and stealthy attacks, such as GPS spoofing, reported in both automotive and aerospace sectors, can subtly manipulate trajectories while evading detection (Eshmawi *et al.*, 2024; Hoffmann *et al.*, 2023; Rathinasamy *et al.*, 2024). Crucially, the compromise of even a single vehicle can destabilize an entire fleet, whether a truck platoon on a highway or a cluster of air taxis navigating crowded urban air corridors (Xie *et al.*, 2023). These threats underscore that CPS-enabled transportation systems are vulnerable not only at the level of individual vehicles but also across the interconnected cyber-physical layers that bind multi-vehicle networks together.

Given the aforementioned cybersecurity challenges, there is an urgent need for methods that go beyond traditional detection-based defenses to enhance the safety and resilience of CPS-based transportation systems. In particular, the severity of stealthy attacks cannot be overstated, as they are deliberately designed to bypass conventional monitoring and anomaly-detection mechanisms (Góes *et al.*, 2017; Kwon *et al.*, 2014; Sui *et al.*, 2020). Left unconsidered, such attacks can compromise mission integrity, cause significant property damage, and endanger human life. To address these limitations, a promising strategy is proactive risk assessment: systematically evaluating and quantifying the potential impacts of stealthy attacks on vehicle performance before they occur. In the transportation domain, this means assessing how adversarial actions could degrade mission objectives, such as formation keeping, trajectory tracking, or collision avoidance, and translating these effects into quantitative security metrics for vehicle operations (Cho *et al.*, 2020; Huang *et al.*, 2022; Sun *et al.*, 2023). Unlike reactive defenses, this proactive approach equips system designers and operators with the ability to anticipate vulnerabilities, prioritize mitigation strategies, and maintain situational awareness even in the absence of explicit attack evidence. This capability is particularly crucial given the sophistication of attackers who can design stealthy strategies to evade onboard monitoring systems and dynamically adapt attack parameters to undermine countermeasures (Tounsi and Rais, 2018).

Consequently, the development of advanced proactive risk assessment frameworks is imperative for ensuring that multi-vehicle transportation systems can withstand stealthy threats and continue to operate safely in mission-antagonistic urban environments.

Building on the above discussions, the primary objective of this chapter is to introduce a risk assessment methodology tailored for transportation systems operating under potential stealthy attacks. At the core of this approach is the concept of reachable sets, which provide a rigorous framework for quantifying how a system's physical states (e.g., position and velocity of a vehicle) may evolve from given initial conditions both in nominal operation and under adversarial influence. In practical terms, reachable set analysis enables the characterization of trajectory deviations induced by stealthy attacks in critical tasks such as reference tracking or coordination. If the reachable set of a vehicle intersects with known static obstacles (e.g., buildings) or overlaps with the reachable sets of nearby vehicles during a mission, it may signal a heightened risk of collisions and potential safety violations. While exact reachable set computation offers the most accurate characterization of such risks under adversarial conditions, it is often computationally intractable due to the curse of dimensionality and the high complexity of adversarial scenarios (Bansal *et al.*, 2017; Herbert *et al.*, 2021; Shafa and Ornik, 2022). These challenges are further exacerbated in multi-vehicle settings, where the number of vehicles and their sophisticated network topology dramatically amplify the computational burden.

Motivated by the above concerns, we develop a Lyapunov-based risk assessment methodology of multi-vehicle systems subject to sensor stealthy attacks. The proposed approach computes ellipsoidal over-approximations of reachable sets that conservatively capture all possible compromised trajectories induced by stealthy attacks, while significantly reducing computational complexity, making it readily applicable to complex transportation systems. These ellipsoidal sets are defined as level sets of a Lyapunov function, obtained by solving a convex optimization problem formulated with the vehicle dynamics, network topology, and norm-bounded residuals representing stealthy attacks. To enable this construction, we derive a set of linear matrix inequality (LMI) conditions that guarantee the existence of a suitable Lyapunov function and its associated level set. The key contribution of our proactive defense strategy lies in establishing a quantifiable security metric at both the vehicle and fleet levels, thereby enabling a comprehensive system-wide risk evaluation. Technically, the fleet-level reachable set is built as the union of the ellipsoidal reachable sets corresponding to individual vehicles, providing a unified representation of the collective system behavior under potential cyber–physical threats. This formulation enables not only the identification of localized risks for individual vehicles but also the assessment of emergent risks arising from inter-vehicle interactions. To demonstrate the effectiveness of the proposed methodology, we present a detailed case study of the UAM scenario, where a multi-vehicle system operates in a 3D urban environment under stealthy attacks. The study illustrates how the proposed strategy facilitates proactive risk assessment and strengthens the resilience of modern transportation systems against sophisticated cyber threats.

The remainder of this chapter is organized as follows: Section 4.2 presents mathematical preliminaries for notations and graph theory. Section 4.3 provides problem formulation for the system model. Section 4.4 discusses methodological foundations for the system stabilization and residual-based attack detector design. Section 4.5 presents the main result of this chapter, namely the Lyapunov-based reachable set analysis and computation using LMIs. Section 4.6 demonstrates the effectiveness of the proposed approach through a UAM case study in an urban-like 3D simulation environment, and Section 4.7 concludes the chapter.

4.2 Preliminaries

This section outlines the essential notations, mathematical definitions, and graph-theoretic background that facilitate better readability of the chapter.

4.2.1 *Definitions and Notations*

Notations are defined as follows: $\mathbb{R}$ and $\mathbb{Z}$ denote the sets of real and integer numbers, with superscript $+$ indicating nonnegative subsets. For a matrix X, its transpose is X^T, and the Hermitian (symmetric) part $He\{X\} = X + X^T$ is used for compactness in block matrices.

The set of $n \times m$ real matrices is denoted by $\mathbb{R}^{n\times m}$. The index set $1, 2, \ldots, N$ is represented as $\mathcal{I}_N$. For any vector $x(k) \in \mathbb{R}^n$, $|x(k)|_2 = \sqrt{x^T(k)x(k)}$ is the Euclidean norm, while $|x(k)|_\infty = \sup_{k\geq 0} |x(k)|_2$ represents the infinity norm of the signal $\{x(k)\}_{k\in\mathbb{Z}^+}$. The symbol $*$ is used to omit symmetric terms in block matrices; I_n is the $n \times n$ identity matrix, and $\mathbf{0}$ represents a zero matrix with matching dimensions. The symbol $\otimes$ denotes the Kronecker product, a widely used operator in the analysis of multi-vehicle systems and frameworks, which forms a larger block matrix by multiplying each entry of one matrix with the entire other matrix.

4.2.2 *Graph Theory*

In this chapter, we study a multi-vehicle system whose interaction topology can be modeled as a graph. Specifically, the communication network is represented by an undirected graph $\mathcal{G} = (\mathcal{V}, \mathcal{E}, \mathcal{A})$, where $\mathcal{V} = \{v_1, \ldots, v_N\}$ denotes the vehicles in the network with $N \in \mathbb{Z}^+$, and $\mathcal{E} \subseteq \mathcal{V}\times\mathcal{V}$ indicates the communication (i.e., network) links. The neighborhood of vehicle i is defined as $\mathcal{N}_i = \{j \in \mathcal{V} \,|\, (v_j, v_i) \in \mathcal{E}\}$. The adjacency matrix $\mathcal{A} = [a_{ij}] \in \mathbb{R}^{N\times N}$ is defined such that $a_{ij} > 0$ if vehicles i and j are connected, i.e., $(v_j, v_i) \in \mathcal{E}$,e and $a_{ij} = 0$ otherwise. The Laplacian matrix $\mathcal{L} = [\mathcal{L}_{ij}] \in \mathbb{R}^{N\times N}$ is defined as $\mathcal{L}_{ii} = \sum_{j\neq i} a_{ij}$ and $\mathcal{L}_{ij} = -a_{ij}$ for $i \neq j$. The Laplacian matrix is a key tool for understanding the connectivity of a multi-vehicle network. Simply put, it tells us how well the vehicles are connected to each other, capturing the overall structure of their communication links. This makes it easier to analyze coordination and interaction among the vehicles.

4.3 Problem Formulation

4.3.1 *Transportation System Design via Multi-Vehicle Systems*

Transportation systems can be modeled within a multi-vehicle system framework, where the overall operation is governed by the dynamics of individual vehicles and their underlying network topology. Such a multi-vehicle system framework is well-suited for the design of modern transportation systems, as vehicles (e.g., autonomous vehicles and UAM AVs) accomplish their missions by exchanging information with neighboring vehicles through sensors and communication modules such as radar, LiDAR, and Vehicle-to-Vehicle (V2V) communication. Given this context, we consider the following discrete linear-time-invariant multi-vehicle system dynamics:

$$\begin{cases} x_i(k+1) = Ax_i(k) + Bu_i(k) + B_w w_i(k), \\ y_i(k) = Cx_i(k) + Fv_i(k) + S_a \delta_{ai}(k), \end{cases} \tag{4.1}$$

where the subscript i denotes the ith vehicle among $N \in \mathbb{Z}^+$ vehicles, where N is the total number of vehicles. We define $x_i(k) \in \mathbb{R}^n$ as the state of vehicle i (e.g., position, velocity), $u_i(k) \in \mathbb{R}^m$ as its control input, and $y_i(k) \in \mathbb{R}^q$ as the corresponding output measurement. The system matrices $A \in \mathbb{R}^{n\times n}$, $B \in \mathbb{R}^{n\times m}$, and $C \in \mathbb{R}^{q\times n}$ denote the system, input, and output matrices, respectively. It is further assumed that the pairs (A, B) and (A, C) satisfy stabilizability and detectability conditions. In addition, the process noise $w_i(k) \in \mathbb{R}^{\bar{w}}$ and measurement noise $v_i(k) \in \mathbb{R}^{\bar{v}}$ enter the system via perturbation matrices $B_w \in \mathbb{R}^{n\times\bar{w}}$ and $F \in \mathbb{R}^{q\times\bar{v}}$, respectively.

Here, it is noted that $\delta_{ai}(k) \in \mathbb{R}^{\bar{\delta}_a}$ with the attack matrix $S_a \in \mathbb{R}^{q\times\bar{\delta}_a}$ represents the cyberattack that stealthily impact the vehicle's sensor, i.e., the measurement output $y_i(k)$. As illustrated in Fig. 4.2, $\delta_{ai}(k)$ can be considered as a potential cyber threat, compromising

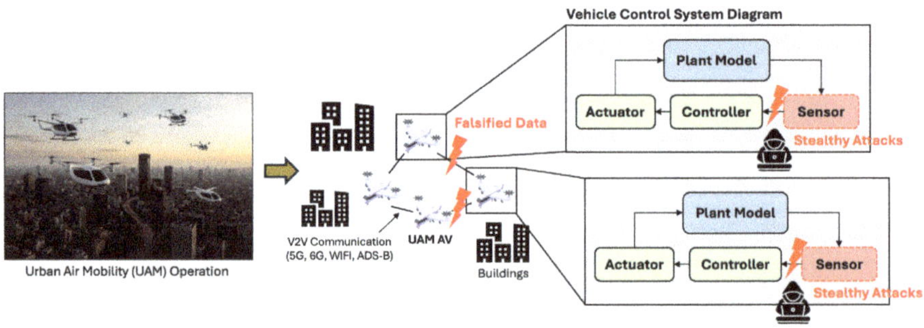

Fig. 4.2. Vulnerability of multi-vehicle systems (e.g., UAM) under stealthy attacks.

the operation of the multi-vehicle system (4.1). A stealthy attack targeting a single vehicle may propagate via the network and adversely affect its neighboring vehicles due to the inter-vehicle communication structure.

In practice, the system dynamics (4.1) are always affected by system noises $w_i(k)$ and $v_i(k)$, which, however, do not grow unbounded. Instead, these noises are typically constrained within a bounded magnitude, which is commonly modeled as follows (Yang *et al.*, 2022):

Assumption 4.1. For the multi-vehicle system (4.1), the process noise $w_i(k)$ and the measurement noise $v_i(k)$ for $\forall i \in \mathcal{I}_N$ are subject to the following norm-bounded constraints:

$$\begin{aligned} |w_i(k)|_2^2 &= w_i^T(k)w_i(k) \leq \tilde{\mathcal{W}}_i, \\ |v_i(k)|_2^2 &= v_i^T(k)v_i(k) \leq \tilde{\mathcal{V}}_i, \end{aligned} \tag{4.2}$$

where the magnitudes are $\tilde{\mathcal{W}}_i \in \mathbb{R}^+$, $\tilde{\mathcal{V}}_i \in \mathbb{R}^+$, and $k \in \mathbb{Z}^+$.

In addition to the norm-bounded conditions for the system noises, the graph $\mathcal{G}$ governing the network connections of the vehicles in (4.1) is assumed to be strongly connected, meaning that every vehicle can reach any other vehicle via network link. This assumption guarantees full information flow across the network and is commonly used to ensure cooperative behaviors in multi-vehicle systems. From a graph-theoretic perspective, $\mathcal{A} = [a_{ij}]$ denotes the adjacency matrix, with entries specified as $a_{ij} = 1$ if a directed link exists from j to i, and $a_{ij} = 0$ in the absence of such a link.

Remark 4.1. In the context of modern transportation systems, such as UAM and CAVs in urban environments (El-Rewini *et al.*, 2020; Hoffmann *et al.*, 2023; Tang, 2021), sensor vulnerabilities to stealthy attacks have been widely documented. Therefore, incorporating compromised measurement outputs caused by these threats into the system modeling in (4.1) is a valid and relevant approach in the context of cybersecurity of modern transportation.

4.3.2 *Target Reference Dynamics*

Here, we assume that the primary objective of the vehicles in transportation missions such as formation control or platooning is to track a designated reference trajectory, thereby ensuring that all vehicles within the network adhere to a common reference signal (Boel *et al.*, 2015; Gu *et al.*, 2023; Lakshmanan *et al.*, 2019). This collaborative behavior allows the vehicles to achieve efficient and safe operation in urban areas (Tang *et al.*, 2021). The target reference, i.e., virtual leader, can be generated by ground control stations or traffic management authorities, whose dynamics can be modeled as follows:

$$\tilde{x}_l(k+1) = A\tilde{x}_l(k) + B\tilde{u}_l(k), \tag{4.3}$$

where the vector $\tilde{x}_l(k) \in \mathbb{R}^n$ represents the virtual leader's state, $\tilde{u}_l(k) \in \mathbb{R}^m$ is its control input. Here, for ease of design, the leader's system matrices are assumed to be identical to those of the vehicle. For the remainder of this chapter, the term target reference is used interchangeably with virtual leader.

4.3.3 *Observer-Based Tracking Controller Design*

Given the above transportation missions, it is crucial to develop a control framework that enables the vehicles in the system (4.1) to accomplish their tracking objectives. Accordingly, we focus on deriving an observer-based tracking control framework based on the dynamics of a multi-vehicle system (4.1) and the virtual leader (4.3). Consider the following synchronization error in terms of vehicle i for $\forall i \in \mathcal{I}_N$:

$$s_i(k) \triangleq x_i(k) - \tilde{x}_l(k).$$

Based on the dynamics of the multi-vehicle system (4.1) and the target reference (4.3), the synchronization error dynamics can be derived as:

$$s_i(k+1) = As_i(k) + B\left(u_i(k) - \tilde{u}_l(k)\right) + B_w w_i(k). \tag{4.4}$$

Within the multi-vehicle framework, the control input of vehicle i can be constructed using an observer approach that accounts for the communication structure with its neighboring vehicles (Xu *et al.*, 2013).

$$u_i(k) = K_c \sum_{j \in \mathcal{N}_i} \left(\hat{x}_i(k) - \hat{x}_j(k)\right) + \alpha_i K_c \left(\hat{x}_i(k) - \tilde{x}_l(k)\right) + \tilde{u}_l(k), \tag{4.5}$$

where $\hat{x}_i(k) \in \mathbb{R}^{n\times n}$ represents the estimated state of vehicle i and $\hat{x}_j(k) \in \mathbb{R}^{n\times n}$ is the estimated state of neighboring vehicle j. The controller gain $K_c \in \mathbb{R}^{m\times n}$ is to be determined, which will be addressed in the subsequent section. Additionally, if the vehicle i is connected to the target reference, the indicator $\alpha_i = 1$, and $\alpha_i = 0$, otherwise.

Next, substituting the control input (4.5) into the synchronization error dynamics (4.4) yields the synchronization error dynamics of the multi-vehicle system (4.1), expressed as follows:

$$s(k+1) = (I_N \otimes A)\, s(k) + \left(\bar{\mathcal{L}} \otimes BK_c\right) \hat{s}(k) + (I_N \otimes B_w)\, w(k), \tag{4.6}$$

where we have the augmented Laplacian matrix $\bar{\mathcal{L}} = \mathcal{L} + \Lambda$ with

$$\Lambda = \mathrm{diag}\left[\alpha_1, \alpha_2, \ldots, \alpha_N\right] \in \mathbb{R}^{N\times N}.$$

This matrix $\bar{\mathcal{L}} = \bar{\mathcal{L}}^T$ indicates the network topology between the vehicles and the virtual leader. Furthermore, the stacked vectors are obtained as follows:

$$\begin{aligned} s(k) &= [s_1^T(k), \ldots, s_N^T(k)]^T, \quad \hat{s}(k) = [\hat{s}_1^T(k), \ldots, \hat{s}_N^T(k)]^T, \\ w(k) &= [w_1^T(k), \ldots, w_N^T(k)]^T, \quad \hat{s}_i(k) = \hat{x}_i(k) - \tilde{x}_l(k). \end{aligned}$$

Lemma 4.1 (Trejo *et al.*, 2021). *The augmented Laplacian matrix $\bar{\mathcal{L}}$ is positive definite, meaning its eigenvalues satisfy the following condition:*

$$0 < \lambda_1(\bar{\mathcal{L}}) < \lambda_2(\bar{\mathcal{L}}) < \cdots < \lambda_N(\bar{\mathcal{L}}).$$

Throughout the remainder of this chapter, the ith eigenvalue of $\bar{\mathcal{L}}$ will, for brevity, be represented as $\bar{\lambda}_i$.

To capture realistic circumstances, we consider the case where the system state $x_i(k)$ of vehicle i is not directly observable (i.e., measurable). In other words, the design of a state observer should account for the possibility that $y_i(k)$ is compromised by $\delta_i(k)$. Therefore, the control input (4.5) utilizes the estimated state, which is obtained from the following state observer (i.e., Luenberger observer):

$$\begin{cases} \hat{x}_i(k+1) = A\hat{x}_i(k) + Bu_i(k) + L_{\mathrm{ob}}\left(y_i(k) - \hat{y}_i(k)\right), \\ \hat{y}_i(k) = C\hat{x}_i(k), \end{cases} \tag{4.7}$$

where $L_{\text{ob}} \in \mathbb{R}^{n\times q}$ and $\hat{y}_i(k) \in \mathbb{R}^q$ denote the state observer gain to be determined, and the estimated output of vehicle i, respectively.

Remark 4.2. In practical systems, not all internal states can be directly measured by sensors. A Luenberger observer can estimate these unmeasurable states by combining a system model with available measurements, effectively acting as a virtual sensor. This allows the system to operate safely and efficiently in scenarios where direct measurements are limited or unavailable. For further technical details, readers are referred to Luenberger (2007).

Next, we define the state estimation error of vehicle i for $\forall i \in \mathcal{I}_N$ as follows:

$$\bar{e}_i(k) \triangleq x_i(k) - \hat{x}_i(k).$$

By combining the dynamics (4.1) and (4.7), we have the following state estimation error dynamics for the multi-vehicle system:

$$\begin{aligned}\bar{e}(k+1) = {} & (I_N \otimes (A - L_{\text{ob}}C))\,\bar{e}(k) + (I_N \otimes B_w)\,w(k) \\ & - (I_N \otimes L_{\text{ob}}F)\,v(k) - (I_N \otimes L_{\text{ob}}S_a)\,\delta_a(k),\end{aligned} \tag{4.8}$$

where the stacked vectors are obtained as:

$$\bar{e}(k) = [\bar{e}_1^T(k), \ldots, \bar{e}_N^T(k)]^T, \quad v(k) = [v_1^T(k), \ldots, v_N^T(k)]^T,$$
$$\delta_a(k) = [\delta_{a1}^T(k), \ldots, \delta_{aN}^T(k)]^T.$$

As a result, we can obtain an augmented closed-loop error dynamics by combining the dynamics (4.6) and (4.8) as follows:

$$\xi(k+1) = \mathcal{B}_1\xi(k) + \mathcal{B}_2 w(k) + \mathcal{B}_3 v(k) + \mathcal{B}_4\delta(k), \tag{4.9}$$

where $\xi(k) \triangleq [s^T(k),\, \bar{e}^T(k)]^T \in \mathbb{R}^{2nN}$, and

$$\mathcal{B}_1 = \begin{bmatrix} I_N \otimes A + \bar{\mathcal{L}} \otimes BK_c & -\bar{\mathcal{L}} \otimes BK_c \\ \mathbf{0} & I_N \otimes (A - L_{\text{ob}}C) \end{bmatrix}, \quad \mathcal{B}_2 = \begin{bmatrix} I_N \otimes B_w \\ I_N \otimes B_w \end{bmatrix},$$

$$\mathcal{B}_3 = \begin{bmatrix} \mathbf{0} \\ -I_N \otimes L_{\text{ob}}F \end{bmatrix}, \quad \mathcal{B}_4 = \begin{bmatrix} \mathbf{0} \\ -I_N \otimes L_{\text{ob}}S_a \end{bmatrix}.$$

Remark 4.3. If the case where the stealthy attack $\delta_a(k)$ is absent, i.e., when the vehicles' sensors are secure, the reference tracking objective of the multi-vehicle system (4.1) can be guaranteed through the appropriate design of the gains K_c and L_{ob}. However, it is noted that the occurrence of a stealthy attack $\delta_a(k)$ has the potential to disrupt consensus and drive the vehicles away from their desired trajectories, which constitutes a significant risk to both individual vehicles and the entire system. For instance, this threat may result in collisions and crashes between obstacles (e.g., buildings) when performing a reference tracking control mission in an urban environment. Considering the potential severity of these threats, the objective of this chapter is to establish a reachability-based risk assessment framework capable of evaluating their impact on the reference tracking control of multi-vehicle transportation systems, thereby improving operational safety.

In preparation for the subsequent sections, we provide the Schur complement and $\mathcal{S}$-procedure (Boyd *et al.*, 1994; VanAntwerp and Braatz, 2000), which will serve as a fundamental linear matrix inequality (LMI) tool in this chapter:

Lemma 4.2 (Schur Complement). *Consider three matrices R_1, R_2, and R_3 with proper dimensions where $R_1 = R_1^T$ and $R_3 = R_3^T$. Then, the following LMIs are equivalent:*

$$\begin{gathered} R_1 < 0, \quad R_3 - R_2 R_1^{-1} R_2^T < 0, \\ R_3 < 0, \quad R_1 - R_2 R_3^{-1} R_2^T < 0, \\ \begin{bmatrix} R_1 & * \\ R_2^T & R_3 \end{bmatrix} < 0. \end{gathered} \tag{4.10}$$

Lemma 4.3 ($\mathcal{S}$-Procedure). *Consider symmetric matrices S_1 and S_2, vectors g_1 and g_2, and scalars h_1 and h_2. Suppose that there exists a vector x_0 such that the following condition is satisfied:*

$$x_0^\top S_1 x_0 + 2 g_1^\top x_0 + h_1 < 0. \tag{4.11}$$

Then, the quadratic implication

$$x^\top S_1 x + 2 g_1^\top x + h_1 \leq 0 \Rightarrow x^\top S_2 x + 2 g_2^\top x + h_2 \leq 0 \tag{4.12}$$

holds, if and only if there exists a scalar $\lambda \in \mathbb{Z}^+$ such that the following LMI condition is satisfied:

$$\lambda \begin{bmatrix} S_1 & g_1 \\ g_1^\top & h_1 \end{bmatrix} - \begin{bmatrix} S_2 & g_2 \\ g_2^\top & h_2 \end{bmatrix} \succeq 0. \tag{4.13}$$

Remark 4.4. The use of LMIs is particularly appealing in the context of multi-vehicle systems. By expressing control and observer design conditions as LMIs, we obtain a systematic method to compute feasible gains that ensure stability and robustness even in the presence of noise, uncertainties, or stealthy attacks. Moreover, the convex nature of LMIs ensures that these computations remain tractable, enabling the development of proactive defense strategies that can be practically implemented in real-world transportation systems. For a comprehensive introduction to the basics and fundamentals of LMIs, we refer readers to VanAntwerp and Braatz (2000) and Boyd *et al.* (1994).

Remark 4.5. The $\mathcal{S}$-procedure provides a systematic way to enforce logical implications between quadratic conditions. To be specific, it allows one to guarantee that the satisfaction of a given quadratic constraint implies the satisfaction of another quadratic condition. By introducing a scalar multiplier, this implication is translated into a tractable LMI, enabling implication-based reasoning within a convex optimization framework.

4.4 Methodological Foundations

As a preliminary step, we design the controller, state estimator, and residual-based attack detector, which form the foundation of the proposed framework, prior to presenting the chapter's main results on reachability-based risk assessment under stealthy attacks.

4.4.1 *Stabilization under Attack-Free Conditions*

Here, we focus on deriving the controller gain K_c, and the state estimator gain L_{ob} that ensure the stability of the multi-vehicle system during tracking control missions in the absence of stealthy attacks (i.e., when $\delta_a(k) = 0$). The following theorem is developed to derive the gain K_c given the augmented Laplacian matrix $\bar{\mathcal{L}}$.

Theorem 4.1. *Consider the synchronization error dynamics of the multi-vehicle system as follows:*

$$s(k+1) = \left(I_N \otimes A + \bar{\mathcal{L}} \otimes BK_c\right) s(k). \tag{4.14}$$

The asymptotic stability of the error dynamics (4.14) *governed by the augmented Laplacian matrix* $\bar{\mathcal{L}}$ *can be accomplished, if there exist matrices* $\mathcal{S} = \mathcal{S}^T \in \mathbb{R}^{n\times n}$ *and* $Z \in \mathbb{R}^{m\times n}$ *such that the following LMIs are satisfied:*

$$\begin{bmatrix} \mathcal{S} & * \\ \mathcal{S}A^T + \bar{\lambda}_i Z^T B^T & \mathcal{S} \end{bmatrix} > 0, \quad \mathcal{S} > 0, \quad \forall i \in \mathcal{I}_N, \tag{4.15}$$

where $\bar{\lambda}_i \in \mathbb{R}^+$ *denotes an* i*th eigenvalue of* $\bar{\mathcal{L}}$ *as provided in Lemma* 4.1. *Finally, the controller gain is computed as* $K_c = Z\mathcal{S}^{-1}$.

Proof. To begin, the Lyapunov candidate is chosen in the form:

$$\mathcal{V}_1(k) = s^T(k)\left(I_N \otimes \mathcal{P}_1\right) s(k),$$

where $\mathcal{P}_1 = \mathcal{P}_1^T \in \mathbb{R}^{n\times n} > 0$. By applying the similarity transformation, we can represent $\bar{\mathcal{L}} = \mathcal{V}_T \Theta \mathcal{V}_T^{-1}$ where we obtain $\mathcal{V}_T \in \mathbb{R}^{N\times N}$ and $\Theta = \text{diag}[\bar{\lambda}_1, \bar{\lambda}_2, \ldots, \bar{\lambda}_N] \in \mathbb{R}^{N\times N}$. By employing the matrix $\mathcal{V}_T$, the coordinate transformation can be applied by the vector $\eta_1(k) \triangleq (\mathcal{V}_T^{-1} \otimes I_n)s(k)$ where $\eta_1(k) = [\eta_{11}^T(k), \ldots, \eta_{1N}^T(k)]^T$. To verify the asymptotic stability of (4.14), the following inequality should be satisfied:

$$\Delta\mathcal{V}_1(k) \triangleq \mathcal{V}_1(k+1) - \mathcal{V}_1(k) < 0.$$

By employing the coordinate transformation with $\eta_1(k)$, we can show the following condition:

$$\Delta\mathcal{V}_1(k) = \sum_{i=1}^{N} \eta_{1i}^T(k) \underbrace{\left[\left(A + \bar{\lambda}_i BK_c\right)^T \mathcal{P}_1 \left(A + \bar{\lambda}_i BK_c\right) - \mathcal{P}_1\right]}_{\Psi_i} \eta_{1i}(k) < 0. \tag{4.16}$$

Based on the Schur complement (refer to Lemma 4.2) and congruence transformation with $\mathcal{S} = \mathcal{P}_1^{-1}$ to Ψ_i, the inequality $\Psi_i < 0$ is equivalent to the conditions in (4.15). Therefore, we can conclude that if LMIs in (4.15) are satisfied, then the synchronization error dynamics is asymptotically stable. This completes the proof for Theorem 4.1. □

As a subsequent step to Theorem 4.1, the observer gain L_{ob} is obtained under the H_∞ control criterion (Burl, 1998; Song *et al.*, 2025). To achieve this, we pose the following control design problem:

Problem 4.1. For some $\gamma \in \mathbb{R}^+$, the primarily objective is to design the observer gain L_{ob} such that the following two conditions are satisfied:

(1) The augmented observer-based error dynamics (4.9) is asymptotically stable when the system noises satisfy $w(k) = 0$ and $v(k) = 0$ for $\forall k \in \mathbb{Z}^+$.
(2) The augmented close-loop error dynamics (4.9) satisfied the following condition under the zero initial condition:

$$\lim_{k\to\infty} \frac{\sum_{i=0}^{k} |\bar{e}(i)|_2^2}{\sum_{i=0}^{k} \left(|w(i)|_2^2 + |v(i)|_2^2\right)} \leq \gamma,$$

where γ is referred to as H_∞ performance index.

Now, we are ready to derive the observer gain L_{ob}.

Theorem 4.2. *Suppose that the LMIs in Theorem* 4.1 *are satisfied, i.e., there exist matrices* $\mathcal{S} = \mathcal{S}^T > 0$ *and* Z*, such that* $\mathcal{P}_1 = \mathcal{S}^{-1}$ *and* $K_c = Z\mathcal{S}^{-1}$*. The augmented error dynamics* (4.9) *is stabilizable by satisfying the* H_∞ *control criterion, if there exist proper matrices* $\mathcal{P}_2 = \mathcal{P}_2^T \in \mathbb{R}^{n\times n}$ *and* $\mathcal{N} \in \mathbb{R}^{n\times q}$ *such that the following optimization is satisfied.*

$$\min_{\gamma} \gamma \quad \text{s.t. } \mathcal{P}_2 > 0, \quad \bar{\mathcal{U}}_i = \left[\begin{array}{c|c} \mathcal{U}_i & * \\ \hline \tilde{\mathcal{U}} & -\mathcal{P}_2 \end{array}\right] < 0 \quad \text{for } \forall i \in \mathcal{I}_N, \qquad (4.17)$$

where

$$\mathcal{U}_i = \begin{bmatrix} \begin{matrix} A^T\mathcal{P}_1A + \bar{\lambda}_i He\{A^T\mathcal{P}_1BK_c\} \\ +\bar{\lambda}_i^2 K_c^T B^T \mathcal{P}_1 BK_c - P_1 \end{matrix} & * & * & * \\ -\bar{\lambda}_i K_c^T B^T \mathcal{P}_1 A - \bar{\lambda}_i^2 K_c^T B^T \mathcal{P}_1 BK_c & \bar{\lambda}_i^2 K_c^T B^T \mathcal{P}_1 BK_c - \mathcal{P}_2 + I_n & * & * \\ B_w^T \mathcal{P}_1 A + \bar{\lambda}_i B_w^T \mathcal{P}_1 BK_c & -\bar{\lambda}_i B_w^T \mathcal{P}_1 BK_c & B_w^T \mathcal{P}_1 B_w - \gamma I_{\bar{w}} & * \\ \mathbf{0} & \mathbf{0} & \mathbf{0} & -\gamma I_{\bar{w}} \end{bmatrix},$$

$$\tilde{\mathcal{U}} = \begin{bmatrix} \mathbf{0} & \mathcal{P}_2 A - \mathcal{N}C & \mathcal{P}_2 B_w & -\mathcal{N}F \end{bmatrix}.$$

Then, we can derive the state observer gain as $L_{\text{ob}} = \mathcal{P}_2^{-1}\mathcal{N}$.

Proof. Consider the following Lyapunov function:

$$\mathcal{V}_2(k) = s^T(k)\left(I_N \otimes \mathcal{P}_1\right)s(k) + e^T(k)\left(I_N \otimes \mathcal{P}_2\right)e(k),$$

where $\mathcal{P}_1 = \mathcal{P}_1^T \in \mathbb{R}^{n\times n} > 0$ is derived from Theorem 4.1 and $\mathcal{P}_2 = \mathcal{P}_2^T \in \mathbb{R}^{n\times n} > 0$. To meet the two conditions in Problem 4.1, the following inequality should be satisfied:

$$\Delta\mathcal{V}_2(k) + |e(k)|_2^2 - \gamma\left(|w(k)|_2^2 + |v(k)|_2^2\right) < 0. \tag{4.18}$$

For the next step, consider the following coordinate transformation for $e(k)$, $w(k)$, and $v(k)$:

$$\eta_2(k) \triangleq (\mathcal{V}_T^{-1} \otimes I_n)e(k), \quad \eta_3(k) \triangleq (\mathcal{V}_T^{-1} \otimes I_{\bar{w}})w(k),$$
$$\eta_4(k) \triangleq (\mathcal{V}_T^{-1} \otimes I_{\bar{v}})v(k),$$

where $\eta_2(k)$, $\eta_3(k)$, and $\eta_4(k)$ are obtained as:

$$\eta_2(k) = \left[\eta_{21}^T(k), \ldots, \eta_{2N}^T(k)\right]^T, \quad \eta_3(k) = \left[\eta_{31}^T(k), \ldots, \eta_{3N}^T(k)\right]^T,$$
$$\eta_4(k) = \left[\eta_{41}^T(k), \ldots, \eta_{4N}^T(k)\right]^T.$$

By employing these vectors, the inequality (4.18) can be represented through the vectorized form:

$$\sum_{i=1}^{N} g_i^T(k)\underbrace{\left[\mathcal{U}_i + \tilde{\mathcal{U}}^T\mathcal{P}_2^{-1}\tilde{\mathcal{U}}\right]}_{\Phi_i} g_i(k) < 0, \tag{4.19}$$

where $g_i(k) = \left[\eta_{1i}^T(k),\, \eta_{2i}^T(k),\, \eta_{3i}^T(k),\, \eta_{4i}^T(k)\right]^T$. To satisfy the inequality (4.19), we need to show $\Phi_i < 0$ for $\forall i \in \mathcal{I}_N$. By applying the Schur complement, the LMI condition $\Phi_i < 0$ is equivalent to the constraint conditions in (4.17), i.e., $\mathcal{P}_2 > 0$ and $\bar{\mathcal{U}}_i < 0$. As a result, we can conclude that if the optimization (4.17) is satisfied, we can derive the observer gain L_{ob} that satisfies the two conditions in Problem 4.1. This completes the proof for Theorem 4.2. □

4.4.2 *Design of Residual-Based Attack Detector*

In this subsection, we present the development of a residual-based monitoring framework to detect stealthy attacks. The underlying principle of the attack detector is intuitive: it identifies potential cyber threats by comparing the measured output with the estimated output. When a significant deviation between these outputs is observed, it can be interpreted as a clear indication that cyberattacks are engaged during system operation. Consider the following residual signal for vehicle i:

$$\begin{aligned} r_i(k) &= y_i(k) - \hat{y}_i(k) \\ &= C\bar{e}_i(k) + Fv_i(k) + S_a\delta_{ai}(k), \end{aligned} \tag{4.20}$$

where $r_i(k) \in \mathbb{R}^q$. Specifically, the attack detector utilizes a distance-based measurement, i.e.,

$$\epsilon_i(k) = r_i^T(k)\mathcal{M}_i r_i(k) \in \mathbb{R}^+, \tag{4.21}$$

where $\mathcal{M}_i = \mathcal{M}_i^T \in \mathbb{R}^{q\times q} \geq 0$ denotes a monitoring matrix to be designed. As shown in Fig. 4.3, the following conditions intuitively show how the attack monitoring mechanism works:

$$\epsilon_i(k) > 1 \Rightarrow \text{ Attack alarm is triggered.} \tag{4.22}$$

$$\epsilon_i(k) \leq 1 \Rightarrow \text{ Attack alarm is not triggered.} \tag{4.23}$$

It is noted that attackers aim to hide their impacts (i.e., stealthiness) by maintaining the condition $\epsilon_i(k) \leq 1$ as shown in (4.23).

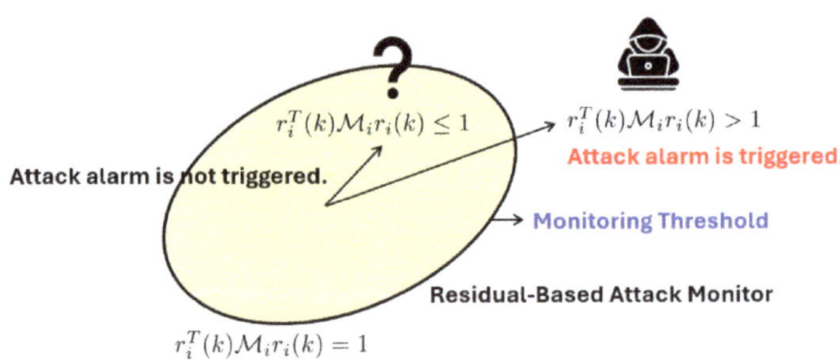

Fig. 4.3. Mechanism of the residual-based attack monitor. The geometry of the ellipsoid is characterized by the design monitoring matrix $\mathcal{M}_i$.

Now, we are ready to determine the attack monitoring matrix $\mathcal{M}_i$ as follows: In an attack-free case when $\delta_{ai}(k) = 0$, the attack monitoring ellipsoid $r_i^T(k)\mathcal{M}_i r_i(k)$ should encompass all possible residuals given the process and measurement noises $w_i(k)$ and $v_i(k)$ for $\forall k \in \mathbb{Z}^+$. To determine the design matrix $\mathcal{M}_i$, the minimal upper bound of the state estimation error $\bar{e}_i(k)$ in the attack-free case should be considered when the dynamics of the state observer reaches a steady state condition. To achieve this, we propose the following theorem:

Theorem 4.3. *Suppose that the LMIs in Theorem 4.2 are satisfied, where there exist matrices $\mathcal{P}_2 = \mathcal{P}_2^T \in \mathbb{R}^{n\times n} > 0$ and $\mathcal{N} \in \mathbb{R}^{n\times q}$ so that we can derive an observer gain $L_{\text{ob}} = \mathcal{P}_2^{-1}\mathcal{N}$. With this observer gain, consider the following state error dynamics in the attack-free case:*

$$\bar{e}_i(k+1) = (A - L_{\text{ob}}C)\,\bar{e}_i(k) + B_w w_i(k) - L_{\text{ob}} F v_i(k), \tag{4.24}$$

where the matrix $(A - L_{\text{ob}}C) \in \mathbb{R}^{n\times n}$ is Schur stable. Let $\alpha \in (0,1)$ be a fixed constant. Suppose there exist positive scalars $\mu_1, \mu_2 \in \mathbb{R}^+$ for which the following optimization problem holds:

$$\min_{\mu_1,\mu_2} \ \mu_1 + \mu_2$$

$$\text{s.t.} \quad \bar{\mathcal{R}} = \left[\begin{array}{c|c} \mathcal{R} & * \\ \hline \tilde{\mathcal{R}} & -\mathcal{P}_2 \end{array}\right] < 0, \quad \mathcal{P}_2 - \mu_2^{-1} I_n \geq 0, \tag{4.25}$$

where

$$\mathcal{R} = \begin{bmatrix} (\alpha-1)\,\mathcal{P}_2 & * & * \\ \mathbf{0} & -\alpha\mu_1 I_{\bar{w}} & * \\ \mathbf{0} & \mathbf{0} & -\alpha\mu_1 I_{\bar{v}} \end{bmatrix},$$

$$\tilde{\mathcal{R}} = [\mathcal{P}_2 A - \mathcal{N}C \ \ \mathcal{P}_2 B_w \ \ \mathcal{N}F],$$

then the upper bound of the state estimation error $\bar{e}_i(k)$ satisfies the following inequality condition:

$$|\bar{e}_i(k)|_2 \leq c\sqrt{\mu_2} + \sqrt{\mu_1\mu_2}\,(\|w_i(k)\|_\infty + \|v_i(k)\|_\infty), \tag{4.26}$$

where

$$c = \sqrt{(1-\alpha)^k \lambda_{\max}(\mathcal{P}_2)|\bar{e}_i(0)|_2},$$

where $\lambda_{\max}(\mathcal{P}_2)$ denotes the maximum eigenvalue of the matrix $\mathcal{P}_2$.

Proof. Consider the following inequality:

$$\Delta\bar{\mathcal{V}}_i(k) + \alpha\bar{\mathcal{V}}_i(k) \leq \alpha\mu_1 \Big(w_i^T(k)w_i(k) + v_i^T(k)v_i(k)\Big), \tag{4.27}$$

where $\bar{\mathcal{V}}_i(k) = \bar{e}_i^T(k)\mathcal{P}_2\bar{e}_i(k)$ and $\Delta\bar{\mathcal{V}}_i(k) = \bar{\mathcal{V}}_i(k+1) - \bar{\mathcal{V}}_i(k)$. It is trivial that the inequality (4.27) results in:

$$\sqrt{\bar{\mathcal{V}}_i(k)} \leq c + \sqrt{\mu_1}\,(\|w_i(k)\|_\infty + \|v_i(k)\|_\infty).$$

If the second LMI in (4.25) is satisfied, then our conclusion (4.26) is obtained. The next question is how to show (4.27) is satisfied. By using a vectorized form, the inequality condition (4.27) now becomes

$$z_i^T(k) \underbrace{\left[\mathcal{R} + \tilde{\mathcal{R}}^T\mathcal{P}_2\tilde{\mathcal{R}}\right]}_{X} z_i(k) \leq 0, \tag{4.28}$$

where $z_i(k) = [\bar{e}_i^T(k),\, w_i^T(k),\, v_i^T(k)]^T$. By employing the Schur complement to the inequality $X < 0$, we can derive the first LMI in (4.25). Finally, if the optimization in (4.25) is satisfied, then we can show that the inequality (4.26) is satisfied. Therefore, the proof for Theorem 4.3 is now completed. □

With the aid of Theorem 4.3 and the $\mathcal{S}$-procedure from Lemma 4.3, the design of the optimal minimal-volume attack monitor reduces to solving the following theorem:

Theorem 4.4. *Provided that Theorem* 4.3 *is satisfied, and considering the upper bound of the estimation error* $\bar{e}_i(k)$ *together with the residual* $r_i(k)$ *from* (4.20), *let there exist* $\kappa_1 \in \mathbb{R}^+$, $\kappa_2 \in \mathbb{R}^+$, *and a symmetric positive semidefinite matrix* $\mathcal{M}_i \in \mathbb{R}^{q\times q}$ *fulfilling the following optimization condition*:

$$\min_{\kappa_1,\kappa_2,\mathcal{M}_i} \quad -\log\det(\mathcal{M}_i)$$

$$\text{s.t.} \begin{bmatrix} \kappa_1 I_n - C^T\mathcal{M}_iC & * & * \\ -F^T\mathcal{M}_iC & \kappa_2 I_{\bar{v}} - F^T\mathcal{M}_iF & * \\ \mathbf{0} & \mathbf{0} & 1-\kappa_1\mu_1\mu_2(\tilde{\mathcal{W}}_i+\tilde{\mathcal{V}}_i)-\kappa_2\tilde{\mathcal{V}}_i \end{bmatrix} \geq 0, \tag{4.29}$$

$$\mathcal{M}_i \geq 0,\ \kappa_1 > 0,\ \kappa_2 > 0.$$

Then, in the absence of attacks, the residual-based monitor in (4.21) *encapsulates all admissible residuals* $r_i(k)$, *provided that the system* *noises $w_i(k)$ and $v_i(k)$ satisfy the norm-bounded constraints, i.e., $w_i^T(k)w_i(k) \leq \tilde{\mathcal{W}}_i$ and $v_i^T(k)v_i(k) \leq \tilde{\mathcal{V}}_i$.*

Proof. Consider the following LMI condition:

$$\Sigma_i := \left[(C\bar{e}_i(k) + Fv_i(k))^T \mathcal{M}_i (C\bar{e}_i(k) + Fv_i(k)) - 1\right] \\ - \kappa_1 \left(\bar{e}_i^T(k)\bar{e}_i(k) - \mu_1\mu_2(\tilde{\mathcal{W}}_i + \tilde{\mathcal{V}}_i)\right) - \kappa_2 \left(v_i^T(k)v_i(k) - \tilde{\mathcal{V}}_i\right) \leq 0.$$

Based on the preliminary results $v_i^T(k)v_i(k) \leq \tilde{\mathcal{V}}_i$, $\bar{e}_i^T(k)\bar{e}_i(k) \leq \mu_1\mu_2\left(\tilde{\mathcal{W}}_i + \tilde{\mathcal{V}}_i\right)$, and $\mathcal{S}$-procedure (Yang *et al.*, 2023), the vectorized form of the LMI condition $\Sigma_i \leq 0$ is same as the condition in (4.29). Finally, this completes the proof for Theorem 4.4. □

Next, by utilizing the residual $r_i(k)$ in (4.20), we can rewrite the stealthy attack $\delta_{ai}(k)$ as follows:

$$\delta_{ai}(k) = S_a^+ (r_i(k) - C\bar{e}_i(k) - Fv_i(k)), \tag{4.30}$$

where S_a^+ represents the Moore-Penrose inverse of S_a^+. Next, by employing (4.30), the augmented error dynamics (4.9) can be reformulated in terms of system noises and residual as follows:

$$\xi(k+1) = \mathcal{C}_1\xi(k) + \mathcal{B}_2 w(k) + \mathcal{C}_2 v(k) + \mathcal{C}_3 r(k), \tag{4.31}$$

where $\xi(k)$ and $\mathcal{B}_2$ are provided in (4.9) and $\mathcal{C}_1$, $\mathcal{C}_2$, and $\mathcal{C}_3$ are

$$\mathcal{C}_1 = \begin{bmatrix} I_N \otimes A + \bar{\mathcal{L}} \otimes BK_c & -\bar{\mathcal{L}} \otimes BK_c \\ \mathbf{0} & I_N \otimes (A - L_{\text{ob}}\mathcal{O}C) \end{bmatrix},$$

$$\mathcal{C}_2 = \begin{bmatrix} \mathbf{0} \\ -I_N \otimes L_{\text{ob}} (F - S_a S_a^+ F) \end{bmatrix},$$

$$\mathcal{C}_3 = \begin{bmatrix} \mathbf{0} \\ -I_N \otimes L_{\text{ob}} S_a S_a^+ \end{bmatrix}, \quad \mathcal{O} = \left(I_q - S_a S_a^+\right),$$

where we have $r(k) = [r_1^T(k), \ldots, r_N^T(k)]^T$, which is the stacked residual vector.

Remark 4.6. A notable aspect of the reformulated dynamics in (4.31) is that the residual $r_i(k)$ substitutes for the stealthy attack $\delta_{ai}(k)$. As a result, the state $\xi(k)$ is perturbed by bounded disturbances $w(k)$, $v(k)$, and $r(k)$. The stealthy attack $\delta_{ai}(k)$, however, is capable of causing considerable estimation error in $\bar{e}_i(k)$, while producing outputs $y_i(k)$ that are nearly indistinguishable from $\hat{y}_i(k)$. This highlights a potential risk of stealthy attacks, namely, their ability to evade detection.

4.5 Risk Assessment via Lyapunov-Based Reachability Analysis

Building on the augmented error dynamics (4.31) under norm-bounded perturbations arising from system noise and residual signals, we propose a reachability-based framework to assess the potential risks associated with stealthy attacks. The proposed methodology leverages LMIs and the Lyapunov stability criterion to obtain the ellipsoidal reachable set. To this end, we provide the following definition and lemma, which serve as the foundation for establishing the main theorem in this section.

Definition 4.1. Consider the discrete linear-time-invariant system as follows:

$$x(k+1) = \tilde{\mathcal{A}}x(k) + \sum_{i=1}^{N_p} \tilde{\mathcal{B}}_i w_i(k), \tag{4.32}$$

where $x(k) \in \mathbb{R}^n$ denotes the system's state perturbed by the external signal $w_i(k) \in \mathbb{R}^m$ with the norm-bounded condition, i.e., $w_i^T(k)\tilde{W}_i w_i(k) \leq 1$ for $\forall i \in \mathcal{I}_{N_p}$. Here, $N_p \in \mathbb{Z}^+$ denotes the number of perturbation signals, and $\tilde{W}_i \in \mathbb{R}^{m\times m}$ represents the ellipsoidal bound of $w_i(k)$. Additionally, $\tilde{\mathcal{A}} \in \mathbb{R}^{n\times n}$ denotes the system matrix satisfying the Schur condition (i.e., a stable matrix) and $\tilde{\mathcal{B}}_i \in \mathbb{R}^{n\times m}$ denotes the perturbation matrix. Starting from the initial state $x(0)$, the collection of all states attainable within k time steps is defined as the reachable set $\mathcal{R}_k^x$. The mathematical description for $\mathcal{R}_k^x$ can be represented as follows:

$$\mathcal{R}_k^x := \left\{ x(k) \in \mathbb{R}^n \,|\, x(k+1) = \tilde{\mathcal{A}}x(k) + \sum_{i=1}^{N_p} \tilde{\mathcal{B}}_i w_i(k), \text{ and } w_i^T(k)\tilde{W}_i w_i(k) \leq 1 \right\}.$$

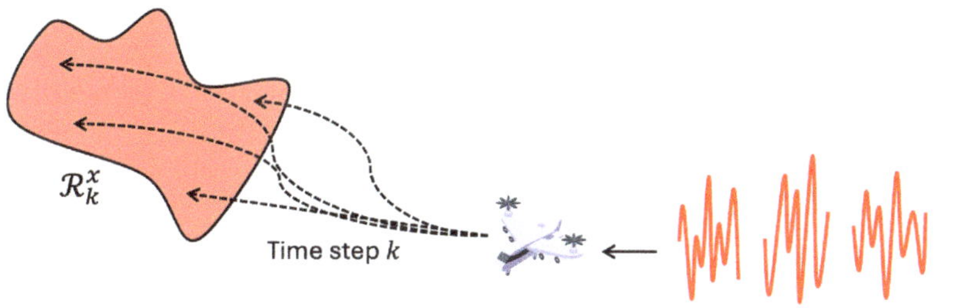

Fig. 4.4. Illustration of the system's reachability under norm-bounded perturbations (see Definition 4.1).

Remark 4.7. Based on Definition 4.1 and the augmented error dynamics (4.31), $\mathcal{R}_k^\xi$ denotes the reachable set for $\xi(k)$, i.e., the set of $\xi(k)$ reachable within k time steps, given the norm-bounded inputs $w(k)$, $v(k)$, and $r(k)$. This concept is illustrated in Fig. 4.4.

Determining the exact reachable set $\mathcal{R}_k^\xi$ is often computationally intractable. Hence, our approach is to construct an over-approximated set that fully contains the actual reachable set. Accordingly, we introduce the following lemma:

Lemma 4.4 (Murguia *et al.*, 2020). *Let $a \in (0, 1)$ be given. Suppose there exists a design parameter $b_i \in (0, 1)$ for $\forall i \in \mathcal{I}_{N_p}$ and a proper function $V : \mathbb{R}^n \to \mathbb{R} > 0$ such that the following inequality holds:*

$$V(x(k+1)) - aV(x(k)) - \sum_{i=1}^{N_p}(1 - b_i)w_i^T(k)\tilde{W}_i w_i(k) \leq 0, \quad (4.33)$$

where $\sum_{i=1}^{N_p} b_i \geq a$, $w_i^T(k)\tilde{W}_i w_i(k) \leq 1$ for $\forall i \in \mathcal{I}_{N_p}$, and $N_p \in \mathbb{Z}^+$ denotes the number of perturbations. Therefore, the following inequality is obtained:

$$V(x(k)) \leq a^k V(x(0)) + \frac{(N_p - a)(1 - a^{k-1})}{(1 - a)}. \quad (4.34)$$

Proof. When the time step $k = 0$, we obtain the following inequality from (4.33):

$$V(x(1)) - aV(x(0)) - \sum_{i=1}^{N_p}(1 - b_i)w_i^T(k)\tilde{W}_i w_i(k) \leq 0. \tag{4.35}$$

Given the conditions $\sum_{i=1}^{N_p} b_i \geq a$ and $w_i^T(k)\tilde{W}_i w_i(k) \leq 1$ for $\forall i \in \mathcal{I}_{N_p}$, the inequality (4.35) becomes

$$V(x(1)) \leq aV(x(0)) + (N_p - a). \tag{4.36}$$

Based on the conditions (4.33) and (4.36), we can obtain the following result:

$$V(x(2)) \leq a^2 V(x(0)) + (N_p - a)(1 + a).$$

Finally, by using a geometric series formula, we can derive the inequality in (4.34). This completes the proof. □

Building upon Lemma 4.4, the subsequent theorem presents an ellipsoid-based over-approximation of $\mathcal{R}_k^{\xi}$ for the dynamics (4.31), which is derived by employing the Lyapunov stability criterion.

Theorem 4.5. *Consider the reformulated augmented error dynamics (4.31) subject to norm-bounded perturbations $w(k)$, $v(k)$, and $r(k)$, and its corresponding reachable set $\mathcal{R}_k^{\xi}$. Given a design constant $a \in (0, 1)$, suppose there exist $b_1, b_2, b_3 \in \mathbb{R}^+$ and a matrix $\mathcal{P}_r = \mathcal{P}_r^T > 0 \in \mathbb{R}^{2nN\times 2nN}$, obtained by solving the following LMI-based optimization problem:*

$$\begin{aligned} &\min_{b_1,b_2,b_3,\mathcal{P}_r} \; -\log\det(\mathcal{P}_r) \\ &\text{s.t.} \begin{bmatrix} a\mathcal{P}_r & * & * \\ \mathbf{0} & \mathcal{E} & * \\ \mathcal{P}_r\mathcal{C}_1 & \mathcal{P}_r\mathcal{D} & \mathcal{P}_r \end{bmatrix} \geq 0, \quad \mathcal{P}_r > 0, \\ &b_1,\; b_2,\; b_3 \in (0,1),\; b_1 + b_2 + b_3 \geq a, \end{aligned} \tag{4.37}$$

where we have

$$\mathcal{E} = \begin{bmatrix} (1-b_1)\mathcal{E}_1 & \mathbf{0} & \mathbf{0} \\ \mathbf{0} & (1-b_2)\mathcal{E}_2 & \mathbf{0} \\ \mathbf{0} & \mathbf{0} & (1-b_3)\mathcal{E}_3 \end{bmatrix}, \quad \mathcal{D} = [\mathcal{B}_2, \mathcal{C}_2, \mathcal{C}_3],$$

$$\mathcal{E}_1 = \frac{1}{N} \times \mathrm{diag}[\tilde{\mathcal{W}}_1^{-1}, \ldots, \tilde{\mathcal{W}}_N^{-1}] \in \mathbb{R}^{N\times N},$$

$$\mathcal{E}_2 = \frac{1}{N} \times \mathrm{diag}[\tilde{\mathcal{V}}_1^{-1}, \ldots, \tilde{\mathcal{V}}_N^{-1}] \in \mathbb{R}^{N\times N},$$

$$\mathcal{E}_3 = \frac{1}{N} \times \mathrm{diag}[\mathcal{M}_1^{-1}, \ldots, \mathcal{M}_N^{-1}] \in \mathbb{R}^{qN\times qN},$$

then we obtain $\mathcal{R}_k^{\xi} \subseteq \mathcal{Q}_k^{\xi} := \left\{\xi(k) \in \mathbb{R}^{2nN} \,|\, \xi^T(k)\mathcal{P}_r\xi(k) \leq \bar{\mathcal{Q}}_k^{\xi}\right\}$, *where the upper bound is calculated as follows:*

$$\bar{\mathcal{Q}}_k^{\xi} := a^k \xi^T(0)\mathcal{P}\xi(0) + \frac{(3-a)(1-a^{k-1})}{1-a}.$$

Proof. We define the Lyapunov function in the following form:

$$\mathcal{V}_r(k) = \xi^T(k)\mathcal{P}_r\xi(k),$$

where $\xi(k)$ is provided in (4.31) and $\mathcal{P}_r = \mathcal{P}_r^T \in \mathbb{R}^{2nN\times 2nN} > 0$. Building upon Lemma 4.4, we assume that the following inequality is satisfied:

$$\begin{aligned} \mathcal{J} := {} & \mathcal{V}_r(\xi(k+1)) - a\mathcal{V}_r(\xi(k)) - (1-b_1)w^T(k)\mathcal{E}_1 w(k) \\ & - (1-b_2)v^T(k)\mathcal{E}_2 v(k) - (1-b_3)r^T(k)\mathcal{E}_3 r(k) \leq 0. \end{aligned}$$

By substituting the dynamics of $\xi(k)$ in (4.31) into the inequality $\mathcal{J} \leq 0$, we obtain the following vectorized form in terms of $\xi(k)$, $w(k)$, $v(k)$, and $r(k)$ as:

$$\mathcal{J} := \theta^T(k) \underbrace{\left[\begin{bmatrix} a\mathcal{P}_r & * \\ \mathbf{0} & \mathcal{E} \end{bmatrix} - \begin{bmatrix} \mathcal{C}_1^T\mathcal{P}_r \\ \mathcal{D}^T\mathcal{P}_r \end{bmatrix} \mathcal{P}_r^{-1} \begin{bmatrix} \mathcal{P}_r\mathcal{C}_1 & \mathcal{P}_r\mathcal{D} \end{bmatrix} \right]}_{\mathcal{H}} \theta(k) \geq 0, \tag{4.38}$$

where $\theta(k) = [\xi^T(k), w^T(k), v^T(k), r^T(k)]^T$. Next, by employing the Schur complement to the LMI condition $\mathcal{H} \geq 0$ in (4.38), we can derive the LMIs in (4.37). Finally, we can obtain $\xi^T(k)\mathcal{P}_r\xi(k) \leq \bar{\mathcal{Q}}_k^{\xi}$, i.e., the disrupted state trajectories $\xi(k)$ are encompassed in the ellipsoid-based over-approximated set $\mathcal{Q}_k^{\xi}$. □

The key contribution of Theorem 4.5 is the construction of the ellipsoid-based set $\mathcal{Q}_k^{\xi}$, serving as an over-approximation of the reachable set with respect to $s(k)$ and $\bar{e}(k)$ across all N vehicles. By applying the ellipsoidal projection to the matrix $\mathcal{P}_r = \mathcal{P}_r^T \in \mathbb{R}^{2nN \times 2nN}$, the reachable set corresponding to vehicle i can be obtained. For more technical details on the ellipsoidal projection, the reader is referred to Corollary 2 in Murguia *et al.* (2020) and Corollary 1 in Yang *et al.* (2023). Based on the over-approximated reachable sets of each vehicle, two geometric operations can then be systematically conducted: (a) union and (b) intersection. These operations specifically facilitate the assessment of potential risks at both the individual-vehicle level and the entire system level. For additional technical details on ellipsoid-based geometric operations, see Chapter 3.7 (Ellipsoidal Approximation) of Boyd *et al.* (1994).

Remark 4.8. The computational complexity of Theorem 4.5, i.e., the complexity associated with solving LMI in (4.37), primarily depends on the number of agents N, the number of states per agent n, and the number of output measurements q. It is known that solving a general semidefinite program (SDP) with a matrix variable of size $\upsilon \times \upsilon$ and ζ affine constraints using a primal–dual interior-point method incurs a per-iteration computational complexity of, $\mathcal{O}(\upsilon^{\omega}\zeta + \zeta^{\omega} + \upsilon^2\zeta^2)$, where ω denotes the exponent of matrix multiplication (Jiang *et al.*, 2020). In theory, the exponent is conservatively bound by $\omega \leq 3$, yielding a per-iteration computational complexity of $\mathcal{O}(\upsilon^3\zeta + \zeta^3 + \upsilon^2\zeta^2)$ (Fujisawa and Endo, 2014). This complexity arises from the cost of forming the Schur complement matrix and performing its Cholesky factorization in each interior-point iteration. Under ideal conditions, assuming strict feasibility and excluding infeasible or ill-conditioned cases, the path-following interior-point method requires $\mathcal{O}(\sqrt{\upsilon}\log\frac{1}{\epsilon})$ iterations to achieve an ϵ-optimal solution (Nesterov and Nemirovskii, 1994). Consequently, the overall computational complexity of solving (4.37) can be upper-bounded by

$$\mathcal{O}(n^{\frac{9}{2}}N^{\frac{9}{2}}\log(1/\epsilon)),$$

under the assumption $m, q = \mathcal{O}(n)$. This result indicates that the runtime required to compute the over-approximated ellipsoidal reachable set grows *polynomially* with respect to both the number of agents and the state dimension. Moreover, the network topology, characterized by the augmented Laplacian matrix $\bar{\mathcal{L}}$, directly affects the sparsity pattern of the resulting LMI matrices. While such sparsity does not alter the theoretical convergence rate of interior-point methods, it can substantially reduce the per-iteration computational cost and memory requirements by enabling more efficient Schur complement formation and sparse Cholesky factorization. As the number of agents increases, exploiting this structural sparsity becomes increasingly important. In this regard, model reduction and network clustering techniques (Monshizadeh *et al.*, 2014) can be employed to further alleviate the computational burden associated with large-scale multi-agent systems.

Remark 4.9. In this chapter, we mainly focus on developing a risk assessment framework for attack scenarios in which vehicles' sensors are subjected to stealthy attacks. However, due to the heavy reliance on inter-vehicle communication links, multi-vehicle system operations may also be vulnerable to other types of cyber threats, such as denial-of-service (DoS) attacks, which disrupt and/or tamper with communication among vehicles. To this end, we proposed a risk assessment strategy for a multi-vehicle system under DoS attacks using the Lyapunov stability criterion and its associated LMIs in Cho *et al.* (2024). Readers are referred to the reference for further technical developments.

Remark 4.10. Regarding the feasibility of the proposed reachability-based risk assessment method on onboard computation, it is important to emphasize that the proposed framework is primarily designed for offline analysis rather than online control synthesis. In particular, the LMI-based optimization problems involved in Theorems 1 and 2 are solved offline, and therefore do not impose significant computational burden on onboard processors during real-time operation. Once the ellipsoidal reachable set is obtained, the online evaluation requires only lightweight algebraic operations, such as matrix–vector multiplications and scalar comparisons, to verify state containment within the pre-computed ellipsoid. As a result, the proposed method can be well suited for safety-critical systems with limited onboard computational resources.

4.6 An Illustrative Example and Discussion

This section presents a UAM case study within a multi-vehicle system framework, illustrating the effectiveness of the proposed Lyapunov-based risk assessment strategy under sensor stealthy attacks. The approach is validated through a simulation in a 3D urban-like environment with multiple obstacles like buildings. In addition, we discuss the practicability of the proposed defense strategy for modern CPS-based transportation systems.

4.6.1 *UAM Mission: Reference Tracking with Formation Control*

In this UAM mission scenario, we consider that five vehicles (e.g., UAM AVs) perform a reference tracking control mission by following a virtual leader, while maintaining formation throughout the operation. This leader-follower setting reflects realistic UAM operations, in which vehicles carry out missions within designated UAM airspace while coordinating their formations (refer to Figure 2 in Murthy *et al.*, 2025). The total simulation time for the UAM case study is set as 45 s, and the system matrices for the vehicles are given as follows:

$$A = \begin{bmatrix} 1 & 0 & 0 & \Delta T_s & 0 & 0 \\ 0 & 1 & 0 & 0 & \Delta T_s & 0 \\ 0 & 0 & 1 & 0 & 0 & \Delta T_s \\ 0 & 0 & 0 & 1 & 0 & 0 \\ 0 & 0 & 0 & 0 & 1 & 0 \\ 0 & 0 & 0 & 0 & 0 & 1 \end{bmatrix}, \quad B = \begin{bmatrix} \Delta T_s^2/2 & 0 & 0 \\ 0 & \Delta T_s^2/2 & 0 \\ 0 & 0 & \Delta T_s^2/2 \\ \Delta T_s & 0 & 0 \\ 0 & \Delta T_s & 0 \\ 0 & 0 & \Delta T_s \end{bmatrix},$$

$$\begin{aligned} C &= I_6, \\ B_w &= 0.1 \times I_6, \quad S_a = \begin{bmatrix} 0.40 & 0.40 & 0.60 & 0.33 & -0.33 & -0.20 \\ 0.07 & -0.16 & -0.33 & 0.07 & 0.07 & 0.13 \end{bmatrix}^T, \\ F &= I_6, \end{aligned}$$

where

$$x(k) = [\mathbf{p}_x^T(k),\, \mathbf{p}_y^T(k),\, \mathbf{p}_z^T(k),\, \mathbf{v}_x^T(k),\, \mathbf{v}_y^T(k),\, \mathbf{v}_z^T(k)].$$

Specifically, $x(k) \in \mathbb{R}^6$ denotes the system state vector, consisting of position and velocity components along the x, y, and z axes. The sampling time is set as $\Delta T_s = 0.5$. Regarding the norm-bounded conditions of the process and measurement noises $w_i(k)$ and $v_i(k)$ for $\forall i \in \mathcal{I}_5$, the following assumptions are satisfied:

$$|w_i(k)|_2^2 = w_i^T(k)w_i(k) \leq 0.01,$$

$$|v_i(k)|_2^2 = v_i^T(k)v_i(k) \leq 0.5, \quad \forall k \in \mathbb{Z}^+.$$

Figure 4.5 presents the augmented Laplacian matrix $\bar{\mathcal{L}}$ (*left*) together with its corresponding formation topology (*right*), offering a more intuitive graphical representation. We assume that the vehicles' formation is arranged in a diamond shape, where vehicles 1, 2, and 3 are connected to the virtual leader (*dotted arrow*). In addition, Fig. 4.6 shows a nominal UAM mission in the absence of stealthy attacks, where the five vehicles successfully track the trajectory of the virtual leader (*blue line*) without any safety violations, while operating in an environment that contains obstacles (e.g., buildings). Building upon Theorems 4.1 and 4.2, the controller gain K_c to stabilize the tracking with formation and the observer gain L_o are calculated

$$\bar{\mathcal{L}} = \mathcal{L} + \Gamma = \begin{bmatrix} 4 & -1 & -1 & -1 & 0 \\ -1 & 4 & -1 & -1 & 0 \\ -1 & -1 & 4 & -1 & 0 \\ -1 & -1 & -1 & 4 & -1 \\ 0 & -1 & -1 & -1 & 3 \end{bmatrix}$$

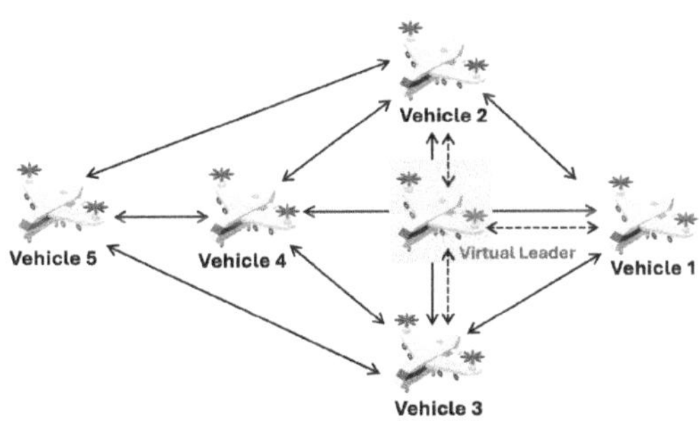

Fig. 4.5. Augmented Laplacian matrix $\bar{\mathcal{L}}$ (left) and the configuration of formation between vehicles and the virtual leader (right).

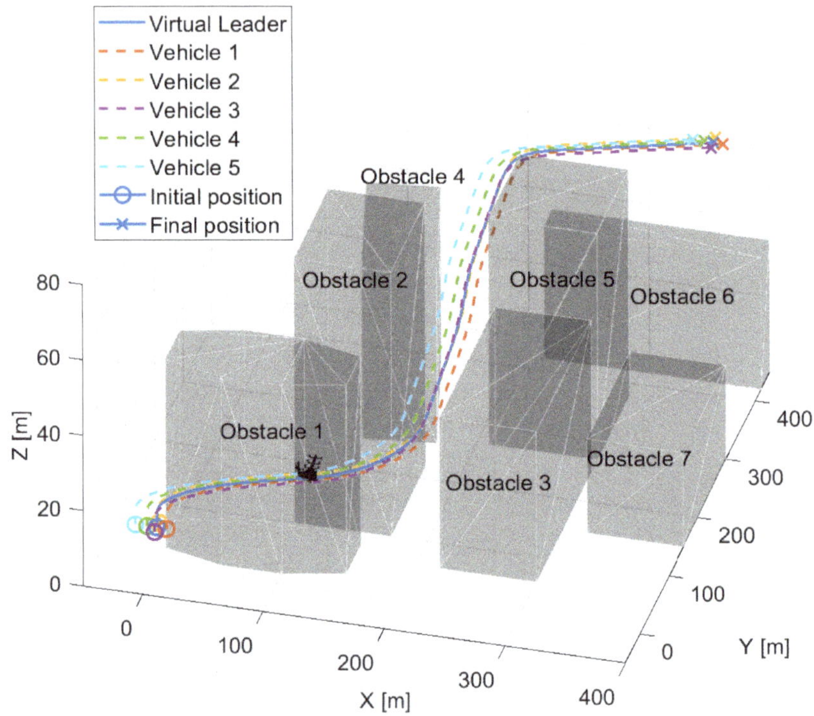

Fig. 4.6. 3D Illustration of a nominal UAM mission profile—reference tracking with formation control in an urban-like environment.

as follows:

$$K_c = \begin{bmatrix} -0.24 & 0 & 0 & -0.69 & 0 & 0 \\ 0 & -0.24 & 0 & 0 & -0.69 & 0 \\ 0 & 0 & -0.24 & 0 & 0 & -0.69 \end{bmatrix},$$

$$L_o = \begin{bmatrix} 1.05 & 0 & 0 & 0.31 & 0 & 0 \\ 0 & 1.05 & 0 & 0 & 0.31 & 0 \\ 0 & 0 & 1.05 & 0 & 0 & 0.31 \\ 0.25 & 0 & 0 & 0.12 & 0 & 0 \\ 0 & 0.25 & 0 & 0 & 0.12 & 0 \\ 0 & 0 & 0.25 & 0 & 0 & 0.12 \end{bmatrix}.$$

Then, Theorems 4.3 and 4.4 provide the upper bound of the state estimation error $\bar{e}_i(k)$ and the attack monitoring matrix, which are given as follows:

$$\mu_1 = 17.23, \quad \mu_2 = 0.71, \quad \kappa_1 = 0.12, \quad \kappa_2 = 0.44,$$
$$\mathcal{M}_i = 0.097 \times I_6, \quad \forall i \in \mathcal{I}_5.$$

4.6.2 *Risk Assessment of UAM Operations under Stealthy Attacks*

As illustrated in Fig. 4.6, complex urban environments can significantly heighten the risk of cyberattacks. In such settings, the presence of dense communication infrastructures and the interaction among multiple vehicles can create expanded surfaces for potential vulnerabilities, thereby offering greater opportunities for malicious intrusions. We assume that the sensor measurements from the output matrix are vulnerable to stealthy attacks via the attack matrix S_a, i.e., such cyber threats may prevent the vehicles from accomplishing the collaborative UAM mission without being detected. In the UAM case study, the attack sequences are generated in the following pattern to meet the stealthy condition (4.23), where they linearly increase over 45 s:

$$\delta_i(k) = [\mathcal{Z}([-0.07, 0.07]) \times k, \quad \mathcal{Z}([-0.07, 0.07]) \times k]^T, \quad \forall i \in \mathcal{I}_5, \tag{4.39}$$

where the function $\mathcal{Z}([a, b])$ provides a random number from the uniform distribution over the interval $[a, b]$ with $a, b \in \mathbb{R}$, and k denotes the discrete time step where $k \in [0, 90]$ with $k \in \mathbb{Z}^+$. Due to the influence of $\delta_i(k)$, UAM aerial vehicles (AVs) may encounter various safety violations, including collisions with buildings or inter-vehicle crashes, potentially compromising the overall safety of UAM operations. From a safety-critical perspective, such incidents should be prevented or proactively detected before the operations.

We now demonstrate how the developed Lyapunov-based risk assessment strategy can be utilized to the UAM case study. Interestingly, two geometric operations—union and intersection—can be conducted based on the reachable sets obtained from individual vehicles. Note that the union of these reachable sets serves as a security metric at the fleet level, while the intersections provide a security metric at the inter-vehicle level. We implement 1000 Monte-Carlo simulations for the UAM case study, and we specifically focus on the UAM operation at two time instances, i.e., $t = 24\,\text{s}$ ($k = 48$) and $t = 26\,\text{s}$ ($k = 52$), where the simulation results are provided in Figs. 4.7 and 4.8, respectively.

Specifically, Fig. 4.7 (left) shows the ellipsoidal over-approximated reachable set in the attack-free case, $\delta_i(k) = 0, \forall i \in \mathcal{I}_5$, where only norm-bounded system noises affect the UAM operation. The reachable set of each vehicle is depicted in a different color—yellow, green, purple, orange, and turquoise—while the blue ellipsoidal set corresponds to the union of the individual reachable sets. For the fleet-level safety, this union can serve as a security metric that quantifies the potential risk during UAM operations. Additionally, Fig. 4.7 (right) demonstrates the effectiveness of the proposed Lyapunov-based risk assessment strategy under stealthy attacks. Although these attacks are articulated by following the form in (4.39) to avoid attack monitors, the developed Lyapunov-based method can effectively handle the potential risk from such an attack by leveraging the norm-bounded condition for residuals, as established in Theorem 4.4. In particular, Fig. 4.7 (right) depicts the ellipsoidal over-approximated reachable sets under stealthy attacks, where sensor compromises result in noticeable trajectory deviations compared to those from nominal operation. Beyond these individual reachable sets, both intersection and union ellipsoids can also be obtained. The intersection ellipsoid, shown in red, serves as an inter-vehicle security metric, indicating whether certain vehicles may be exposed to potential risks relative to one another. Similar to the attack-free case, the union ellipsoid functions as a fleet-level security metric. At time $t = 24\,\text{s}$ ($k = 48$) overlaps are observed between Obstacles 2 and 3, and the union ellipsoid, indicating that the fleet may be at risk of collision at that time in the presence of stealthy attacks. The interpretation of the simulation results in Fig. 4.8 can be carried out in a similar manner to those in Fig. 4.7, where potential risks are also identified at both the inter-vehicle and fleet levels.

Moreover, it is worth noting that the objective of the present UAM case study is not to benchmark detection accuracy or post-attack recovery performance against existing frameworks. Instead, the obtained ellipsoidal over-approximated reachable set provides a

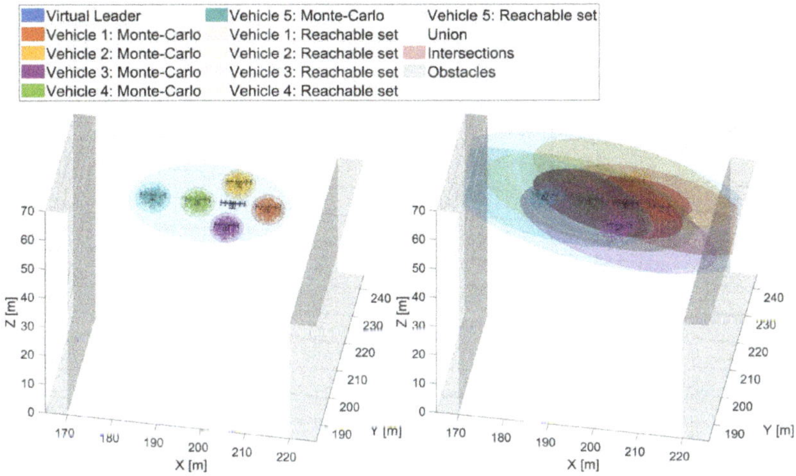

Fig. 4.7. Illustration of the ellipsoidal over-approximated reachable set at $t = 24$ s ($k = 48$): attack-free case (left) and under stealthy attacks (right).

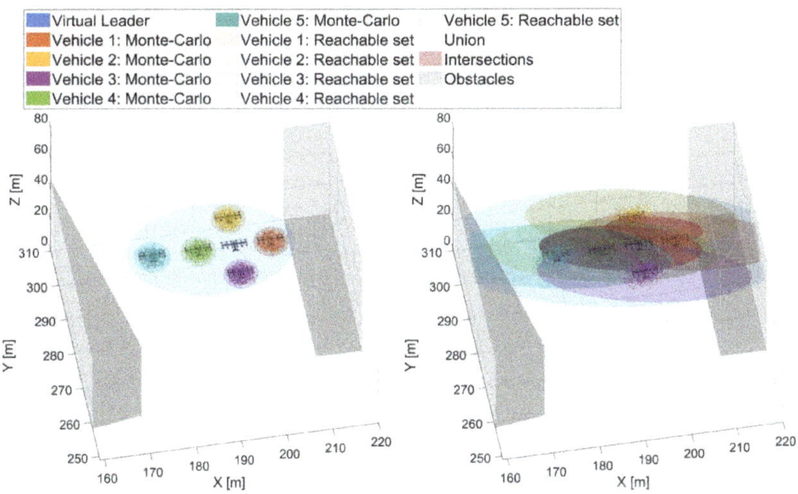

Fig. 4.8. Illustration of the ellipsoidal over-approximated reachable set at $t = 26$ s ($k = 52$): attack-free case (left) and under stealthy attacks (right).

proactive risk characterization, quantifying a worst-case safety envelope under bounded system noises, as well as the impact from stealthy attacks, even prior to the activation of detection or mitigation mechanisms. In this sense, the proposed approach complements detection- and resilience-oriented methods by offering risk bounds that can inform potential safety margins, monitoring thresholds, and contingency planning in UAM operations.

4.6.3 *Discussion of the Proposed Defense Strategy*

4.6.3.1 *Applicability of the Proposed Reachability-Based Risk Assessment Method under Stealthy Attacks*

From a safety-critical perspective, the proposed Lyapunov-based reachability framework offers a rigorous and quantitative security metric for evaluating the impact of stealthy attacks on system operations. Unlike conventional qualitative assessments, this metric systematically captures how adversarial perturbations may propagate through system dynamics and compromise mission performance. As demonstrated in the UAM case study with multiple UAM AVs, when the computed reachable set intersects with critical sets corresponding to hazardous or unsafe conditions, the system can be formally inferred as being at risk of potential hazardous behavior. This characterization not only uncovers vulnerabilities in the presence of hidden attacks but also quantifies their safety implications. In this regard, the proposed method serves as a proactive defense strategy against cyber threats, capable of anticipating risks even when attacks evade traditional detection mechanisms.

A key strength of the proposed method lies in its wide applicability across different transportation domains. Since the framework relies on general system models, Lyapunov functions, and convex optimization, the method can be seamlessly integrated into navigation, guidance, and path planning modules and enhance overall operational safety and resilience for diverse platforms—including urban air mobility, autonomous ground vehicles, and maritime fleets, without being restricted to a specific vehicle type or mission profile. Furthermore, by leveraging reachability-based security metrics, mission operators are able to dynamically adapt mission profiles or adjust control strategies in response to identified potential risks. Such adaptability enables real-time safety assurance and contributes to long-term system robustness, thereby improving the overall reliability and trustworthiness of safety-critical operations across a broad spectrum of modern transportation systems.

4.6.3.2 *Applicability of the Proposed Reachability-Based Risk Assessment Method to Real-World Transportation Platforms*

It is acknowledged that the proposed risk assessment framework is developed based on linear system models. In real-world operations, however, the vehicle dynamics of modern transportation systems (e.g., UAM platforms) may exhibit strong nonlinear behaviors. Nevertheless, linear system models are widely used to characterize the system dynamics within local operating regions, where linearization remains valid and is commonly adopted for safety- and certification-oriented analyses. Importantly, the objective of the proposed reachability-based method is not to provide an exact prediction of nonlinear trajectories, but rather to compute a conservative safety envelope that upper-bounds all possible system evolution under stealthy attacks and bounded uncertainties.

Although the present chapter primarily focuses on linear system models, the proposed LMI-based reachability framework can be naturally extended to nonlinear systems through well-established system modeling techniques. One promising approach is the Takagi–Sugeno (T–S) fuzzy modeling framework (Nguyen *et al.*, 2019), in which the nonlinear dynamics is represented as a convex combination of multiple local linear subsystems with the corresponding real-valued membership (weight) functions, while explicitly accounting for norm-bounded modeling approximation errors. Within this setting, the ellipsoidal reachability analysis and LMI-based conditions developed in this chapter can be effectively applied to fuzzified nonlinear systems, thereby enabling conservative yet tractable risk assessment for nonlinear dynamics under stealthy attacks.

4.7 Summary

In this chapter, we examined the critical cybersecurity challenges confronting modern transportation systems, particularly in smart vehicles like urban air mobility (UAM), where the dense, dynamic, and complex operating environment substantially amplifies the risks posed by stealthy attacks. These attacks are especially concerning because they can evade conventional monitoring while significantly disrupting vehicle dynamics and mission objectives. In light of these challenges, we proposed a novel Lyapunov-based framework to assess the risks associated with multi-vehicle systems. By leveraging linear matrix inequality (LMI) techniques and ellipsoidal reachable set computation, our framework enables a conservative yet systematic characterization of state deviations under sensor stealthy attacks. This approach transforms qualitative notions of vulnerability into quantifiable security metrics, allowing risk evaluation at both the vehicle and fleet levels. Such metrics are central to proactive defense strategies, as they empower vehicles and operators to anticipate and mitigate risks before adversarial actions escalate into safety-critical failures. The proposed methodology captures how stealthy adversaries may undermine essential mission objectives, including formation control, collision avoidance, and reference tracking, functions that underpin the safety and efficiency of next-generation transportation networks. By quantifying the potential for mission compromise, vehicles and operators are empowered to anticipate, assess, and mitigate risks, even in the absence of explicit attack detection. Our UAM case study conducted in a 3D urban-like environment demonstrated the practicality of the approach, illustrating how the proposed reachability-based risk assessment can uncover vulnerable operational states and guide system adaptive responses in real time. To sum up, the developed framework offers a principled and scalable foundation for proactive risk assessment in CPS-enabled transportation. It paves the way for the integration of security-aware control and planning into autonomous vehicles, thereby strengthening public trust and ensuring the reliable operation of safety-critical transportation infrastructures in increasingly complex and adversarial environments.

Bibliography

Bansal, S., Chen, M., Herbert, S., and Tomlin, C. J. (2017). Hamilton-Jacobi reachability: A brief overview and recent advances, in *2017 IEEE 56th Annual Conference on Decision and Control (CDC)* (IEEE), pp. 2242–2253.

Boel, R. K., Marinică, N. E., and Sarlette, A. (2015). Leader-follower cooperative control paradigm, with applications to urban traffic coordination control, in *2015 European Control Conference (ECC)* (IEEE), pp. 2208–2215.

Boyd, S., El Ghaoui, L., Feron, E., and Balakrishnan, V. (1994). *Linear Matrix Inequalities in System and Control Theory* (SIAM, Philadelphia, PA, USA).

Burl, J. B. (1998). *Linear Optimal Control: H (2) and H (Infinity) Methods* (Addison-Wesley Longman Publishing Co., Inc., Reading, MA, USA).

Cho, J.-H., Sharma, D. P., Alavizadeh, H., Yoon, S., Ben-Asher, N., Moore, T. J., Kim, D. S., Lim, H., and Nelson, F. F. (2020). Toward proactive, adaptive defense: A survey on moving target defense, *IEEE Communications Surveys & Tutorials* **22**, 1, pp. 709–745.

Cho, M., Hwang, S., and Hwang, I. (2024). Risk assessment of multi-agent system under denial-of-service cyberattacks using reachable set synthesis, in *2024 American Control Conference (ACC)* (IEEE), pp. 1293–1298.

Deka, L., Khan, S. M., Chowdhury, M., and Ayres, N. (2018). Transportation cyber-physical system and its importance for future mobility, in *Transportation Cyber-Physical Systems* (Elsevier), pp. 1–20.

Dibaji, S. M., Pirani, M., Flamholz, D. B., Annaswamy, A. M., Johansson, K. H., and Chakrabortty, A. (2019). A systems and control perspective of cps security, *Annual Reviews in Control* **47**, pp. 394–411.

Duo, W., Zhou, M., and Abusorrah, A. (2022). A survey of cyber attacks on cyber physical systems: Recent advances and challenges, *IEEE/CAA Journal of Automatica Sinica* **9**, 5, pp. 784–800.

El-Rewini, Z., Sadatsharan, K., Sugunaraj, N., Selvaraj, D. F., Plathottam, S. J., and Ranganathan, P. (2020). Cybersecurity attacks in vehicular sensors, *IEEE Sensors Journal* **20**, 22, pp. 13752–13767.

Eshmawi, A., Umer, M., Ashraf, I., and Park, Y. (2024). Enhanced machine learning ensemble approach for securing small unmanned aerial vehicles from GPS spoofing attacks, *IEEE Access* **12**, pp. 27344–27355.

Freeman, K. and Lewis, T. D. (2022). NASA's secured airspace for urban air mobility (UAM), in *Secured Airspace for UAM Workshop.*

Fujisawa, K. and Endo, T. (2014). Peta-scale general solver for semidefinite programming: Extremely large-scale parallel Cholesky solver, *Transactions of the Operations Research Society of Japan* **2014**, pp. 104–105.

Góes, R. M., Kang, E., Kwong, R., and Lafortune, S. (2017). Stealthy deception attacks for cyber-physical systems, in *2017 IEEE 56th Annual Conference on Decision and Control (CDC)* (IEEE), pp. 4224–4230.

Gu, Z., Huang, X., Sun, X., Xie, X., and Park, J. H. (2023). Memory-event-triggered tracking control for intelligent vehicle transportation systems: A leader-following approach, *IEEE Transactions on Intelligent Transportation Systems* **25**, 5, pp. 4021–4031.

Herbert, S., Choi, J. J., Sanjeev, S., Gibson, M., Sreenath, K., and Tomlin, C. J. (2021). Scalable learning of safety guarantees for autonomous systems using Hamilton-Jacobi reachability, in *2021 IEEE International Conference on Robotics and Automation (ICRA)* (IEEE), pp. 5914–5920.

Hill, B. P., DeCarme, D., Metcalfe, M., Griffin, C., Wiggins, S., Metts, C., Bastedo, B., Patterson, M. D., and Mendonca, N. L. (2020). UAM vision concept of operations (ConOps) UAM maturity level (UML) 4.

Hoffmann, R., Pereira, D. P., and Nishimura, H. (2023). Security viewpoint and resilient performance in the urban air mobility operation, *IEEE Open Journal of Systems Engineering* **1**, pp. 123–138.

Huang, H., Wlazlo, P., Mao, Z., Sahu, A., Davis, K., Goulart, A., Zonouz, S., and Davis, C. M. (2022). Cyberattack defense with cyber-physical alert and control logic in industrial controllers, *IEEE Transactions on Industry Applications* **58**, 5, pp. 5921–5934.

Islam, M. M., Newaz, A. A. R., Song, L., Lartey, B., Lin, S.-C., Fan, W., Hajbabaie, A., Khan, M. A., Partovi, A., Phuapaiboon, T., *et al.* (2023). Connected autonomous vehicles: State of practice, *Applied Stochastic Models in Business and Industry* **39**, 5, pp. 684–700.

Jiang, H., Kathuria, T., Lee, Y. T., Padmanabhan, S., and Song, Z. (2020). A faster interior point method for semidefinite programming, in *2020 IEEE 61st Annual Symposium on Foundations of Computer Science (FOCS)* (IEEE), pp. 910–918.

Kayan, H., Nunes, M., Rana, O., Burnap, P., and Perera, C. (2022). Cybersecurity of industrial cyber-physical systems: A review, *ACM Computing Surveys (CSUR)* **54**, 11s, pp. 1–35.

Kwon, C., Liu, W., and Hwang, I. (2014). Analysis and design of stealthy cyber attacks on unmanned aerial systems, *Journal of Aerospace Information Systems* **11**, 8, pp. 525–539.

Lakshmanan, S., Yan, Y., Baek, S., and Alghodhaifi, H. (2019). Modeling and simulation of leader-follower autonomous vehicles: Environment effects, in *Unmanned Systems Technology XXI*, Vol. 11021 (SPIE), pp. 116–123.

Lee, E. A. (2015). The past, present and future of cyber-physical systems: A focus on models, *Sensors* **15**, 3, pp. 4837–4869.

Liu, Y., Peng, Y., Wang, B., Yao, S., and Liu, Z. (2017). Review on cyber-physical systems, *IEEE/CAA Journal of Automatica Sinica* **4**, 1, pp. 27–40.

Luenberger, D. G. (2007). Observing the state of a linear system, *IEEE Transactions on Military Electronics* **8**, 2, pp. 74–80.

Meng, H. (2023). Vulnerability of unmanned aerial vehicles under stealthy perception-based attacks, Master's Thesis, Duke University.

Monshizadeh, N., Trentelman, H. L., and Camlibel, M. K. (2014). Projection-based model reduction of multi-agent systems using graph partitions, *IEEE Transactions on Control of Network Systems* **1**, 2, pp. 145–154.

Murguia, C., Shames, I., Ruths, J., and Nešić, D. (2020). Security metrics and synthesis of secure control systems, *Automatica* **115**, p. 108757.

Murthy, S. K., Ingebrand, T., Smith, S., Topcu, U., Wei, P., and Neogi, N. A. (2025). Separation assurance in urban air mobility systems using shared scheduling protocols, in *AIAA SCITECH 2025 Forum*, p. 2116.

Nesterov, Y. and Nemirovskii, A. (1994). *Interior-Point Polynomial Algorithms in Convex Programming* (SIAM, Philadelphia, PA, USA).

Nguyen, A.-T., Taniguchi, T., Eciolaza, L., Campos, V., Palhares, R., and Sugeno, M. (2019). Fuzzy control systems: Past, present and future, *IEEE Computational Intelligence Magazine* **14**, 1, pp. 56–68.

Patil, A. K., Punugupati, B., Gupta, H., Mayur, N. S., Ramesh, S., and Honnavalli, P. B. (2025). Building the future of transportation: A comprehensive survey on AV perception, localization, and mapping, *Sensors* **25**, 7, p. 2004.

Picone, M., Busanelli, S., Amoretti, M., Zanichelli, F., Ferrari, G., *et al.* (2015). Advanced technologies for intelligent transportation systems.

Rani, S., Kataria, A., Kumar, S., and Karar, V. (2025). A new generation cyber-physical system: A comprehensive review from security perspective, *Computers & Security* **148**, p. 104095.

Rathinasamy, D., Sivaramakrishnan, R., Pavithra, R., and Kaliappan, V. K. (2024). Safety and security challenges in implementing urban air mobility, in *Urban Air Mobility* (River Publishers, Aalborg, Denmark), pp. 211–238.

Safwat, N. E.-D., Thangavel, K., Hussain, K. F., Gardi, A., and Sabatini, R. (2024). Intelligent cyber-physical system for advanced air mobility and UAS traffic management, in *2024 AIAA DATC/IEEE 43rd Digital Avionics Systems Conference (DASC)* (IEEE), pp. 1–10.

Shafa, T. and Ornik, M. (2022). Reachability of nonlinear systems with unknown dynamics, *IEEE Transactions on Automatic Control* **68**, 4, pp. 2407–2414.

Song, G. I., Park, H. Y., and Kim, J. H. (2025). The H infinity robust stability and performance conditions for uncertain robot manipulators, *IEEE/CAA Journal of Automatica Sinica* **12**, 1, pp. 270–272.

Sui, T., Mo, Y., Marelli, D., Sun, X., and Fu, M. (2020). The vulnerability of cyber-physical system under stealthy attacks, *IEEE Transactions on Automatic Control* **66**, 2, pp. 637–650.

Sun, N., Ding, M., Jiang, J., Xu, W., Mo, X., Tai, Y., and Zhang, J. (2023). Cyber threat intelligence mining for proactive cybersecurity defense: A survey and new perspectives, *IEEE Communications Surveys & Tutorials* **25**, 3, pp. 1748–1774.

Tang, A. C. (2021). A review on cybersecurity vulnerabilities for urban air mobility, in *AIAA Scitech 2021 Forum*, p. 0773.

Tang, Q., Cheng, Y., Hu, X., Chen, C., Song, Y., and Qin, R. (2021). Evaluation methodology of leader-follower autonomous vehicle system for work zone maintenance, *Transportation Research Record* **2675**, 5, pp. 107–119.

Tounsi, W. and Rais, H. (2018). A survey on technical threat intelligence in the age of sophisticated cyber attacks, *Computers & Security* **72**, pp. 212–233.

Trejo, J. A. V., Rotondo, D., Medina, M. A., and Theilliol, D. (2021). Robust observer-based leader-following consensus for a class of nonlinear multi-agent systems: Application to UAV formation control, in *2021 International Conference on Unmanned Aircraft Systems (ICUAS)* (IEEE), pp. 1565–1572.

VanAntwerp, J. G. and Braatz, R. D. (2000). A tutorial on linear and bilinear matrix inequalities, *Journal of Process Control* **10**, 4, pp. 363–385.

Xie, Y., Gardi, A., and Sabatini, R. (2023). Cybersecurity risks and threats in avionics and autonomous systems, in *2023 IEEE International Conference on Dependable, Autonomic and Secure Computing, International Conference on Pervasive Intelligence and Computing, International Conference on Cloud and Big Data Computing, International Conference on Cyber Science and Technology Congress (DASC/PiCom/CBDCom/CyberSciTech)* (IEEE), pp. 0814–0819.

Xiong, G., Zhu, F., Liu, X., Dong, X., Huang, W., Chen, S., and Zhao, K. (2015). Cyber-physical-social system in intelligent transportation, *IEEE/CAA Journal of Automatica Sinica* **2**, 3, pp. 320–333.

Xu, X., Chen, S., Huang, W., and Gao, L. (2013). Leader-following consensus of discrete-time multi-agent systems with observer-based protocols, *Neurocomputing* **118**, pp. 334–341.

Yang, H., Zhang, Y., Gu, W., Yang, F., and Liu, Z. (2022). Set-membership filtering for automatic guided vehicles with unknown-but-bounded noises, *Transactions of the Institute of Measurement and Control* **44**, 3, pp. 716–725.

Yang, T., Murguia, C., and Lv, C. (2023). Risk assessment for connected vehicles under stealthy attacks on vehicle-to-vehicle networks, *IEEE Transactions on Intelligent Transportation Systems* **24**, 12, pp. 13627–13638.

https://doi.org/10.1142/9789819831548_0005

Chapter 5

Misdirected Guidance and Urban Traffic Network Resilience

Eunhan Ka* **and Satish V. Ukkusuri,*,†**

**Lyles School of Civil and Construction Engineering*
Purdue University, West Lafayette, IN, USA
†Department of Computer Science (Courtesy)
Purdue University, West Lafayette, IN, USA

Abstract
This chapter examines how compromised navigation services can affect the resilience of urban traffic networks. It first reviews the threat landscape of modern navigation systems, with an emphasis on positioning attacks, traffic data manipulation, digital map poisoning, and route guidance attacks. It then models how drivers respond to deceptive travel-time information under perfect rationality, logit-based stochastic choice, and bounded rationality. These behavioral responses are coupled with the Generalized Bathtub Model to translate individual route choices into network-level congestion, throughput, travel time, and resilience losses. The chapter shows that the same cyber manipulation can produce different network impacts depending on driver compliance and tolerance to perceived travel-time differences. The findings motivate defense strategies that combine signal authentication, cross-source data validation, anomaly detection, and human-centered guidance design to reduce the effect of deceptive routing information on network operations.

Keywords: Navigation systems, Route guidance attacks, Traffic network resilience, Bounded rationality, Generalized bathtub model.

5.1 Introduction

Modern *navigation systems* have become an indispensable part of daily transportation, guiding billions of drivers through real-time traffic conditions. For instance, Google Maps alone serves over 1 billion monthly users, and roughly 90% of drivers report using navigation at least weekly (Harutyunyan, 2025; Tires, 2024). By dynamically rerouting around incidents and congestion, these services can improve travel efficiency. One case study found that real-time routing reduced average travel times by about 6.5% and emissions by 1.7% in a city network (Arora *et al.*, 2021). Navigation apps can also enhance traffic resilience: for example, smartphone guidance diverted up to 22% of vehicles away from a suddenly closed freeway, speeding up crash recovery and preventing secondary traffic jams (Guin *et al.*, 2021). However, the growing reliance on navigation systems has introduced new cyber-physical vulnerabilities. Recent incidents and research demonstrate that attacks on navigation can have serious real-world consequences for safety, efficiency, and network resilience.

Security researchers in 2014 and 2018 successfully generated *ghost traffic jams* on the Waze app by using Sybil accounts (fake virtual cars), leading unwitting drivers to reroute onto less efficient routes (Tufnell, 2014; Wang *et al.*, 2018). In 2020, a person towing 99 smartphones could fool Google Maps into showing a phantom traffic jam and divert drivers away from an empty road (Weckert, 2020). Figure 5.1 illustrates a data poisoning attack on a navigation service. Figure 5.1(a) shows an attacker pushing a cart loaded with smartphones that submit clustered low-speed reports, thereby fabricating a congestion signal. Across the map panels, the dashed ellipse highlights the targeted bridge segment where normal flow (green) in Fig. 5.1(b) is turned into apparent heavy congestion (red) in Fig. 5.1(d). In a recent real-world incident in 2025, dozens of residents in Zandvoort, Netherlands, coordinated to fake road closures on Google Maps in order to redirect tourist traffic away from their neighborhood (NL Times, 2025). Their coordinated disinformation campaign succeeded in rerouting drivers, prompting the municipality to issue warnings advising drivers to turn off navigation on specific days. These examples illustrate that both physical-layer attacks (e.g., Global Navigation Satellite System (GNSS) jamming/spoofing) and data-layer attacks (like map or traffic data poisoning) can significantly impair the reliability of navigation systems and disrupt traffic.

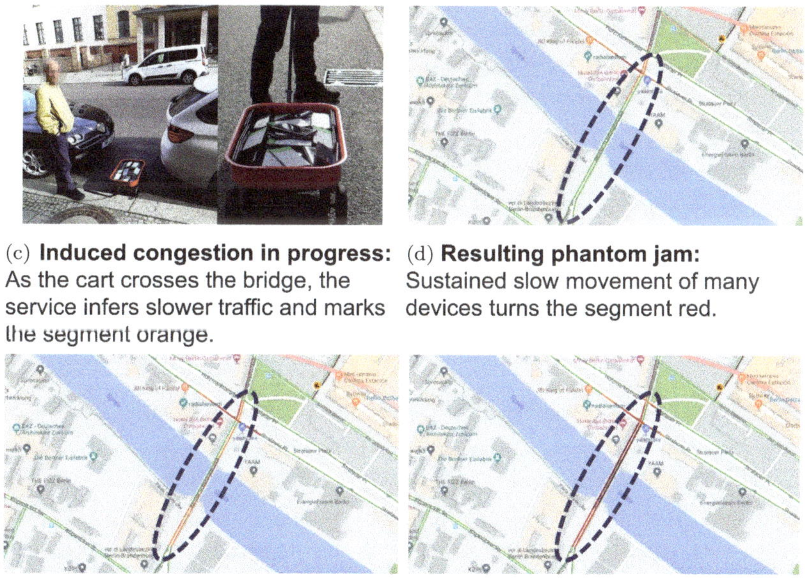

(a) **Attack setup:** A handcart carrying 99 smartphones is moved at walking speed to create fake traffic jam.

(b) **Nominal scenario before the attack:** The highlighted bridge segment shows free flow conditions.

(c) **Induced congestion in progress:** As the cart crosses the bridge, the service infers slower traffic and marks the segment orange.

(d) **Resulting phantom jam:** Sustained slow movement of many devices turns the segment red.

Fig. 5.1. Demonstration of false traffic data injection into navigation systems: (a) Attack setup. (b) Benign scenario before the attack. (c) Induced congestion. (d) Resulting phantom jam.

Source: Images adapted from Simon Weckert (2020).

The impacts of compromised navigation can be quantified. Field reports and simulations show that malicious navigation signals or data can induce substantial delays and congestion. For instance, an adversarial manipulation of live traffic reports was found to increase total travel times across an urban network by over 20% compared to normal conditions (Eryonucu and Papadimitratos, 2022; Waniek *et al.*, 2021). In a controlled study, a stealthy GPS spoofing attack misled 95% of test drivers into following an incorrect route without noticing the deception (Zeng *et al.*, 2018), highlighting the safety risk if drivers are led into dangerous or unauthorized areas. On the other hand, navigation failures (whether accidental or due to attack) degrade efficiency. If drivers lose guidance or trust in the system, they may react with sudden stops, missed turns, or suboptimal route choices that compound congestion. Even the resilience of transportation networks (i.e., the ability to recover from disruptions) diminishes when navigation information cannot be trusted. Drivers might flock to a few perceived safe routes and overload them, or conversely avoid certain areas entirely, either of which can exacerbate the spatial concentration of traffic beyond what an unbiased system would produce. In summary, navigation system attacks have the potential not only to inconvenience individual drivers, but also to trigger network-wide traffic breakdowns or safety incidents.

Research Question and Contributions. This chapter asks how route guidance attacks propagate from cyber manipulation to network-level congestion and resilience, and how driver response moderates that propagation. We make three contributions. First, we synthesize the navigation-system threat landscape and define route guidance attacks (RGAs) as a cross-layer threat that links data manipulation with human route choice. Second, we formalize three driver response models (i.e., perfect rationality, logit-based stochastic choice, and bounded rationality) to represent different levels of compliance under deceptive travel-time information. Third, we couple these micro-level choices with the Generalized Bathtub Model to quantify system-wide efficiency and resilience losses across attack intensities. A key message is that driver response is a first-order determinant of resilience under deceptive guidance.

Overview of This Chapter. Section 5.2 surveys the threat landscape of modern navigation systems by separating physical layer and data layer attacks and by motivating RGAs as the link between cyber manipulation and human decision making. Building on that, Section 5.3 models driver response to deceptive travel-time information under three behavioral paradigms and formalizes the attacker's goal, capability, and knowledge. Section 5.4 then translates these micro-level route choices into macro-level network dynamics using the Generalized Bathtub Model (GBM), whose state variables and conservation laws track the evolution of network-wide congestion and throughput. Section 5.5 describes the demand generation, the attack mechanism, the model discretization, and the evaluation metrics used to measure efficiency and resilience. Section 5.6 quantifies how attack intensity and behavior interact to shape travel times, throughput, and resilience, and explains the mechanisms behind the observed patterns. Figure 5.2 shows an overview of this chapter. In this figure, a scenario is shown in which a malicious actor injects falsified

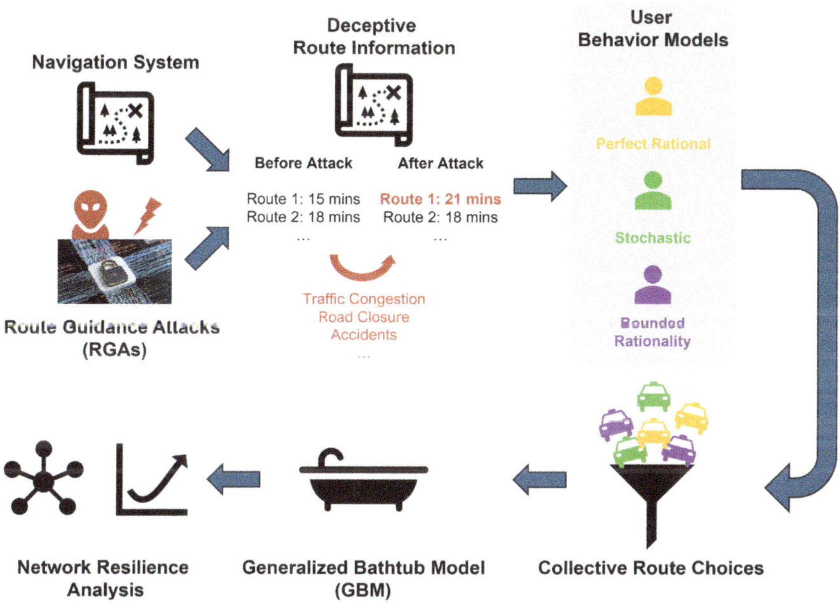

Fig. 5.2. Overview of the analysis pipeline for traffic network resilience under RGAs. The sequence corresponds to (i) deceptive travel-time information (Section 5.2), (ii) behavioral response models (Section 5.3), (iii) GBM network dynamics (Section 5.4), and (iv) efficiency and resilience metrics (Sections 5.5–5.6).

routing data into the navigation system, causing users to receive inaccurate travel time guidance (for example, *Route 1: 21 min*). The analysis considers three behavioral models, perfect rationality (PR), stochastic logit (LOGIT), and bounded rationality (BR), which lead to different route splits and traffic patterns. Section 5.7 translates these findings into defense and mitigation strategies that combine signal security, data validation, and human-centered route guidance. Lastly, Section 5.8 concludes all findings and discussions.

5.2 Threat Landscape of Navigation Systems

Modern vehicle navigation systems are complex cyber-physical platforms composed of several interconnected components. At the foundation are the **GNSS signals** (e.g., GPS, GLONASS, Galileo) broadcast from satellites, which vehicles use to determine position and time. This positioning feeds into **on-board navigation devices** or smartphone apps that combine it with a digital **map database** (containing road layouts, speed limits, etc.) to compute optimal routes. Route computation can be done on the device or via cloud servers, often using real-time inputs. One crucial input is **real-time traffic data**, which may come from centralized infrastructure (sensors, cameras) but increasingly from decentralized crowdsourcing: the aggregate speed and incident reports from many user devices on the road (Reece, 2020). Many navigation services also integrate user-generated content such as reports of accidents, hazards, road closures, and map edits contributed by volunteers. In advanced settings, navigation systems interface with **vehicle-to-everything (V2X)** communications (e.g., receiving safety messages or sharing probe data with traffic management centers). Finally, these systems interact with the human driver: the guidance module recommends turn-by-turn directions, which drivers may accept or override based on their own judgment.

Each of these components is a potential attack surface. To better understand and mitigate risks, it is useful to classify threats into layers based on which part of the system they target:

- **Physical Layer (Positioning and Timing):** Attacks that manipulate the signals or sensors providing position, velocity, and time to the navigation system. The primary target here is GNSS signals, which are vulnerable to jamming and spoofing due to their low power and lack of authentication (Psiaki and Humphreys, 2016a; Tippenhauer *et al.*, 2011). Other sensors, such as odometers, inertial units, or compasses, could also be manipulated (e.g., via electromagnetic interference) in principle.
- **Data Layer (Digital Maps and Traffic Data):** Attacks on the digital information that the routing engine uses. This includes static map data (road network topology and attributes) as well as dynamic data (traffic congestion levels, incidents, and travel times). By falsifying or tampering with this data (e.g., inserting fake road closures or feeding incorrect speed information), an attacker can indirectly influence route calculations. Unlike physical attacks, data-layer attacks do not require proximity and can often be executed remotely via the network or user-contributed channels (Chen *et al.*, 2018; Wang *et al.*, 2018).

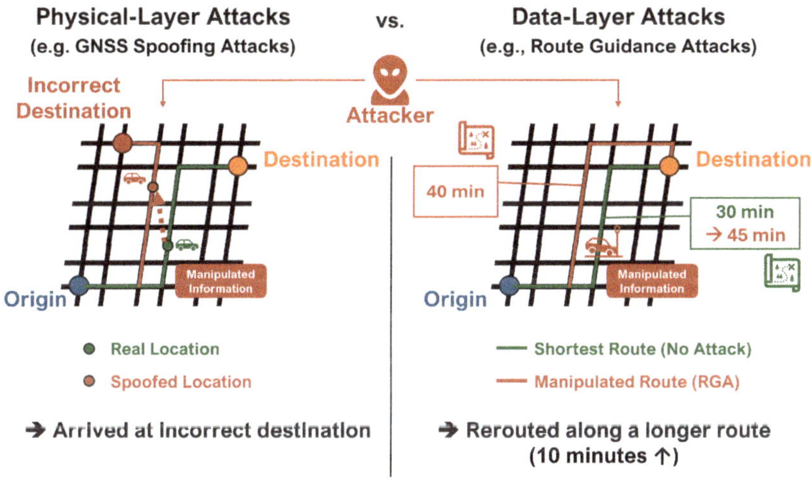

Fig. 5.3. Comparison of physical layer and data layer attacks.

At the interface of the physical and data layers are *Route Guidance Attacks* (RGAs): attacks that target the decision-making process of the navigation service and the behavior of users following it. We use the term *Route Guidance Attacks* (RGAs) to describe attempts to maliciously bias routing outcomes, effectively tricking the algorithm or the user into choosing suboptimal or attacker-desired paths (Waniek *et al.*, 2021; Zeng *et al.*, 2018). RGAs often build on data-layer manipulations (since the easiest way to change routing advice is to feed false inputs), but they warrant separate analysis because they directly exploit how routing algorithms and human drivers respond to information. Here, the attacker's focus is on shaping global traffic patterns (e.g., creating or alleviating jams at specific locations) rather than just corrupting data for its own sake. Figure 5.3 left depicts a transmitter that spoofs the GNSS-based position, shifting it from the true route to a fabricated one and misleading the driver without clear signs of malfunction. Figure 5.3 illustrates a route guidance attack on the right, in which the adversary inflates the reported travel time on the actually fastest route. The service then infers a major jam and recommends a longer, manipulated alternative.

In the following sections, we examine each layer in detail, discussing the attackers' techniques, the conditions required for success, the observed or potential impacts on traffic flow, and known mitigation approaches. We devote special attention to RGAs, because they represent a convergence of cyber and human factors and pose complex challenges for defense.

5.2.1 *Physical-Layer Attacks*

Physical-layer attacks on navigation primarily involve *interfering with GNSS signals*, which form the backbone of positioning and timing in most vehicular navigation systems. GNSS signals are remarkably weak by the time they reach receivers on Earth (on the order of 10^{-16} W); they are also unencrypted in civilian use. This makes them inherently vulnerable: an attacker with a radio transmitter can overshadow or mimic the real satellite signals with relatively little power or cost (Psiaki and Humphreys, 2016a). The two fundamental threat modes are *jamming* and *spoofing*.

Jamming is conceptually simple: the attacker broadcasts noise or a strong continuous wave on the GPS frequency bands to drown out the authentic signals. A cheap jammer (sometimes called a GPS blocker) can be bought on the gray market and plugged into a vehicle's cigarette lighter; it will cause all nearby GPS receivers to lose their lock on satellite signals (Psiaki and Humphreys, 2016b). Jamming does not falsify data, but it creates a denial of service: the navigation device may report a lost GPS signal or fall back to dead reckoning. In one real incident, a truck driver in New Jersey used a GPS jammer to evade fleet tracking; the interference was so strong that it repeatedly disrupted the GPS-based landing system at Newark Airport until authorities located the source (Matyszczyk, 2013). The immediate impact of jamming is loss of navigation and timing. For drivers, this might mean a sudden loss of turn-by-turn directions. If it happens at scale (as reported in aviation contexts), safety is at risk because systems relying on GPS for timing (like some communication protocols or collision avoidance) may malfunction. Fortunately, GPS jamming is usually easy to detect (i.e., the receiver typically alerts when signals are unavailable). Jamming can be mitigated by using directional or high-gain antennas and filtering, though these measures are more common in military receivers. Another approach is leveraging signals from multiple constellations (GPS, Galileo, etc.): an attacker must jam all frequencies simultaneously, which is harder (though still feasible with a broadband jammer).

Spoofing is a more insidious attack where the adversary sends counterfeit GNSS signals that the target receiver locks onto, thereby outputting a bogus position or time. A spoofer typically imitates a set of satellites, gradually overpowering the real signals and causing the receiver to transfer its lock from the authentic signals to the counterfeit signals (Tippenhauer *et al.*, 2011). This is technically challenging but has been demonstrated in practice. A classic example is the 2013 experiment by University of Texas researchers, who spoofed a yacht's GPS and caused it to deviate off course without triggering an alarm (Kerns *et al.*, 2014). In a vehicular context, researchers have shown it is possible to spoof a car's GPS and trick navigation apps into mislocating the vehicle on a nearby road (Zeng *et al.*, 2018). More deviously, an attacker can create matched spoofing by forging signals that place a group of vehicles at mirror-image positions (e.g., each vehicle appears a few hundred meters from its true location). All those vehicles might then be instructed by their navigation systems to, say, turn around or take detours, potentially causing chaos on the road. One documented spoofing incident occurred in the Black Sea in 2017, where the GPS readings of over 20 ships showed them to be inland at a nearby airport; analysts suspect it was a state-sponsored spoofing attack, given the sophistication (Vaas, 2017).

If a physical-layer attack is successful, what are the broader traffic impacts? For isolated vehicles, spoofing could mislead a driver down incorrect routes (e.g., navigating to a fake destination or into a wrong-way entry if the map alignment is off). The greater danger is if many vehicles are affected simultaneously over an area. Jamming in a city center during rush hour could disable navigation for thousands of drivers, possibly causing slower responses to incidents and generally more erratic route choices (as drivers fall back on memory or smartphone maps without GPS). One analysis by European aviation agencies noted that GPS jamming events correlated with increased air traffic control workload and minor reroutings (Topham, 2024); by analogy on roads, one could expect increased travel times and potentially accidents (if, say, truck fleet guidance systems fail). Spoofing could be even more disruptive: an attacker might, for example, spoof all vehicles on a highway to believe they are a few hundred meters ahead of their actual position. This could confuse not only human drivers but also any connected/autonomous vehicle systems relying on geofencing or map data. If synchronized, such an attack could reroute multiple drivers simultaneously. Consider an attacker spoofing the GPS in an entire suburb to convince navigation apps that a main bridge is 5 km farther than it is. The apps might then direct all drivers to take a longer alternate route to reach that bridge location, effectively detouring all traffic and leaving the real bridge underutilized. The literature on multi-target spoofing shows that if done cleverly, attackers can indeed cause *coordinated misrouting* without triggering obvious alarms (Zeng *et al.*, 2018). Traffic-flow simulations have indicated that even a 10–20% subset of vehicles receiving incorrect GPS position data can increase average network delays, especially if those vehicles occupy critical links or intersection approaches (Chen *et al.*, 2018). Physical attacks can also be a precursor to data attacks: by feeding a vehicle incorrect coordinates, the vehicle's reported traffic data (speed and location) to the server becomes misinformation that can distort traffic maps for others.

In summary, physical-layer attacks exploit the unprotected nature of positioning signals. They can be detected by vigilant systems (multi-antenna checks, clock consistency checks, etc.), but many consumer navigation devices today lack such robust defenses. Techniques like signal authentication, adaptive filtering, and cooperative detection (e.g., vehicles sharing anomaly alerts) are active areas of development (Jansen *et al.*, 2018; Sathaye *et al.*, 2022). As vehicles increasingly rely on precise positioning (for example, in automated driving or platooning), securing the physical layer becomes critical to prevent cascading failures in higher-layer navigation functions.

5.2.2 *Data-Layer Attacks*

At the data layer, attackers manipulate the digital map content or real-time traffic information that navigation systems use to compute routes. These attacks do not require direct interference with radio signals; instead, they exploit the openness and connectivity of modern mapping platforms. A defining feature of current navigation services is their heavy reliance on user-contributed and third-party data. Services like Waze, Google Maps, and OpenStreetMap leverage millions of user inputs (e.g., reports of traffic slowdowns, accident notifications, and map edits for new roads) to provide up-to-date navigation. While this crowdsourcing provides rich, timely information, it also introduces vulnerabilities because malicious actors can pose as contributors.

One form of data-layer attack is **map database poisoning**. This involves inserting false information or deleting or altering correct information in the base map. OpenStreetMap (OSM), a widely used open mapping platform, has experienced such vandalism. For example, during the popularity of Pokémon GO, some pranksters edited OSM to add nonexistent footpaths or to delete roads, in order to influence the game or simply cause mischief (Juhász *et al.*, 2020). In a transportation context, an attacker might add a fake road that creates a shortcut in the map. Navigation apps could then route drivers onto a nonexistent path, which could lead to dead ends or dangerous off-road situations. Conversely, marking a legitimate road as closed or private could cause the navigation app to avoid that road, thereby overloading alternate routes. A notable incident occurred in 2020, when TomTom accused an OSM edit of introducing errors that affected its routing, highlighting the tension between open data and reliable navigation (Geere, 2012). Map poisoning can also occur via compromised official updates: connected cars receive periodic map updates over-the-air; if an attacker infiltrates that supply chain, they could deliver malicious map firmware (e.g., swap the speed limit signs or turn restrictions in the data). Although no major such case has been publicly recorded, the possibility has been raised by industry and standards bodies (Intellias, 2025).

Another prevalent attack category is **traffic data spoofing** through crowdsourcing channels. Navigation apps continuously collect anonymous speed and location data from users to infer traffic conditions. They also allow users to actively report incidents (crashes, police, hazards) and road status (closures, construction). Attackers can abuse these features by submitting false reports or simulating ghost vehicles. In the Waze attacks conducted by researchers, bots were programmed to simulate numerous virtual cars crawling along roads at slow speeds, fooling the app into detecting a nonexistent traffic jam (Tufnell, 2014; Wang *et al.*, 2018). Similarly, false accident reports can cause the system to mark a road as partially closed or delay-prone, influencing route choice for many drivers. In 2019, researchers Eryonucu and Papadimitratos demonstrated Sybil attacks on Google Maps, showing that even with some defenses in place, an attacker controlling a few dozen virtual clients could significantly alter the overall traffic conditions in a city (Eryonucu and Papadimitratos, 2022). The Zandvoort case in 2025, as previously mentioned, illustrates an incident of synchronized misinformation: a large number of residents collectively labeled their streets as closed on the map, thus creating a digital blockade (NL Times, 2025). Waze's developers have had to patch and improve their system after students repeatedly exposed such vulnerabilities; for example, after the 2014 experiment, Waze introduced stricter checks on user reports and capped the influence of new users to mitigate Sybil attacks (Matei, 2016; Tufnell, 2014).

A more sophisticated data-layer attack might target the **back-end algorithms or APIs** of the navigation service. For example, if an adversary hacks into the traffic data provider's system (imagine compromising a traffic camera feed or the city's open traffic API), they could inject fictitious data that propagates to all user devices. An attacker could also exploit any lack of authentication in the communication between the app and server. For example, if navigation instructions or traffic updates are not encrypted, a man-in-the-middle could alter them. There have been academic proofs of concept in which researchers intercepted navigation app communications and modified the route or map data on the fly (Cao *et al.*, 2019). Such attacks blur into route guidance attacks, since they effectively tamper with the control plane of the service.

Route Guidance Attacks (RGAs) refer to a category of cyber-attacks aimed at systematically misdirecting vehicles by exploiting the way navigation systems generate and disseminate route advice. Unlike a generic data attack, an RGA is typically strategic: the attacker has a specific objective in terms of traffic outcome (for example, maximize congestion on Route A, or prevent any traffic from reaching Location X) and tailors the misinformation to achieve that goal, while remaining undetected or plausible to users. To understand RGAs, it is essential to recall how route guidance typically works. Navigation algorithms (like those based on Dijkstra's or A* shortest path algorithms) seek to minimize some cost for each user, typically travel time or distance. Many services also incorporate predictive or real-time traffic data, effectively solving a shortest path problem on a time-varying weighted graph. In some cases, they attempt to achieve user equilibrium (no driver can improve by unilaterally changing route) or even system optimum (minimize total travel time) by nudging drivers with appropriate routes. However, the algorithm's output is ultimately advisory. Drivers may or may not follow it. This opens two dimensions for an attacker to exploit: the algorithm's **cost calculations** and the **users' response behavior**. Navigation services often provide alternative route suggestions or distribute drivers among multiple routes when those routes are nearly equal. An attacker can manipulate this by introducing small asymmetries. For example, if two routes are normally equal, the service might split drivers between them. An attacker could slightly degrade one route's apparent time, causing the algorithm to send every driver along the other route and potentially overloading that route. This tactic is referred to in some research as an alternative route attack (Fontaine *et al.*, 2022). It leverages the algorithm's deterministic nature: small input changes can cause a large shift in outcome if many users switch routes simultaneously (a threshold effect).

The consequences of data-layer attacks can be wide-ranging. By poisoning data, attackers indirectly cause the routing engine to produce *misleading or suboptimal routes* for potentially thousands of users. For instance, a fake traffic jam on a major highway might lead the app to reroute drivers through secondary streets; this can clog the smaller roads (a *spillover congestion*) while the highway actually remains underutilized. A real-world parallel occurred unintentionally in many cities as navigation apps became popular: drivers avoided main roads and crowded residential streets, prompting complaints. An attacker can deliberately induce that effect in a concentrated way. A recent study in *Scientific Reports* modeled disinformation attacks on traffic networks and found that even a small fraction of drivers receiving false congestion alerts could dramatically worsen overall congestion by causing cascades of rerouting (Waniek *et al.*, 2021). They distinguished divergence attacks (steering drivers away from a location) and convergence attacks (luring drivers to a location). In a divergence scenario, falsifying a road closure at a critical bridge can distribute traffic to longer detours (overall delay increases, and other bridges get jammed). In a convergence scenario, an attacker might, for example, create a fake event or incentive (say, a false message: 50% off fuel at Exit 20 today) that causes many drivers to converge on one place, creating a jam there (Waniek *et al.*, 2021). Although convergence attacks may require more creativity (people are less likely to follow a random lure than to heed a warning), the study found a significant portion of people would still alter their route based on such disinformation if it appeared credible.

Data attacks can also cause dangerous situations: Consider if an attacker removes a weight limit or height restriction from a bridge in the map data—trucks following navigation might be routed over a bridge that can't handle them or into tunnels they don't fit, leading to accidents. Or false info could route traffic through sensitive areas (e.g., residential zones or even military bases) intentionally to cause disturbance. There was a case where navigation apps sent drivers through wildfire zones due to incorrect data, raising concerns that data integrity is also a safety issue (although that case was likely not an intentional attack but rather a data lag).

Therefore, data-layer attacks turn the strengths of connected navigation into weaknesses. By attacking the information inputs, adversaries can effectively achieve outcomes similar to physical attacks (vehicles misrouted or delayed), but possibly on a larger scale and with remote reach. The defense requires rigorous data validation, trust management for user inputs, and fail-safes (if unsure, the system might revert to more conservative routing rather than aggressively rerouting on unverified info). Users also play a role: encouraging driver feedback (Is this road really closed? Report if not.) can create a self-correcting crowd to some extent. However, a knowledgeable adversary can still slip through, which is why the next section addresses the culmination of such efforts: when attackers orchestrate data-layer manipulations specifically to hijack the *route guidance* logic of the system.

5.2.3 *Summary*

Table 5.1 summarizes the navigation system threat landscape across these two layers, listing the key assets and typical attack methods at each, along with example impacts and possible indicators of attack. Importantly, these layers are interdependent. For example, a GNSS spoofing attack (physical layer) could cause a vehicle to appear at an incorrect location in the map database (data-layer issue), which in turn might result in bizarre routing instructions (route guidance anomaly). Likewise, a cleverly executed data-layer attack that falsifies traffic speeds can effectively become a route guidance attack if it leads many drivers to reroute. In practice, attackers may combine methods across layers to achieve their goals, so defenders need a

Table 5.1. Navigation system threats by layer. Each layer's assets can be targeted by specific attack techniques, leading to characteristic impacts.

Layer	Key Assets & Targets	Attacks & Impacts
Physical (Positioning & Timing)	GNSS signals (GPS, etc.); on-board sensors (IMU, speedometer); time synchronization	*Jamming*: overwhelm receivers with noise, causing loss of GPS lock. *Spoofing/Meaconing*: broadcast counterfeit GNSS signals to mislead the receiver's calculated location or time (Psiaki and Humphreys, 2016a; Tippenhauer *et al.*, 2011). **Impacts:** Vehicles may think they are at false locations or lose navigation function entirely; coordinated spoofing could detour multiple vehicles off-course or into unsafe areas. Time spoofing can disrupt networked systems (e.g., V2X communications timing). Detection indicators include sudden jumps in reported location, many users in an area losing satellite lock simultaneously, or impossible movements (as if teleporting).
Data (Maps & Traffic Information)	Digital map database (road geometry, speed limits, etc.); real-time traffic feeds (speeds, incidents); user reports and map edits; routing algorithms (cost functions, optimization criteria)	*Map database poisoning*: alter road data (e.g., close/open roads, change road connectivity or attributes) (Juhász *et al.*, 2020). *Crowdsourced traffic spoofing*: inject fake congestion reports or Sybil vehicles sending false slow-speed data (Eryonucu and Papadimitratos, 2022; Wang *et al.*, 2018). *Incident/closure falsification*: falsely reporting accidents or roadblocks (as in the Zandvoort case) (NL Times, 2025).

(*Continued*)

Table 5.1. *(Continued)*

Layer	Key Assets & Targets	Attacks & Impacts
		Backend compromise: hack the routing server or map update API to feed malicious data. *Route Guidance Attacks (RGAs)*: strategically manipulate inputs or the algorithm to steer drivers toward or away from specific routes (Chen *et al.*, 2018; Zeng *et al.*, 2018). Methods include subtle congestion data falsification to bias route costs, creating fake road closures (so traffic diverges), or even manipulating user perception via misleading prompts. **Impacts:** Navigation algorithms compute suboptimal or dangerous routes (e.g., sending drivers into heavy actual traffic or down inappropriate roads). Attackers can create virtual traffic jams that divert vehicles and cause congestion elsewhere or completely block navigation through certain areas (Waniek *et al.*, 2021). Indicators include contradictory data sources (e.g., user reports of a jam with no corroboration from road sensors), anomalous surges in incident reports, or frequent map changes by untrusted accounts (potential vandalism, Truong *et al.*, 2018).

holistic view of the system. Section 5.3 next formalizes how drivers respond when RGAs distort perceived route costs.

5.3 Modeling Driver Behavior under Deceptive Routing

Modern navigation systems greatly influence drivers' route choices, so a malicious actor can exploit them by providing deceptive guidance (incorrect travel times or incidents) to misdirect traffic. In this section, we develop models of driver decision-making under such attacks, considering three increasingly realistic paradigms: perfect rationality, logit-based stochastic choice, and bounded rationality. These models capture how drivers respond to route suggestions, ranging from idealized utility optimizers to humans with limited perception and varying levels of compliance. They form the basis for analyzing how a *route guidance attack* can perturb individual choices and, ultimately, network conditions. We align our terminology with prior discussions of driver compliance thresholds and attack-induced perturbations, examining how each behavioral assumption affects the efficacy of deceptive routing.

5.3.1 *Perfect Rationality*

Under the perfect rationality paradigm, drivers are assumed to be fully informed, utility-maximizing agents who always choose the route that minimizes their travel cost (typically travel time) (Yu *et al.*, 2020). This is analogous to Wardrop's First Principle of traffic equilibrium, wherein all drivers select the shortest (or least costly) path and no one can unilaterally improve their travel time by switching routes (Wardrop and Whitehead, 1952). Mathematically, if C_r denotes the perceived cost (e.g., travel time) of route r, a perfectly rational driver will choose route r^* such that

$$C_{r^*} = \min_{s \in \mathcal{R}} C_s \tag{5.1}$$

where $\mathcal{R}$ is the set of all routes. At equilibrium, all utilized routes have equal minimal cost. This classical *homo economicus* model corresponds to drivers having perfect information and obeying deterministic user-optimal behavior (Wardrop and Whitehead, 1952).

In the context of deceptive routing, perfect rationality implies that drivers follow guidance without hesitation if they believe it yields the shortest travel time. The navigation system effectively dictates the route choice by altering the perceived costs. A perfectly rational (but misinformed) driver will treat the falsified travel times as truth and thus may willingly take a longer or indirect route if the guidance suggests it is optimal. For example, if an attacker broadcasts that the direct highway route is heavily congested (high cost) and a circuitous route is *faster*, all rational drivers will abandon the highway and flock to the suggested detour. This represents a worst-case scenario from a vulnerability standpoint: 100% of drivers comply with the malicious advice, immediately creating a large-scale rerouting. Indeed, prior studies on route guidance attacks often made this simplifying assumption of full compliance. Under perfect rationality, the induced traffic pattern can be analyzed using conventional user equilibrium tools, but with the equilibrium computed based on the *perceived* costs rather than the true costs. The result is generally a distorted traffic assignment, where some routes are overutilized due to fake low costs, and others are underutilized due to fake incidents, potentially leading to severe congestion on the misdirected paths (Waniek *et al.*, 2021). Therefore, perfectly rational drivers represent a highly vulnerable case: any sufficiently convincing disinformation (e.g., false traffic jam reports) would be accepted by all drivers, causing a cascade of rerouting. This assumption provides an upper bound on attack impact, but it neglects the variability and skepticism observed in real driver behavior.

5.3.2 *Logit-Based Stochastic Choice*

Real-world drivers do not all make identical route choices even under the same information; there is inherent heterogeneity and randomness in decision-making. Discrete choice theory models this by treating route choice as a stochastic process driven by random utility (Ben-Akiva and Lerman, 1985). One widely used formulation is the multinomial logit model, in which each route r has an associated utility $U_r = V_r + \tilde{\epsilon}_r$. Here V_r is a measurable part (e.g., negative travel time or cost) and $\tilde{\epsilon}_r$ is a random error term reflecting unobserved factors and idiosyncrasies. A driver chooses route r with probability

$$P(r) = \frac{\exp(\beta V_r)}{\sum_{s \in \mathcal{R}} \exp(\beta V_s)}, \tag{5.2}$$

where $\beta > 0$ is the scale (sensitivity) parameter. Equation (5.2) is the multinomial logit choice function. As $\beta \to \infty$, the choice becomes nearly deterministic (approaching the perfectly rational case where the highest V_r is chosen with probability 1). As $\beta \to 0$, choices become random (all routes chosen equally). For intermediate β, shorter routes are chosen more often, but occasionally a longer route may be taken due to unmodeled preferences or information discrepancies. This model yields a *stochastic user equilibrium* when extended to many drivers, meaning that no driver can improve their *expected* utility by unilaterally changing their route (Daganzo and Sheffi, 1977).

The logit-based model is well-suited to capturing partial compliance with routing advice. Suppose an adversary falsifies the travel time on a certain link or route to lure drivers. In a logit model, this decreases the perceived cost V_r of that route, thereby increasing its choice probability $P(r)$ but not to 100%. Instead, a fraction of drivers will still choose other routes due to the random utility terms. In effect, route choice becomes *soft*: a malicious signal influences many drivers but leaves some unconvinced. For example, if a normally unattractive route is reported (falsely) as only slightly longer than the usual route, the logit model might predict, say, 30% of drivers divert to it and 70% stay, depending on the β parameter. This heterogeneity can blunt the impact of an attack compared to the all-or-nothing rational case.

From the attacker's perspective, a stochastic choice model means the outcome of a disinformation attack is probabilistic. An adversary can no longer guarantee all drivers will avoid a targeted road; instead, they can only increase the probability of avoidance. Consequently, some level of network resilience emerges: not every driver falls for the false information, so the induced congestion might be less than intended. On the other hand, stochasticity also means an attacker could try repeated or randomized tactics to gradually shift probabilities in their favor without detection. For instance, by subtly exaggerating travel times on a certain corridor, an attacker can incrementally divert a fraction of drivers. Each driver independently makes a stochastic choice via Eq. (5.2), so the attack's effect is spread across the population. Empirical evidence supports using such models: drivers' compliance with navigation advice is far from uniform. Some drivers are highly responsive to reported delays, while others exhibit inertia or a lack of trust. In a recent city-scale simulation, only about half of drivers followed fake congestion alerts on average (Waniek *et al.*, 2021). This variability can be captured by calibrating β (or using mixture models) to match observed compliance rates. Thus, logit-based models provide a more realistic baseline for analyzing deceptive routing, as they incorporate randomness and partial compliance, highlighting that not all drivers can be steered like *perfect optimizers.*

5.3.3 *Bounded Rationality*

While logit models introduce randomness, they still assume drivers evaluate all options in a compensatory manner. Real drivers are also *boundedly rational*: they satisfice rather than optimize perfectly, and they often use heuristics with thresholds for decision-making. One key concept here is the **indifference band** introduced by Mahmassani and Chang (1987). An indifference band means that a traveler is unwilling to change routes for a time saving below a certain threshold. In other words, if the travel time difference between two routes is small (within ε minutes or a certain percentage), the driver perceives them as essentially equal and may stick with the familiar route (Di *et al.*, 2017; Mahmassani and Chang, 1987). Only when the difference exceeds this threshold will the driver consider switching. Bounded rationality acknowledges limited perception and inertia: drivers do *not* always act on minute differences in travel time, especially under uncertainty or habit. They require a sufficiently large benefit to be enticed to switch, and conversely, will tolerate a certain amount of delay or extra distance without feeling misled.

Formally, a boundedly rational route-choice model can be expressed as an ε-*equilibrium.* In an ε-user equilibrium, no driver can improve their travel time by more than ε by unilaterally changing their routes (Di *et al.*, 2017). This relaxes Wardrop's condition to account for insensitivity to small differences. Empirical studies have quantified indifference bands. Mahmassani and Chang's experiments indicated that typical commuters had a time threshold on the order of a few minutes (around 5–10 min) or about 15–20% of travel time (Mahmassani and Chang, 1987). For instance, one experiment found that drivers would not switch to a route unless it saved roughly 19% of travel time pre-trip (or 18% on-route). In practical terms, a commuter might be indifferent between a 50-min route and a 53-min route, considering them effectively the same. Only a difference larger than $\varepsilon \approx 3$ min would prompt reconsideration. Bounded rationality also entails that drivers may distrust routes that appear too circuitous or deviate significantly from their usual experience. They possess a qualitative sense of what constitutes a reasonable detour, and any advice beyond that threshold is often ignored or viewed with skepticism.

Under deceptive routing attacks, bounded rationality imposes a critical constraint on the adversary: the false guidance must be plausibly within drivers' acceptance range, otherwise many drivers will simply reject it. If an attacker exaggerates a delay by an outrageous amount or suggests a detour that is clearly far longer than the usual route, boundedly rational drivers will likely not comply. For example, if a navigation app suddenly recommends a 20-km longer route that would add 30 min, many users will perceive this as suspicious or clearly suboptimal (outside their indifference band) and will stick to their known route. On the other hand, if the app suggests a minor detour with only a 5-min longer travel time, drivers might accept it, as this difference falls within their tolerance. In essence, the attacker faces a trade-off: the fake information must lie within the narrow range of deviation that users will tolerate enough to influence route choice, but not so large as to trigger disbelief. Recent research formalized this in an attack framework by incorporating a bounded rationality model, where the adversary constrained the magnitude of travel-time perturbation to stay within typical daily variability, thereby remaining credible to users. This approach, dubbed *plausibility preservation,* recognizes that boundedly rational users impose natural limits on what an attack can achieve before it is self-defeating.

In practical modeling, bounded rationality can be implemented by decision rules or modified logit functions. One simple rule-based model is to *switch to an alternate route if and only if it improves travel time by at least ε.* This leads to hysteresis effects and more stable route choices, as minor fluctuations in reported congestion will not cause massive rerouting. Alternatively, a softened approach can be employed, such as a stepwise logit model, in which perceived cost differences below ε are heavily dampened (drivers remain nearly indifferent) (Di *et al.*, 2017). Either way, the model better matches observed driver compliance behavior. Surveys indicate that not all drivers follow navigation advice blindly; overall adherence to route guidance is roughly 40–50%, dropping even further when the advice suggests a significant detour from one's normal route. Previous studies reported that while approximately 88% of drivers followed the guidance when it aligned with their habitual route, only about 31% did so when the recommended route deviated substantially from their usual commute (Ringhand and Vollrath, 2018; van Essen *et al.*, 2020). Such heterogeneity underscores the importance of bounded rationality in modeling. It captures both the existence of a compliance threshold and the fact that different drivers have different thresholds or trust levels.

To summarize, bounded rationality introduces a realistic nuance: drivers require a *credible* benefit to change behavior. In a cyber-physical attack scenario, this means minor fake congestion reports may have little effect (drivers shrug them off as normal variation), whereas carefully calibrated disinformation that just exceeds typical thresholds can cause significant, yet still believable, re-routing. The defender of the system can exploit this too: by ensuring any anomalous routing suggestions would require implausibly large deviations, one can reduce the fraction of drivers who fall victim to an attack. Bounded rationality thus provides a form of behavioral resilience, which is a filter that not all malicious signals can penetrate if they are beyond drivers' acceptance range. The next section will build on these behavioral models to examine how deceptive routing at scale impacts the overall network, using aggregate flow models and considering performance and resilience metrics.

5.3.4 *Threat Model*

We consider RGAs that inject false traffic information into a navigation system. RGAs aim to (i) inflate the perceived travel time of the currently shortest route and (ii) redirect drivers onto a suboptimal route while remaining inconspicuous. Our framework assumes the adversary can perform real-time data manipulation at a sufficient scale to noticeably affect the route-planning algorithm's output. The attacker must continuously counteract legitimate traffic reports to maintain the fake traffic jam, and modern navigation services may employ basic anomaly filters to flag or ignore suspicious congestion reports. We focus on a worst-case scenario where the attacker's manipulations are not filtered out, reflecting documented real-world vulnerabilities in crowdsourced navigation systems. Prior work has demonstrated that even a single cleverly deployed Sybil device can trigger false congestion and reroute traffic on Waze, and that deployed navigation apps historically lacked robust security mechanisms (Barrett, 2020; Tufnell, 2014).

Adversary's Goal. The adversary's primary objective is to induce drivers to abandon the true shortest path in favor of a slower, attacker-chosen path. This goal hinges on converting the driver's perceived travel time gain into an actual delay in the physical world. Merely altering the app's display does not suffice; the driver must genuinely follow the deceptive route and experience a measurable increase in travel time. Such delay attacks can have tangible real-world consequences, including forcing strategic detours that create congestion or that benefit the attacker in location-based fraud scenarios (Wang *et al.*, 2018; Waniek *et al.*, 2021). Consequently, the adversary's challenge is to maximize the real-world delay by manipulating the travel time to a critical point that is persuasive enough to trigger a detour, yet subtle enough to remain credible.

Adversary's Capabilities. An adversary can inject false congestion data into the navigation service's data stream. This can be achieved by creating or compromising multiple traffic-reporting clients, known as Sybil devices, which collectively submit false traffic incident reports (Barrett, 2020; Essers, 2013; Tufnell, 2014; Wang *et al.*, 2018). While achieving the necessary scale for such an attack is a significant requirement, similar data-poisoning exploits have been demonstrated in practice with moderate resources, proving their feasibility (Barrett, 2020; Eryonucu and Papadimitratos, 2022;

Essers, 2013; Tufnell, 2014). We assume the attacker's capabilities are bounded. The adversary cannot modify the reported travel times of alternative routes, directly control traffic signals, or alter the physical road infrastructure (Barrett, 2020; Tufnell, 2014). Instead, the attack leverages the trust-based process by which navigation systems aggregate user-submitted data to compute real-time route recommendations (Essers, 2013; Waniek *et al.*, 2021).

Adversary's Knowledge. We model an adversary who understands the operational logic of the navigation system. We assume the navigation system treats incoming data reports from its clients as trustworthy for generating a ranked list of routes (Chen *et al.*, 2018; Raponi *et al.*, 2021; Sinai *et al.*, 2014; Wang *et al.*, 2018; Zeng *et al.*, 2017). While modern services like Google Maps or Waze likely employ anomaly detection to filter out spurious data, we explore the vulnerability that exists if these defenses can be bypassed or are not tuned to detect subtle, localized manipulations. The attack's design aims to generate perturbations that fall within plausible traffic variations to remain below detection thresholds.

On the user side, we model boundedly rational drivers who trust their navigation app (Di *et al.*, 2017; Mahmassani and Chang, 1987; Ramirez *et al.*, 2021). A driver will not switch from their current path for a negligible time-saving. However, if the manipulated data makes the original route appear sufficiently slower, exceeding the driver's indifference threshold, they will accept the recommended detour. This behavioral assumption is supported by empirical findings, which indicate that a perceived travel time saving of approximately 20% is needed to induce a route change (Di *et al.*, 2017; Mahmassani and Liu, 1999).

The adversary must balance the inflation of travel times to be persuasive enough to trigger this switch, yet plausible enough to avoid raising driver suspicion or being flagged by the system as an anomaly. By operating within these constraints, the adversary can covertly misguide travelers and effectively prolong their journey. Our model focuses on the direct impact on the targeted user's routing decision, acknowledging that the dynamic reactions of other, non-targeted users in the network present a complex factor that warrants further investigation.

5.4 Generalized Bathtub Model for Network Traffic Dynamics

Deceptive routing attacks do not merely affect individual drivers; their ultimate impact is felt at the network level in the form of altered congestion patterns and overall traffic flows. To analyze system-wide effects and vulnerabilities, we adopt a macroscopic modeling approach. In particular, we use the *Generalized Bathtub Model* (GBM) framework, a recent analytical model for city-scale traffic dynamics (Jin, 2020). The GBM treats an urban road network as a single reservoir of traffic, abstracting away the spatial details and focusing on the aggregate evolution of trips in the network (Jin, 2020).

In our framework, the route-choice component in Section 5.3 determines the realized trip distances of entering travelers under a given attack intensity and behavioral assumption. These realized distances define the time-dependent entering trip-length distribution $\tilde{\varphi}(t, x)$ (or its survivor $\tilde{\Phi}(t, x)$), which is a key GBM input together with the inflow rate $f(t)$. The GBM then propagates these inputs into the evolution of active trips, mean speed, outflow, and resilience metrics.

The bathtub model analogy, originally attributed to Vickrey (1991), envisions the entire road network as a bathtub through which vehicles flow in and out (Vickrey, 2020). Inflow (vehicles entering the network) is like water from the faucet, outflow (vehicles completing their trips) is like water draining, and the accumulation of vehicles in the network corresponds to the water level. Vickrey's seminal concept was based on three premises:

- The congested road network is treated as a single undifferentiated movement area (a single bathtub or reservoir), ignoring specific origin-destination pairs or route topologies.
- A network-level speed-density relationship (also known as a network fundamental diagram) governs traffic: as the density (vehicles per unit network) increases, the average speed decreases. There is a well-defined critical density at which flow (drainage rate) is maximized, and beyond which additional vehicles cause throughput to decline (the onset of hyper-congestion or gridlock).

- Trip distances are assumed to follow a fixed negative exponential distribution (memoryless) that does not change over time. This implies most trips are short but a few are very long, and importantly, that the probability a trip is not completed is proportional to its remaining distance (a property of the exponential distribution).

To overcome the restrictive assumption of constant trip length in traditional bathtub models, Jin (2020) introduced the Generalized Bathtub Model (GBM), which does not constrain trip lengths to a specific distribution function (Jin, 2020). Unlike the traditional bathtub model, the GBM assigns each vehicle an individual trip length while regulating its speed according to the mean speed of the reservoir derived from the speed–density relationship, similar to the trip-based modeling framework. However, the GBM extends beyond the trip-based model by allowing trip lengths to follow a variety of distributions and by explicitly accounting for the remaining trip distance over time. In contrast, trip-based models typically track vehicle trajectories using a fixed trip length, L_i, and the mean speed determined from the Macroscopic Fundamental Diagram (MFD). By relaxing assumptions on trip length, the GBM captures flow evolution more accurately under conditions of sudden changes in inflow. As a result, the GBM provides a dynamic representation of not only individual trip lengths but also the aggregate number of vehicles and their corresponding remaining travel distances across the network.

5.4.1 *Variables*

We use five classes of variables (Jin, 2020): *supply (network) variables*, *inflow variables*, *outflow variables*, *traffic state variables*, and *active trip variables*.

Supply Variables (Network Variables). We model the road network as a single reservoir. Let L_{net} be the total road length, T the total time, and $\mathbf{T} = [0, T]$ the time horizon. The speed–density relation follows the MFD:

$$v(t) = V(\rho(t)), \tag{5.3}$$

where $v(t)$ is the average network speed, $V(\cdot)$ is the MFD mapping, and $\rho(t)$ is the average density per unit road length. The flow–density relation is

$$q(t) = \rho(t)\, v(t) = \rho(t)\, V(\rho(t)), \tag{5.4}$$

where $q(t)$ is the average network flow.

Inflow Variables (Demand Variables). Let $f(t)$ denote the entering trip rate and

$$F(t) = \int_0^t f(s)\, ds \tag{5.5}$$

the cumulative inflow. Let $\mathbf{X} = [0, X_{\max}]$ be the distance domain. The probability density of entering trip distance $x \in \mathbf{X}$ at time t is $\tilde{\varphi}(t, x)$ and the survival distribution is

$$\tilde{\Phi}(t, x) = \int_x^\infty \tilde{\varphi}(t, s)\, ds, \tag{5.6}$$

with $\int_0^\infty \tilde{\varphi}(t, x)\, dx = 1$, $\tilde{\varphi}(t, x) \geq 0$, $\tilde{\Phi}(t, 0) = 1$, and $\tilde{\Phi}(t, \infty) = 0$. The mean entering trip distance is

$$\tilde{B}(t) = \int_0^\infty x\, \tilde{\varphi}(t, x)\, dx. \tag{5.7}$$

Traffic State Variables. Inside the reservoir, let $\varphi(t, x)$ be the probability density of remaining trip distance x and

$$\Phi(t, x) = \int_x^\infty \varphi(t, s)\, ds, \tag{5.8}$$

the survival distribution. Then $\int_0^\infty \varphi(t, x)\, dx = 1$, $\varphi(t, x) = -\partial\Phi(t, x)/\partial x$, $\Phi(t, 0) = 1$, and $\Phi(t, \infty) = 0$. The mean remaining distance is

$$B(t) = \int_0^\infty x\, \varphi(t, x)\, dx. \tag{5.9}$$

Active Trip Variables. Let $\lambda(t)$ be the number of active trips and $\rho(t) = \lambda(t)/L_{\text{net}}$ the average density. The density of active trips with remaining distance x is

$$k(t,x) = \lambda(t)\,\varphi(t,x), \qquad \int_0^\infty k(t,x)\,dx = \lambda(t). \tag{5.10}$$

The number of active trips with remaining distance at least x is

$$K(t,x) = \lambda(t)\,\Phi(t,x), \tag{5.11}$$

which summarizes the distribution of remaining distances and is a key GBM state. The boundary condition is $\lambda(t) = K(t,0)$.

Outflow Variables. Let $g(t)$ be the outflow rate of trips that exit during $[t, t+\Delta t]$ and $G(t)$ its cumulative counterpart.[1] When a remaining distance reaches zero, the trip exits the network (Jin, 2020).[2]

$$g(t)\,\Delta t = \int_0^{v(t)\Delta t} k(t,x)\,dx \approx k(t,0)\,v(t)\,\Delta t, \tag{5.12}$$

so that

$$g(t) = k(t,0)\,v(t) = \lambda(t)\,\varphi(t,0)\,v(t), \qquad G(t) = \int_0^t g(s)\,ds. \tag{5.13}$$

5.4.2 *Governing Equations*

Conservation of Total Trip Miles.

$$\lambda(0)B(0) + \int_0^t f(s)\,\tilde{B}(s)\,ds - \int_0^t \lambda(s)\,v(s)\,ds = \lambda(t)B(t). \tag{5.14}$$

Equation (5.14) is a stock and flow balance written in units of vehicle miles. The left hand side comprises the initial stock of remaining trip miles, plus the cumulative miles injected by new arrivals, minus the cumulative miles traveled by the active fleet. The right hand side is

[1] $q(t)$ is the within-network flow, while $g(t)$ is the exit flow from the reservoir at time t.

[2] Negative remaining distances in computation arise from time discretization and indicate that a vehicle has effectively completed its trip.

the remaining miles at time t. Define the total remaining miles as $M(t) \equiv \lambda(t)B(t)$. Differentiating (5.14) gives the pointwise law

$$\frac{d}{dt}[\lambda(t)B(t)] = f(t)\,\tilde{B}(t) \;-\; \lambda(t)\,v(t),$$

which states that the rate of change of remaining miles equals the inflow of intended miles minus the miles drained by motion at the average network speed $v(t)$. Units are consistent: $f\,\tilde{B}$ has units of veh·mi/h and λv has the same units.

This conservation law follows directly from the survivor representation of remaining distances. Note that $\int_0^\infty \Phi(t,x)\,dx = B(t)$, hence $\int_0^\infty K(t,x)\,dx = \lambda(t)B(t)$. Integrating the continuous transport equation in (5.16) over $x \in [0,\infty)$ and using the boundary conditions $K(t,\infty) = 0$ and $K(t,0) = \lambda(t)$ yields

$$\frac{d}{dt}\left[\int_0^\infty K(t,x)\,dx\right] = -\,v(t)\,K(t,0) \;+\; f(t)\int_0^\infty \tilde{\Phi}(t,x)\,dx,$$

that is, $\frac{d}{dt}[\lambda B] = -v\,\lambda + f\,\tilde{B}$, which integrates in time to (5.14) (Jin, 2020).

Conservation of the Number of Active Trips.

$$G(t) = \lambda(0) + F(t) - \lambda(t). \tag{5.15}$$

Equation (5.15) is the stock balance for the population of active trips. It states that cumulative departures $G(t)$ equal the initial stock plus cumulative arrivals minus the current stock. Differentiating gives the instantaneous form

$$\frac{d\lambda(t)}{dt} = f(t) \;-\; g(t),$$

where $g(t) = dG(t)/dt$ is the exit flow. This links the head count of vehicles in the reservoir to the input and output streams without reference to distance. The exit flow consistent with the GBM is $g(t) = k(t,0)\,v(t) = \lambda(t)\,\varphi(t,0)\,v(t)$, which counts the mass that crosses the boundary $x = 0$ per unit time.

Consistency with the transport equation is immediate. Evaluating (5.16) at $x = 0$ and recalling $K(t,0) = \lambda(t)$ and

$\partial K/\partial x|_{x=0} = -k(t,0)$ yields

$$\frac{d\lambda(t)}{dt} = -\,v(t)\,\frac{\partial K}{\partial x}(t,0) \;+\; f(t)\,\tilde{\Phi}(t,0) = f(t) \;-\; k(t,0)\,v(t),$$

since $\tilde{\Phi}(t,0) = 1$. Integrating in time recovers (5.15). Thus, the head count balance is the zero moment of the K transport.

Conservation by Remaining Distance. The evolution of $K(t,x)$ admits a continuous and a discrete form:

$$\begin{aligned}\frac{\partial}{\partial t}K(t,x) &= v(t)\,\frac{\partial}{\partial x}K(t,x) + f(t)\,\tilde{\Phi}(t,x)\\ &= V\left(\frac{K(t,0)}{L_{\text{net}}}\right)\,\frac{\partial}{\partial x}K(t,x) + f(t)\,\tilde{\Phi}(t,x), \qquad (5.16)\\ K(t+\Delta t,x) &= K(t,\, x + v(t)\Delta t) + f(t)\,\tilde{\Phi}(t,x)\,\Delta t, \qquad (5.17)\end{aligned}$$

Equations (5.16)–(5.17) are transport balances in the remaining distance domain. The advection term $v(t)\,\partial K/\partial x$ shifts the survivor K toward smaller x as vehicles consume distance at the common speed $v(t)$, while the source term $f(t)\,\tilde{\Phi}(t,x)$ injects mass at all thresholds x for which entering trips have total length at least x. Boundary conditions are $K(t,\infty) = 0$ and $K(t,0) = \lambda(t)$; the latter connects the survivor to the head count. Initial data $K(0,x) = \lambda(0)\,\Phi(0,x)$ completes the problem.

The discrete form (5.17) is a first order upwind scheme that is conservative by construction: the shift $x \mapsto x + v\Delta t$ transports existing mass, and the additive term accounts for new entries. A standard Courant condition $v(t)\,\Delta t/\Delta x \le 1$ ensures stability and nonnegative updates. Taking moments of (5.16) reproduces the two aggregate conservations above: integrating over x gives (5.14) in differential form through the identity $\int_0^\infty K\,dx = \lambda B$, and evaluating at $x = 0$ recovers (5.15). Hence, the transport equation is the fundamental statement from which the total miles and the head count balances follow (Jin, 2020).

5.5 Experimental Setup

We design controlled experiments to quantify how RGAs reshape the GBM under three driver responses: perfectly rational (PR), logit-based stochastic choice (LOGIT), and bounded rationality (BR).

5.5.1 *Demand Generation and RGA Mechanism*

We generate $N_{\text{trav}} = 2400$ trips. Entry times are sampled on a 1-min grid from a trapezoidal baseline inflow $f(t)$ so that $\mathbb{E}[\text{arrivals}] = f(t)\,\Delta t$. Each traveler draws an intended distance

$$x \sim \mathrm{U}(0,\, 2\tilde{B}(t)), \quad \tilde{B}(t) = 2 + \max\{0, \min\{7.5t,\, 3,\, 7.5(1-t)\}\}\ \text{mi},$$

which induces a time varying mean trip length (Fig. 5.4(c)).

At decision time the navigation service presents two candidates: the shortest route before attack (*route 1*) and the second shortest route before attack (*route 2*). To isolate the role of attack strength in a transparent way, we use a stylized two-route setting parameterized by an attack intensity $\alpha \in [0, 1]$. When $\alpha = 0$ (no attack), the two candidates are identical so that all behavioral models share the same baseline. As α increases, the attacker inflates the *perceived* travel time of route 1 while route 2 becomes progressively more sub-optimal in reality, representing more severe misguidance. Specifically, the navigation service advertises perceived travel times

$$d_1^p = x(1+\alpha), \quad d_2^p = x(1+\alpha/2),$$

while the true travel times remain

$$d_1^r = x, \quad d_2^r = x(1+\alpha/2).$$

Behavioral responses are:

(1) **PR:** choose the smaller perceived cost, which yields d_2^r.
(2) **LOGIT:** choose d_2^r with probability

$$P_{\text{alt}} = \frac{\exp(-\beta d_2^p)}{\exp(-\beta d_1^p) + \exp(-\beta d_2^p)}, \quad \beta = 1.0.$$

(3) **BR:** remain on d_1^r unless the reported inflation on route 1 exceeds the indifference threshold τ (here $\tau = 0.20$), after which the same logit rule applies.

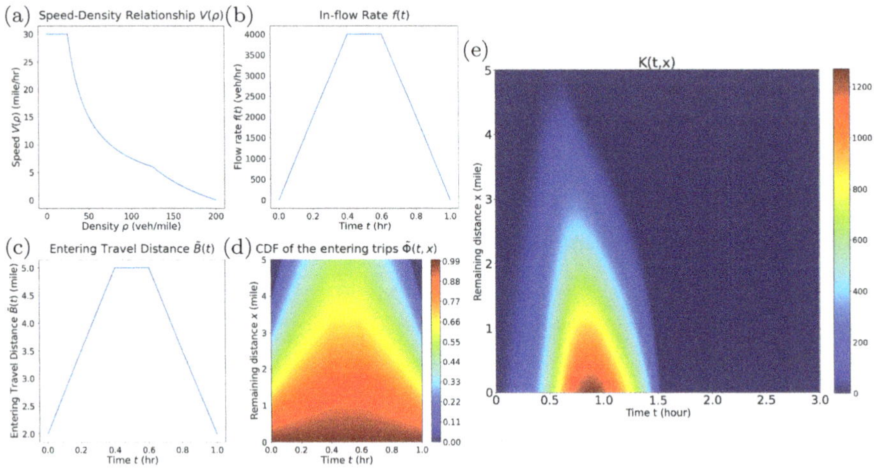

Fig. 5.4. Experimental setting: (a) speed–density relation $V(\rho)$; (b) inflow $f(t)$; (c) average entering trip distance $\tilde{B}(t)$; (d) time-dependent entry distribution $\tilde{\Phi}(t,x)$; (e) survivor of remaining distance $K(t,x)$.

Source: Ka and Ukkusuri (2026).

The entry process $f(t)$ is held fixed across scenarios. For each $(\alpha, \text{behavior})$ pair, the behavioral rule maps the intended distance x_i to a realized distance $x_i^{\text{act}} \in \{d_1^r, d_2^r\}$. The realized distances determine the entering trip-length distribution $\tilde{\varphi}_\alpha(t,x)$ and survivor $\tilde{\Phi}_\alpha(t,x)$ used by the GBM. This is the interface that couples the human decision layer with network dynamics; identical attack intensities can therefore induce different congestion and recovery paths across PR, LOGIT, and BR.

5.5.2 *GBM Discretization and Parameters*

For notational convenience, we define a normalized time $\hat{t} = t/T \in [0,1]$ when specifying $f(\cdot)$ and $\tilde{B}(\cdot)$. The simulation horizon is $T = 3$ h with $\Delta t = 1$ min. Following Jin (2020), the horizon $[0, T]$ with $T = 3$ h is discretized into $J = 180$ steps of $\Delta t = 1$ min. The distance range $[0, X_{\max}]$ with $X_{\max} = 10$ mi is partitioned into $I = 320$ cells of $\Delta x = 2^{-6}$ mi ($\approx$82.5 ft). The corridor length is $L_{\text{net}} = 10$ lane miles. The piecewise MFD used for Fig. 5.4(a) is

$$V(\rho) = \min\left\{30,\ \frac{750}{\rho},\ 10\left(\frac{200}{\rho} - 1\right)\right\}\ \text{mi} \cdot \text{h}^{-1},$$

and the boundary inflow in Fig. 5.4(b) is

$$f(t) = \max\{0,\ \min\{10{,}000\,t,\ 4000,\ 10{,}000(1-t)\}\}.$$

At each time index j the scheme updates active trips λ^j, speed v^j, and the characteristic shift in distance. The time-dependent entry distribution in Fig. 5.4(d) is

$$\tilde{\Phi}(t,x) = \max\left\{0,\ 1 - \frac{x}{2\tilde{B}(t)}\right\}.$$

We test eleven attack intensities $\alpha \in \{0, 0.1, \ldots, 1.0\}$ to sweep from the no attack baseline to a fully attacked system. The three hour horizon captures the lag between peak demand and peak congestion. In particular, the concentration of $K(t,x)$ peaks between 0.75 and 1 hour (Fig. 5.4(e)).

5.5.3 *Evaluation Metrics*

We assess efficiency and resilience with compact, model-based metrics evaluated over $[0, T]$ with $\Delta t = 1$ min. Network efficiency reflects systemwide operating cost; static resilience compares cumulative outcomes at the end of the disturbance; dynamic resilience tracks their time evolution. All metrics are summarized in Table 5.2.

5.6 Results and Discussion

5.6.1 *Impact on Travel Times and Congestion*

We first establish the baseline ($\alpha = 0$). All three behaviors produce the same totals: TSTT $= 944.7$ veh $\cdot$ h and $\bar{\tau} \approx 0.394$ h (about 23.6 min), confirming identical efficiency in the absence of manipulation.

Under small attacks, PR and LOGIT already deteriorate while BR remains unchanged at the outset. At $\alpha = 0.1$, PR yields TSTT $=$ 1067.3 and $\bar{\tau} = 0.445$; LOGIT is milder with TSTT $= 1014.5$ and $\bar{\tau} = 0.423$; BR matches the baseline. This pattern persists through $\alpha = 0.2$: bounded rationality absorbs small distortions, while full compliance magnifies them and probabilistic compliance partially mitigates them.

Table 5.2. Network efficiency and resilience metrics (evaluated over $[0, T]$). The attack intensity is α.

Category	Metric [Unit]	Formula	Meaning
Network efficiency	Total System Travel Time [veh h]	$\mathrm{TSTT}(\alpha) = \int_0^T \lambda_\alpha(t)\, dt$	Cumulative vehicle presence
Network efficiency	Average Travel Time [h]	$\bar{\tau}(\alpha) = \dfrac{\mathrm{TSTT}(\alpha)}{\int_0^T f_\alpha(t)\, dt}$	Mean time per completed trip
Static resilience	Total Performance Loss [veh]	$L(\alpha) = \int_0^T \max\big(g_0(t) - g_\alpha(t), 0\big)\, dt$	Throughput shortfall vs. baseline
Static resilience	Resilience Index [–]	$R(\alpha) = 1 - \dfrac{L(\alpha)}{\int_0^T g_0(t)\, dt}$	Fraction of baseline output retained
Dynamic resilience	Cumulative Loss to t [veh]	$C_\alpha(t) = \int_0^t \max\big(g_0(\tau) - g_\alpha(\tau), 0\big)\, d\tau$	Loss accumulated up to t
Dynamic resilience	Time-varying Resilience [–]	$R_\alpha(t) = 1 - \dfrac{C_\alpha(t)}{\int_0^t g_0(\tau)\, d\tau}$	Baseline output retained by t

Source: Reprinted from Ka and Ukkusuri (2026).

As α increases, congestion rises in every regime but at different rates. Around $\alpha \approx 0.3$, BR begins to degrade (TSTT $= 1200.7$, $\bar{\tau} = 0.501$), yet PR is already worse (TSTT $= 1303.9$, $\bar{\tau} = 0.544$). For intermediate intensities, LOGIT and BR track closely and consistently outperform PR. At $\alpha = 0.5$, PR averages about 0.664 h per trip versus ≈ 0.59 h for LOGIT and BR. At $\alpha = 1.0$, gridlock-like conditions push PR to TSTT $= 2725.8$ (about 2.9 times baseline), while LOGIT and BR are lower (about 2.5 times baseline). In short, higher attack intensities harm all driver behavior scenarios, but the perfect-compliance case produces the largest delays; even modest heterogeneity helps moderate the cascade of congestion at low to moderate intensities.

5.6.2 *Impact on Network Throughput and Resilience*

Resilience metrics confirm these patterns. At $\alpha = 0.1$ the Resilience Index falls to 0.826 for PR, 0.873 for LOGIT, and remains 1.000 for BR. BR preserves full performance through $\alpha = 0.2$, revealing a behavioral threshold below which small distortions do not propagate. Beyond $\alpha = 0.3$, all regimes lose resilience but PR declines fastest. For example, at $\alpha = 0.5$ the retained performance is about 0.64 for BR versus 0.58 for PR. At $\alpha = 1.0$ all fall below one half, again with PR lowest.

Throughput loss tells a complementary story. For small attacks, BR loses virtually nothing and LOGIT loses little; PR already shows a nontrivial shortfall. Losses grow with intensity across all regimes, peaking near two to three dozen equivalent vehicles at $\alpha = 1.0$. Thus, the principal effect of strong attacks is prolonged travel rather than a complete failure to serve demand. Dynamic resilience curves $C_\alpha(t)$ and $R_\alpha(t)$ emphasize the same ordering: PR accumulates loss fastest and to the greatest extent, LOGIT rises more gradually, and BR is immune up to the tolerance threshold and then behaves like LOGIT thereafter (Fig. 5.5).

5.6.3 *Discussion*

Our experiments show that driver response is a first-order determinant of resilience under deceptive guidance. When all travelers behave as perfectly rational agents, even small distortions in perceived travel time trigger herding onto purportedly faster paths, concentrate flow, and create large delays. Heterogeneous responses mitigate this amplification. With logit-based stochastic choice, compliance is partial, and dispersion across alternatives absorbs a portion of the shock. With bounded rationality, a tolerance band serves as a behavioral filter, blocking small perturbations from propagating through the network. These findings connect to informational paradoxes in routing, where individually rational reactions to information can degrade collective performance (Acemoglu *et al.*, 2018; Ukkusuri and Patil, 2007). They also align with prior attack studies that assume full compliance and find large network losses (Waniek *et al.*, 2021), while our results quantify how losses shrink once realistic noncompliance is acknowledged (Di *et al.*, 2017; Ramirez *et al.*, 2021; van Essen *et al.*, 2020).

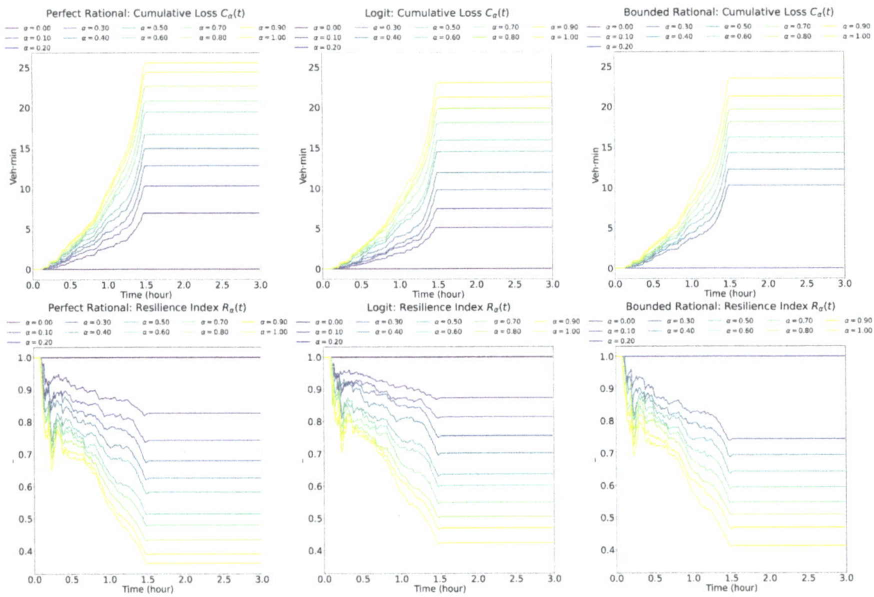

Fig. 5.5. Time-varying cumulative loss $C_\alpha(t)$ and resilience index $R_\alpha(t)$ for PR, LOGIT, and BR under $\alpha \in [0,1]$. The BR tolerance is $\tau = 0.20$.

A clear threshold emerges for bounded rational users. Performance remains close to the no-attack baseline until the perceived advantage exceeds the indifference band, after which outcomes converge toward the probabilistic case. This suggests an actionable principle for resilient information design: discourage overreaction to marginal differences and cultivate mild route diversity so that no single path becomes a coordination focal point under erroneous signals. In practice, resilience depends not only on secured infrastructure and robust algorithms but also on the distribution of user trust, attention, and tolerance. Systems that surface uncertainty and promote measured responses can prevent small misinformation from cascading into large-scale congestion.

5.7 Defense and Mitigation Strategies

We organize defenses by layer and emphasize mechanisms that reduce both the probability of successful manipulation and the impact when it occurs.

5.7.1 *Physical Layer Defenses*

Satellite Signal Authentication. One strategy to preempt spoofing attacks is the authentication of navigation signals at the source. Researchers have long proposed cryptographic methods to authenticate civil GNSS broadcasts, allowing receivers to verify the origin and integrity of the signals (Lo and Enge, 2010; Scott, 2003). For example, Lo and Enge (2010) outlined a scheme to digitally sign augmentation messages in aviation navigation, foreshadowing later developments in authenticated GNSS. In recent years, these concepts have materialized: the European Galileo system now transmits Open Service Navigation Message Authentication (OSNMA), enabling receivers to cryptographically validate satellite data in real time. Field tests of OSNMA demonstrate that cryptographic authentication can significantly enhance resilience against spoofed or replayed signals (Rusu-Casandra and Lohan, 2025). In a 2025 study, Rusu-Casandra and Lohan (2025) demonstrated that an OSNMA-enabled receiver maintained robust positioning even under interference, underscoring the preventive value of authenticated GNSS signals. By making it computationally infeasible for an adversary to forge valid navigation messages, signal authentication serves as a proactive defense that prevents many spoofing attacks from succeeding (O'Hanlon *et al.*, 2013; Rusu-Casandra and Lohan, 2025). Early experimental work and patents proposed similar ideas (e.g., encrypted spreading codes or TESLA-based delayed key disclosure) to authenticate GPS signals (O'Hanlon *et al.*, 2013; Scott, 2003). However, such measures began to see practical deployment only in the last decade. This progress marks a significant preventive improvement at the physical layer: modern vehicular navigation receivers can be equipped to reject unauthenticated or inconsistent satellite signals before they ever mislead the vehicle.

Spoofing Detection and Resilient Positioning. Complementary to source authentication, extensive research has focused on detecting spoofing or jamming events and maintaining navigation when they occur. Multi-antenna receiver techniques provide one line of defense: by comparing signals at multiple antennas, a receiver can detect anomalous wavefronts or direction-of-arrival inconsistencies indicative of a spoofing device. Montgomery *et al.* (2009)

demonstrated an early multi-antenna spoofing detector that autonomously flagged and rejected counterfeit GPS signals by recognizing that genuine satellite signals arrive from widely spaced directions, unlike a spoofing source. Such spatial processing can be both preventive (when it enables automatic blocking of spoofer signals) and mitigative (by alerting the system to revert to backup navigation). Similarly, monitoring physical signal metrics provides detection cues: sudden shifts in carrier-to-noise ratio, signal power, or Doppler that deviate from normal satellite behavior can raise alarms (Broumandan *et al.*, 2016; Psiaki and Humphreys, 2016a).

Modern approaches increasingly fuse auxiliary sensor data to improve spoofing detection sensitivity. For instance, inertial measurement unit (IMU) readings and odometry can be continuously cross-checked against GNSS outputs; any growing divergence suggests a possible spoof. Gu *et al.* (2021) presents a coupled visual-inertial-GNSS scheme that triggers an alarm when the GNSS position drifts from the IMU/visual odometry solution beyond a statistically determined threshold (Gu *et al.*, 2021). Once a spoofing attack is detected, their system locks out the corrupted GNSS data and relies on the inertial/visual subsystem to carry on positioning, thereby mitigating the attack's impact (Gu *et al.*, 2021). In general, multi-sensor fusion provides a robust mitigative strategy: an attack that succeeds in deceiving one sensor is likely to leave telltale inconsistencies with readings from independent sensors (IMU, camera, map). Researchers have developed attack-resilient state estimators that combine redundant sensor inputs and can tolerate or isolate a certain degree of corrupted data (Gu *et al.*, 2021). Such techniques do not necessarily prevent spoofing outright, but they mitigate its effect by enabling the vehicle to detect the deception and remain navigationally inertial or use alternative references until authentic signals are restored (Montgomery *et al.*, 2009). Notably, the last decade has seen considerable progress in this area: from early single-antenna detectors that detected only extreme anomalies, to sophisticated machine-learning-based spoofing detectors that leverage side-channel data (e.g., vision or even satellite imagery, Gu *et al.*, 2021; O'Hanlon *et al.*, 2013) to reliably recognize subtle spoofing attempts. Modern vehicles thus can be equipped with an immune system at the physical layer that flags suspect signals and sustains navigation via trusted sensors, exemplifying a strong mitigative strategy against physical-layer attacks.

5.7.2 *Data Layer Security Strategies*

Anomaly Detection and Data Validation. Even with authentication, there remains the risk of an insider adversary or a compromised but valid unit broadcasting misleading information (for example, a hacked car sending out false traffic congestion alerts with a valid certificate). To handle such cases, vehicles and infrastructure employ misbehavior detection and data validation algorithms as a mitigative strategy. These systems continuously monitor incoming navigation-related data for inconsistencies, implausibilities, or statistically anomalous patterns that suggest malicious manipulation. For instance, a vehicle receiving crowd-sourced traffic reports will check if the data makes sense: Are multiple trusted sources confirming the same road blockage, or is there just a single reporter? Does the reported congestion align with the vehicle's own sensors (e.g., its speed) or other context (time of day, known road capacity)? Lin *et al.* (2018) studied data integrity attacks on dynamic route guidance and implemented a forged data filtering mechanism. Essentially, the navigation system cross-checks each piece of traffic state data against its neighbors and expected traffic models, filtering out reports that are aberrant or that would lead to physically inconsistent network states. Their evaluations showed that such filtering can mitigate the impact of an attacker injecting false traffic readings, preserving efficient routing for honest drivers (Lin *et al.*, 2018).

Another example is Sybil attack detection in crowd-sourced mapping services. An attacker might create numerous phantom vehicles (Sybil nodes) that all report a fake traffic jam to trick navigation systems. To counter this, researchers have devised techniques where vehicles use local wireless communication to verify the presence of other reporters (plausibility whispers or identity verification exchanges) (So *et al.*, 2019). If a cluster of traffic messages appears to originate from distinct sources but no actual vehicles are detected at those locations, the system can flag a likely Sybil attack and disregard those messages. So *et al.* (2019) demonstrated a physical-layer plausibility check where the vehicle's radio measures the signal characteristics of alleged nearby transmitters; if one transmitter claims to be multiple vehicles (Sybil devices), the radio-fingerprint similarity can expose the fraud. While it doesn't prevent the false data from being sent, it detects it and stops it from affecting route guidance.

In the past few years, machine learning methods have been applied to learn what normal navigation data looks like, so that deviations can be caught in real time. Indeed, security surveys emphasize that misbehavior detection is key in vehicular networks to identify attacks that pure cryptography cannot catch (e.g., an authenticated yet dangerous command) (Bouchouia *et al.*, 2023; Campos *et al.*, 2024). Over the last decade, the automotive research community has developed a toolbox of validation techniques, ranging from simple rule-based checks (e.g., speed reports exceeding physical limits) to complex reputation systems, where each data source is assigned a trust score that decreases if it sends information that later proves to be false. When an irregularity is detected, navigation systems may discard the data or switch to a fail-safe mode (for example, if the traffic info is deemed unreliable, the system might default to a routing that avoids major changes based on that info, or at least warn the user). The mitigative impact is clear: even if an attacker injects malicious data, the system strives to limit the effect on route calculation by catching the lie in the act (Lin *et al.*, 2018; So *et al.*, 2019).

Redundancy and Cross-Source Verification. A final data-layer strategy is the use of redundant data sources and cross-verification across them, which can serve both preventive and mitigative purposes. Modern vehicular navigation does not rely on a single stream of information; instead, it can aggregate data from GNSS, inertial sensors, vehicle-to-vehicle (V2V) alerts, digital maps, and cellular connectivity to cloud services. By comparing these sources against each other, the system can often detect discrepancies that signal an attack. For example, a navigation unit might receive traffic speed data from an internet service and simultaneously observe the spacing and speed of nearby connected vehicles via V2V beacons. If the cloud service says heavy traffic (20 km/h) on a segment but all nearby cars are broadcasting much higher speeds, the system infers the cloud data may be erroneous or malicious. It could then ignore that outlier or seek confirmation from yet another source (e.g., roadside sensors or a different provider) before altering the route. This kind of cross-source verification was highlighted by early work from the crowd-sourced navigation domain: Wang *et al.* (2016) (as referenced by later studies) proposed that mapping services incorporate user reputation and multiple input modalities (GPS probes vs. user reports) to validate incident reports.

In practice today, major navigation platforms do use multiple confirmation mechanisms (for instance, Google Waze flags a reported road closure as unconfirmed until enough distinct users or an authoritative data feed confirm it). From an academic perspective, redundancy is often discussed in terms of Byzantine fault tolerance or data fusion trust models. So *et al.* (2019) and others have noted that requiring agreement between independent channels can drastically raise the effort for an attacker: instead of compromising one system, they would need to compromise several different ones simultaneously. Thus, redundancy acts as a preventive measure by raising the bar for successful attacks (multiple systems need to be fooled simultaneously), and as a mitigative measure by allowing fallback. If one data source becomes untrustworthy, the system can downweight it and still function using the trustworthy sources that remain. A clear example appeared in navigation map update security: vehicles increasingly receive map corrections (e.g., new speed limits, road geometry changes) over-the-air.

Another instance is in route guidance algorithms that incorporate real-time data: Lin *et al.* (2018) and others recommend that the algorithm be robust to isolated bad data. For example, compute the nominal optimal route from the available data, and then also compute the second-best route and a set of routes that each omit one data point. Comparing these solutions shows whether any single observation exerts undue influence on the outcome. If removing a particular incident dramatically changes the chosen route, the system might treat that incident as a suspect. This approach, akin to influence diagnostics, provides a way to mitigate the impact of any single false datum. In summary, exploiting redundancy and performing cross-checks is a strategy that both preempts certain attacks and reduces harm from those that occur. It embodies the classic principle of never trust any single point of information, which is a principle that the automotive domain has increasingly embraced in the last decade to secure the data layer of navigation.

5.7.3 *Route Guidance and Human-Factors Mitigations*

Context-Aware and Safe Route Guidance. Beyond securing signals and data, there is a growing recognition that the route guidance algorithms themselves should be designed to anticipate and avoid potential dangers—whether those dangers arise accidentally or due to malicious manipulation. One preventive strategy is to incorporate richer context (such as safety, legality, and driver preferences) into the route selection logic, rather than blindly optimizing for the shortest travel time. Recent research has shown that conventional navigation apps, by focusing on efficiency, can sometimes lead drivers onto more hazardous roads (e.g., those with poor lighting or higher crash rates). Sohrabi and Lord (2022) found that in Texas, taking the fastest route (just 8% shorter in time) could increase crash risk by over 23%, because the shortest path often traversed local roads with safety issues. In response, they propose a new routing paradigm that explicitly factors in road safety metrics, such as historical accident data, road geometry, and weather conditions, to identify the safest route rather than just the fastest (Sohrabi and Lord, 2022). This work highlights a preventive approach: if a navigation system can proactively steer users away from objectively risky routes, it also thwarts attackers who might try to exploit those vulnerabilities (for example, an attacker who cuts power to streetlights on one road cannot lure drivers there if the nav system already tends to avoid dark, unsafe roads).

In the last decade, industry and academia have started acknowledging such human-centric route guidance. Modern systems may include options like avoiding unsafe areas or using driver profiles to prevent routes through, say, neighborhoods with high crime late at night (a concern not directly cyber-related but important for holistic safety). From a human-factors perspective, navigation guidance is being improved to present more context to drivers about the route. Instead of just saying take Exit 23, systems now might warn of heavy rain reported along this route, drive with caution, or that this route saves 5 min but involves winding mountain roads. Providing this additional information serves a security purpose: it helps drivers make informed decisions rather than blindly following turn-by-turn instructions. Brown and Laurier (2012) observed that drivers are not mere automatons. They engage in active interpretation of GPS instructions and often face troubles when the system's guidance is unclear or lacks critical context (like legal restrictions or road surface conditions).

One implication is that navigation systems should be designed to better fit the real-world complexities of driving, thereby preventing confusion or dangerous situations that could be exacerbated by malicious interference. For example, a malicious actor might try to divert a truck down a small rural road by falsely indicating a highway closure. A context-aware system, however, might prevent this by noting the suggested detour goes via a narrow road unsuited for trucks and either avoiding it or alerting the driver. In essence, building in knowledge of road rules (like height clearances, turn restrictions), environmental factors, and vehicle capabilities can prevent many ill-advised routing decisions, including those that an attacker might try to induce. Indeed, Lin *et al.* (2017)'s analysis of Death by GPS incidents (cases where drivers followed nav directions into peril) concluded that many catastrophes occurred due to missing data (e.g., the system not knowing a road was impassable or dangerous) and poor communication of route hazards. Their recommendations urge navigation services to advance space usage rule mapping (better data on where vehicles are allowed and safe) and to incorporate real-time environmental information (weather, road closures) into routing decisions (Lin *et al.*, 2017). Following these recommendations inherently acts as a safeguard against malicious data as well: if the system already cross-checks routes against authoritative road condition sources and avoids ones that are marked closed or hazardous, an attacker's false suggestion to take a closed road would be ignored.

In summary, enhancing route guidance to be context-aware, safety-optimized, and transparent is a preventive strategy that reduces the system's susceptibility to both benign errors and targeted exploits. By guiding drivers along routes that have safety margins and by informing them of relevant conditions, modern nav systems create fewer opportunities for an attacker to trick the driver into a truly dangerous situation (Brown and Laurier, 2012; Lin *et al.*, 2017; Sohrabi and Lord, 2022). This marks a notable progress in the past decade: the navigation function is evolving from a pure time/distance optimizer to a more nuanced co-pilot that prioritizes overall safety and reliability, thereby preemptively countering certain classes of route manipulation attacks.

Driver Alerts, Transparency, and Control. No matter how well-designed an automated navigation system is, there remains the possibility that it will at times be fed incorrect or malicious inputs, leading to questionable guidance. At this point, the last line of defense is the human driver. Therefore, a crucial set of mitigation strategies revolves around keeping the driver in the loop, maintaining appropriate trust, and designing the human-machine interface such that the driver can detect and correct anomalies in navigation. One strategy is to make the system's decision-making more transparent to the user. Instead of silently rerouting the driver with no explanation, a transparent system might display, Rerouting to avoid a reported accident ahead, or show a comparison of the current route with an alternate. This gives the attentive driver a chance to judge whether the change is sensible. If an attacker had injected a fake accident report to divert traffic (a data-layer attack), a savvy driver might notice, for instance, that local radio or their own observation doesn't corroborate any accident, and could question the reroute. Many navigation interfaces now allow users to see the reason for delays or route changes (such as construction or an accident), implementing this transparency as a standard feature. Aylward *et al.* (2020) provides evidence from the maritime domain that greater information sharing can paradoxically increase trust to a point of over-reliance: professional navigators who were given a new route-exchange tool trusted it so much that they were more likely to violate certain right-of-way rules. The lesson for road navigation is that while transparency is important, designers must also caution against blind trust.

Mitigative measures in this vein include periodic reminders or training to users that navigation suggestions are advisory. Some systems now deliberately include slight friction for major deviations. For example, if the proposed route is substantially longer or passes through areas with elevated safety risk, the system should display a confirmation prompt, "Do you want to accept this reroute?", rather than automatically confirming the change. This gives the driver an opportunity (albeit brief) to reflect on whether the new route seems reasonable, acting as a human-mediated checkpoint against erroneous or malicious changes. Another human-factor defense is providing alerts when the system itself is unsure or detects anomalies. If GPS signal quality drops or sensor conflicts arise (potentially due to spoofing), the navigation system might display GPS signal unreliable and proceed with caution rather than silently switching to a possibly degraded mode. This warning can spur the driver to pay more attention to road signs or use their own judgment until the system recovers. It mitigates the risk that the driver follows incorrect guidance from a malfunctioning system.

Driver familiarity and training also play a role: over the past decade, there has been increasing emphasis on educating drivers not to over-trust their devices. Just as driver's education now often covers the limitations of GPS (some safety organizations release advisories like Don't follow GPS into a lake if it looks wrong!), Car manufacturers have begun to integrate user instructions about how their navigation and driver-assist systems should be monitored. The concept of calibrated trust is key: drivers should trust the system when it is likely correct, but remain vigilant for when it might err. Research in automated driving indicates that too high a trust can reduce a driver's vigilance, whereas too little trust means the system's benefits are not realized (Aylward *et al.*, 2020). To calibrate trust, mitigative interface designs have been tried, such as varying the display of confidence. For instance, if traffic data is sparse or conflicting, the system could present the route with a disclaimer (traffic estimates uncertain) shown in a different color. The driver, seeing this, might be less surprised by an unusual route and more prepared to adapt if needed. In extreme cases where a potential attack is detected (say, misbehavior detection flags that a traffic jam report is likely false), the system could even inform the driver: Unverified traffic incident reported. Navigation will double-check or may ignore if not confirmed. Although such honesty might reduce the user's perception of system infallibility, it engages the driver as an ally in security. Ultimately, giving the driver final control is perhaps the strongest mitigator.

Navigation systems have begun to allow easy overrides—one can tap "Avoid this road" or "Reject reroute" if something seems off. Earlier systems often auto-rerouted without asking; now, user-initiated reroute avoidance is common, reflecting the recognition that drivers should be empowered to correct the course. Brown and Laurier (2012) argued for designing GPS interfaces for the active driver who is always interpreting and can disagree with the device. In practice, this means interfaces should make it easy to compare alternate routes, return to a previous route, or simply ignore guidance for a time without the system incessantly recalculating in a confusing manner. Over the last ten years, navigation apps have indeed become more user-friendly in this respect: they often show multiple route options by default, and if the driver deviates intentionally, they gracefully adapt. This user autonomy is a defense against malicious or erroneous instructions. The driver is not forced down any path and can leverage their own instincts or local knowledge.

In conclusion, human-factors strategies act as a mitigation layer that catches what technical measures might miss. By keeping the driver informed, aware, and in control, the system reduces the chances that a spoofed or misleading instruction actually results in harm. The academic and industry consensus emerging is that cybersecurity for navigation must include the human: designing systems that enhance a driver's situational awareness and encourage appropriate skepticism can prevent many worst-case outcomes of navigation system attacks or failures (Aylward *et al.*, 2020; Brown and Laurier, 2012; Lin *et al.*, 2017).

5.8 Conclusion

This chapter introduces a behavior-aware framework to analyze RGAs and their network-level impacts, which integrates three driver response models with the GBM to connect deceptive information, individual route choice, and aggregate congestion. The experiments show that perfect rationality produces the largest losses because small distortions trigger coordinated rerouting. Logit-based choice moderates this effect by dispersing flow, and bounded rationality introduces a threshold that filters minor misinformation. Across metrics of efficiency and resilience, heterogeneity in behavior improves performance relative to the full compliance case, especially at low and moderate attack intensities.

The results suggest practical design principles. Navigation services should validate and cross-check traffic inputs, limit the influence of unverified reports, surface uncertainty to users, and encourage mild route diversity so that no single path becomes a focal point under erroneous signals. Future work can calibrate behavior to field data, extend the model to multiple regions with richer demand feedback, and study adaptive attackers and defenders that learn over time. By combining technical safeguards with behavior-sensitive guidance, systems can reduce both the likelihood and the impact of deceptive routing at scale.

Bibliography

Acemoglu, D., Makhdoumi, A., Malekian, A., and Ozdaglar, A. (2018). Informational Braess' paradox: The effect of information on traffic congestion, *Operations Research* **66**, 4, pp. 893–917, doi:10.1287/opre.2017.1712.

Arora, N., Cabannes, T., Ganapathy, S., Li, Y., McAfee, P., Nunkesser, M., Osorio, C., Tomkins, A., and Tsogsuren, I. (2021). Quantifying the sustainability impact of Google Maps: A case study of Salt Lake City, *arXiv preprint* arXiv:2111.03426.

Aylward, K., Weber, R., Man, Y., Lundh, M., and MacKinnon, S. N. (2020). "Are you planning to follow your route?" The effect of route exchange on decision making, trust, and safety, *Journal of Marine Science and Engineering* **8**, 4, p. 280.

Barrett, B. (2020). An artist used 99 phones to fake a Google Maps traffic jam, https://www.wired.com/story/99-phones-fake-google-maps-traffic-jam/, accessed: August 24, 2025.

Ben-Akiva, M. E. and Lerman, S. R. (1985). *Discrete Choice Analysis: Theory and Application to Travel Demand*, Vol. 9 (MIT Press, Cambridge, MA, United States).

Bouchouia, M. L., Labiod, H., Jelassi, O., Monteuuis, J.-P., Jaballah, W. B., Petit, J., and Zhang, Z. (2023). A survey on misbehavior detection for connected and autonomous vehicles, *Vehicular Communications* **41**, p. 100586.

Broumandan, A., Jafarnia-Jahromi, A., Daneshmand, S., and Lachapelle, G. (2016). Overview of spatial processing approaches for GNSS structural interference detection and mitigation, *Proceedings of the IEEE* **104**, 6, pp. 1246–1257.

Brown, B. and Laurier, E. (2012). The normal natural troubles of driving with GPS, in *Proceedings of the SIGCHI Conference on Human Factors in Computing Systems*, pp. 1621–1630.

Campos, E. M., Hernandez-Ramos, J. L., Vidal, A. G., Baldini, G., and Skarmeta, A. (2024). Misbehavior detection in intelligent transportation systems based on federated learning, *Internet of Things* **25**, p. 101127.

Cao, Y., Luo, Q., and Liu, J. (2019). Road navigation system attacks: A case on GPS navigation map, in *IEEE International Conference on Communications (ICC)*.

Chen, Q. A., Yin, Y., Feng, Y., Mao, Z. M., and Liu, H. X. (2018). Exposing congestion attack on emerging connected vehicle based traffic signal control, in *Network and Distributed System Security (NDSS) Symposium*, San Diego, California, USA.

Daganzo, C. F. and Sheffi, Y. (1977). On stochastic models of traffic assignment, *Transportation Science* **11**, 3, pp. 253–274.

Di, X., Liu, H. X., Zhu, S., and Levinson, D. M. (2017). Indifference bands for boundedly rational route switching, *Transportation* **44**, 5, 1169–1194.

Eryonucu, C. and Papadimitratos, P. (2022). Sybil-based attacks on Google Maps or how to forge the image of city life, in *Proceedings of the 15th ACM Conference on Security and Privacy in Wireless and Mobile Networks*.

Essers, L. (2013). Researcher: Hackers can cause traffic jams by manipulating real-time traffic data, https://tinyurl.com/bddehz6v, accessed: August 24, 2025.

Fontaine, S. L., Muralidhar, N., Clifford, M., Eliassi-Rad, T., and Nita-Rotaru, C. (2022). Alternative route-based attacks in metropolitan traffic systems, in *Proceedings of 52nd IEEE/IFIP International Conference on Dependable Systems and Networks Workshops (DSN-W)*, pp. 20–27, doi: 10.1109/DSN-W54100.2022.00014.

Geere, D. (2012). Tomtom launches assault on open mapping data, https://www.wired.com/story/tomtom-attacks-openstreetmap/, accessed: August 24, 2025.

Gu, N., Xing, F., and You, Z. (2021). GNSS spoofing detection based on coupled visual/inertial/GNSS navigation system, *Sensors* **21**, 20, p. 6769.

Guin, A., Hadi, M., Wang, F., Watkins, K., Hunter, M. P., Iqbal, M. S., Kiriazes, R., Bu, L., Tariq, M. T., Arafat, M., *et al.* (2021). Impact of smartphone applications on trip routing, Technical Report, Southeastern Transportation Research, Innovation, Development and Education Center.

Harutyunyan, M. (2025). Essential Google Maps statistics & trends to watch in 2025, https://www.loopexdigital.com/blog/google-maps-statistics#Section-2, accessed: August 24, 2025.

Intellias (2025). Over-the-air map updates for connected cars, https://intellias.com/air-map-updates-connected-cars/, accessed: August 24, 2025.

Jansen, K., Schäfer, M., Moser, D., Lenders, V., Pöpper, C., and Schmitt, J. (2018). Crowd-GPS-Sec: Leveraging crowdsourcing to detect and localize GPS spoofing attacks, in *2018 IEEE Symposium on Security and Privacy (SP)*.

Jin, W.-L. (2020). Generalized bathtub model of network trip flows, *Transportation Research Part B: Methodological* **136**, pp. 138–157.

Juhász, L., Novack, T., Hochmair, H. H., and Qiao, S. (2020). Cartographic vandalism in the era of location-based games—The case of OpenStreetMap and Pokémon GO, *ISPRS International Journal of Geo-Information* **9**, 4, 197.

Ka, E. and Ukkusuri, S. V. (2026). Resilience of traffic networks to route guidance attacks: The role of driver behaviour heterogeneity. *Transportmetrica B: Transport Dynamics* **14**, 1, Article 2674257.

Kerns, A. J., Shepard, D. P., Bhatti, J. A., and Humphreys, T. E. (2014). Unmanned aircraft capture and control via GPS spoofing, *Journal of Field Robotics* **31**, 4, 617–636.

Lin, A. Y., Kuehl, K., Schöning, J., and Hecht, B. (2017). Understanding "death by GPS" a systematic study of catastrophic incidents associated with personal navigation technologies, in *Proceedings of the 2017 CHI Conference on Human Factors in Computing Systems*, pp. 1154–1166.

Lin, J., Yu, W., Zhang, N., Yang, X., and Ge, L. (2018). Data integrity attacks against dynamic route guidance in transportation-based cyber-physical systems: Modeling, analysis, and defense, *IEEE Transactions on Vehicular Technology* **67**, 9, 8738–8753.

Lo, S. C. and Enge, P. K. (2010). Authenticating aviation augmentation system broadcasts, in *IEEE/ION Position, Location and Navigation Symposium* (IEEE), pp. 708–717.

Mahmassani, H. S. and Chang, G.-L. (1987). On boundedly rational user equilibrium in transportation systems, *Transportation Science* **67**, 9, 8738–8753.

Mahmassani, H. S. and Liu, Y.-H. (1999). Dynamics of commuting decision behaviour under advanced traveller information systems, *Transportation Research Part C: Emerging Technologies*, **7**, 2–3, 91–107.

Matei, M. (2016). Waze addresses "ghost drivers" exploit; Wazers are safe, https://www.androidheadlines.com/2016/04/waze-addresses-ghost-drivers-exploit-wazers-safe.html, accessed: August 11, 2025.

Matyszczyk, C. (2013). Truck driver has GPS jammer, accidentally jams Newark airport, https://www.cnet.com/culture/truck-driver-has-gps-jammer-accidentally-jams-newark-airport/, accessed: August 24, 2025.

Montgomery, P. Y., Humphreys, T. E., and Ledvina, B. M. (2009). Receiver-autonomous spoofing detection: Experimental results of a multi-antenna receiver defense against a portable civil GPS spoofer, in *Proceedings of the 2009 International Technical Meeting of the Institute of Navigation*, pp. 124–130.

NL Times (2025). Zandvoort residents trick Google Maps to block tourists from their streets, https://nltimes.nl/2025/07/20/zandvoort-residents-trick-google-maps-block-tourists-streets.

O'Hanlon, B. W., Psiaki, M. L., Bhatti, J. A., Shepard, D. P., and Humphreys, T. E. (2013). Real-time GPS spoofing detection via correlation of encrypted signals, *Navigation* **60**, 4, pp. 267–278.

Psiaki, M. L. and Humphreys, T. E. (2016a). GNSS spoofing and detection, *Proceedings of the IEEE* **104**, 6, 1258–1270.

Psiaki, M. L. and Humphreys, T. E. (2016b). Protecting GPS from spoofers is critical to the future of navigation, *IEEE Spectrum* **10**.

Ramirez, H. G., Leclercq, L., Chiabaut, N., Becarie, C., and Krug, J. (2021). Travel time and bounded rationality in travellers' route choice behaviour: A computer route choice experiment, *Travel Behaviour and Society* **22**, 59–83.

Raponi, S., Sciancalepore, S., Oligeri, G., and Di Pietro, R. (2021). Road traffic poisoning of navigation apps: Threats and countermeasures, *IEEE Security & Privacy* **20**, 3, 71–79.

Reece, K. (2020). Google Maps 101: How contributed content makes a more helpful map, https://tinyurl.com/mwtamaca, accessed: August 11, 2025.

Ringhand, M. and Vollrath, M. (2018). Make this detour and be unselfish! Influencing urban route choice by explaining traffic management, *Transportation Research Part F: Traffic Psychology and Behaviour* **53**, 99–116.

Rusu-Casandra, A. and Lohan, E. S. (2025). Experimental assessment of Osnma-enabled GNSS positioning in interference-affected rf environments, *Sensors* (Basel, Switzerland) **25**, 3, p. 729.

Sathaye, H., LaMountain, G., Closas, P., and Ranganathan, A. (2022). SemperFi: Anti-spoofing GPS receiver for UAVs, in *Network and Distributed Systems Security (NDSS) Symposium.*

Scott, L. (2003). Anti-spoofing & authenticated signal architectures for civil navigation systems, in *Proceedings of the 16th International Technical Meeting of the Satellite Division of The Institute of Navigation (ION GPS/GNSS 2003)*, pp. 1543–1552.

Sinai, M. B., Partush, N., Yadid, S., and Yahav, E. (2014). Exploiting social navigation, *arXiv preprint* arXiv:1410.0151.

So, S., Petit, J., and Starobinski, D. (2019). Physical layer plausibility checks for misbehavior detection in V2X networks, in *Proceedings of the 12th Conference on Security and Privacy in Wireless and Mobile Networks*, pp. 84–93.

Sohrabi, S. and Lord, D. (2022). Navigating to safety: Necessity, requirements, and barriers to considering safety in route finding, *Transportation Research Part C: Emerging Technologies* **137**, p. 103542.

Tippenhauer, N. O., Pöpper, C., Rasmussen, K. B., and Capkun, S. (2011). On the requirements for successful GPS spoofing attacks, in *Proceedings of the 18th ACM Conference on Computer and Communications Security*, 75–86.

Tires, U. (2024). Study reveals where drivers are most reliant on their GPS, https://www.utires.com/articles/where-drivers-need-gps-the-most/?srsltid=AfmBOook-eUTEFHN-WeAuaeq0UZc7N0Sh0_BcrvTmE6wfLZm2YQ55TX2, accessed: August 24, 2025.

Topham, G. (2024). Thousands of flights to and from Europe affected by suspected Russian jamming, https://tinyurl.com/52hwtee2, accessed: August 24, 2025.

Truong, Q. T., Touya, G., and de Runz, C. (2018). Towards vandalism detection in OpenStreetMap through a data driven approach, in *International Conference on Geographic Information Science (GIScience).*

Tufnell, N. (2014). Students hack Waze, send in army of traffic bots, https://www.wired.com/story/waze-hacked-fake-traffic-jam/, accessed: August 24, 2025.

Ukkusuri, S. V. and Patil, G. R. (2007). Exploring user behavior in online network equilibrium problems, *Transportation Research Record* **2029**, 1, pp. 31–38.

Vaas, L. (2017). Suspected mass-spoofing of ships' GPS in the black sea, https://news.sophos.com/en-us/2017/09/26/suspected-mass-spoofing-of-ships-gps-in-the-black-sea/, accessed: August 24, 2025.

van Essen, M., Thomas, T., van Berkum, E., and Chorus, C. (2020). Travelers' compliance with social routing advice: Evidence from SP and RP experiments, *Transportation* **47**, 3, 1047–1070.

Vickrey, W. (2020). Congestion in midtown Manhattan in relation to marginal cost pricing, *Economics of Transportation* **21**, 100152.

Wang, G., Wang, B., Wang, T., Nika, A., Zheng, H., and Zhao, B. Y. (2016). Defending against Sybil devices in crowdsourced mapping services, in *Proceedings of the 14th Annual International Conference on Mobile Systems, Applications, and Services*.

Wang, G., Wang, B., Wang, T., Nika, A., Zheng, H., and Zhao, B. Y. (2018). Ghost riders: Sybil attacks on crowdsourced mobile mapping services, *IEEE/ACM Transactions on Networking* **26**, 3, 1123–1136

Waniek, M., Raman, G., AlShebli, B., Peng, J. C.-H., and Rahwan, T. (2021). Traffic networks are vulnerable to disinformation attacks, *Scientific Reports* **11**, 1, 5329.

Wardrop, J. G. and Whitehead, J. I. (1952). Correspondence. Some theoretical aspects of road traffic research, *Proceedings of the Institution of Civil Engineers* **1**, 5, pp. 767–768.

Weckert, S. (2020). Google Maps hacks, https://simonweckert.com/googlemapshacks.html, accessed: August 24, 2025.

Yu, Y., Han, K., and Ochieng, W. (2020). Day-to-day dynamic traffic assignment with imperfect information, bounded rationality and information sharing, *Transportation Research Part C: Emerging Technologies* **114**, pp. 59–83.

Zeng, K. C., Liu, S., Shu, Y., Wang, D., Li, H., Dou, Y., Wang, G., and Yang, Y. (2018). All your {GPS} are belong to us: Towards stealthy manipulation of road navigation systems, in *27th USENIX Security Symposium*, 1527–1544.

Zeng, K. C., Shu, Y., Liu, S., Dou, Y., and Yang, Y. (2017). A practical GPS location spoofing attack in road navigation scenario, in *Proceedings of the 18th International Workshop on Mobile Computing Systems and Applications*, 85–90.

https://doi.org/10.1142/9789819831548_0006

Chapter 6

End-to-End Co-Simulation Testbed for Cybersecurity Research and Development in Intelligent Transportation Systems

Minhaj Uddin Ahmad, Akid Abrar, Sagar Dasgupta, and Mizanur Rahman

Department of Civil, Construction and Environmental Engineering University of Alabama, Tuscaloosa, USA

Abstract

Intelligent Transportation Systems (ITS) have been widely deployed across major metropolitan regions worldwide to improve roadway safety, optimize traffic flow, and reduce environmental impacts. These systems integrate advanced sensors, communication networks, and data analytics to enable real-time traffic monitoring, adaptive signal control, and predictive maintenance. However, such integration significantly broadens the cyber attack surface, exposing critical infrastructures to cyber threats that jeopardize safety, data integrity, and operational resilience. Ensuring robust cybersecurity is therefore essential, yet comprehensive vulnerability assessments, threat modeling, and mitigation validations are often cost-prohibitive and time-intensive when applied to large-scale, heterogeneous transportation systems. Simulation platforms offer a cost-effective and repeatable means for cybersecurity evaluation, and the simulation platform should encompass the full range of ITS dimensions—mobility, sensing, networking, and applications. This chapter discusses an integrated co-simulation testbed that links CARLA for 3D environment and sensor modeling, SUMO for microscopic traffic simulation and control, and OMNeT++ for Vehicle to Everything (V2X) communication simulation. The co-simulation testbed enables end-to-end experimentation, vulnerability identification, and mitigation, providing practical insights for developing secure, efficient, and resilient ITS infrastructures. The chapter also presents a case study of a Cellular V2X (C-V2X) proactive safety alert system enhanced with post-quantum cryptography, demonstrating end-to-end cryptographic integration within the testbed, highlighting its role in advancing secure, efficient, and resilient ITS infrastructures.

Keywords: Co-simulation, Simulation testbed, OpenCAMS, Post quantum cryptography, Intersection collision warning, CV2X, Cybersecurity, Cyberattack.

6.1 Introduction

Intelligent Transportation Systems (ITS), now deployed in major metropolitan regions across Europe, Asia, and the United States, play a vital role in enhancing roadway safety, optimizing traffic flow, and reducing environmental impact. The integration of advanced sensors, communication networks, and data analytics enables real-time traffic monitoring, adaptive signal control, and predictive maintenance. However, these capabilities also significantly expand the ITS cyber-attack surface, exposing systems to increasingly sophisticated cyber threats from malicious actors. One of the major concerns in ITS security lies in the security of the roadside infrastructure. Components such as traffic signals, sensors, and variable message signs often operate on outdated or poorly maintained systems, making them attractive targets for attackers due to weak authentication mechanisms and unpatched software. In addition, denial-of-service and jamming attacks pose serious operational challenges (Islam *et al.*, 2024).

These vulnerabilities are not limited to roadside infrastructure. Inside the vehicle, traditional in-vehicle networks were not designed with cybersecurity as a primary consideration. For example, the Controller Area Network (CAN) bus lacks native encryption and authentication, meaning that a compromised node can inject arbitrary messages (Bozdal *et al.*, 2018). Likewise, physical access points such as the On-Board Diagnostics (OBD) port can provide attackers with direct control over critical vehicle functions if not properly secured (Ammar *et al.*, 2020). As vehicles continue to evolve toward greater connectivity and automation, such weaknesses are no longer confined to local physical attacks but can be exploited remotely once the external network perimeter is breached.

Communication channels introduce yet another dimension of risk. In the absence of resilient authentication and encryption mechanisms, V2X messages are vulnerable to interception, manipulation, or spoofing. For example, attackers may exploit unsecured wireless links to inject false information, such as fabricated hazard alerts or manipulated signal phase data, that can mislead drivers and compromise traffic safety (Intelligent Transportation Society of America, 2024; SAE International, 2024). The use of multiple communication technologies, including Cellular V2X (C-V2X), 4G/5G/6G, Wi-Fi, and Bluetooth, further broadens the attack surface, multiplying opportunities for cyber intrusions into both vehicles and infrastructure (5G Automotive Association, 2019). As time-sensitive ITS applications rely heavily on wireless communication, attacks such as spectrum flooding through jamming devices can degrade or completely block message delivery. Signal spoofing techniques can also inject malicious data that masquerades as legitimate information. In safety-critical scenarios where split-second decisions are required, the loss or corruption of these time-sensitive messages can create hazardous roadway conditions. In addition, the continuous exchange of data that underpins ITS raises significant privacy concerns. Vehicles routinely transmit sensitive information, including location, identity, and travel patterns. If inadequately protected, this data can be exploited for surveillance, tracking, or manipulation. Beyond endangering individual users, such intrusions risk eroding public confidence in ITS technologies and their broader adoption (SAE International, 2024).

Implementing cybersecurity measures across ITS infrastructures is of utmost importance to ensure continuous Availability, preserve Integrity, and maintain Confidentiality (CIA), the core principles of cybersecurity whose compromise could endanger human safety and operational resilience. The vast scale and heterogeneity of transportation infrastructure and cybersecurity domains, combined with the safety-critical nature of real-world deployments, render comprehensive vulnerability assessments, threat model development, intrusion detection evaluations, and mitigation strategy validations both cost-prohibitive and time-intensive. Simulation platforms provide a safe, repeatable, and cost-effective environment for conducting comprehensive cybersecurity evaluations of ITS infrastructures; however, no single simulator integrates ITS applications, detailed sensor models, realistic mobility patterns, and network connectivity components to represent the whole transportation ecosystem. For example, CARs Learning to Act (CARLA) (Dosovitskiy *et al.*, 2017) delivers high-fidelity 3D world modeling and diverse sensor simulation, Simulation of Urban MObility (SUMO) (Lopez *et al.*, 2018) excels at microscopic traffic dynamics and signal controller modeling, and Objective Modular Network Testbed in C++ (OMNeT++) specializes in detailed network and V2X communication simulations. Individually, these simulators capture only subsets of the critical dimensions required to emulate a complete ITS ecosystem, including mobility, sensing, network connectivity, and application logic. The gap is bridged by integrating these individual simulators into an end-to-end co-simulation platform that couples high-fidelity environment and sensor modeling, comprehensive traffic and signal control simulation, and realistic network and V2X communication simulation to enable rigorous cybersecurity research and development.

This book chapter explains how a co-simulation platform can be constructed and applied to ITS cybersecurity research. The platform comprises of three open-source simulators: (i) CARLA for 3D world modeling and sensor simulation, (ii) SUMO for microscopic traffic simulation and traffic controller modeling, and (iii) OMNeT++ for V2X communication simulation. The development of the co-simulation platform, OpenCAMS, is presented in detail by Ahmad *et al.* (2025). This chapter demonstrates the platform's operation through a case study of a C-V2X-based proactive safety alert system that employs post-quantum cryptography (PQC) to strengthen V2X messaging and communication security across domains, enabling cohesive, realistic interactions among mobility, sensing, and networking components. Such co-simulation testbeds empower researchers and practitioners to identify cybersecurity vulnerabilities before deployment, benchmark and refine mitigation strategies, and perform rigorous performance evaluations, providing practical guidance for developing secure, efficient, and resilient ITS solutions that meet evolving transportation cybersecurity needs.

6.2 Simulation Platforms for ITS

Simulation is a well-established approach for studying and analyzing complex systems. Instead of experimenting directly in the real world, which is often costly, risky, or infeasible, researchers build simulators to replicate a system's behavior in a controlled virtual environment. A simulator creates a model of reality, abstracting away unnecessary details while preserving the dynamics most relevant to the problem at hand. By doing so, it enables safe, repeatable, and large-scale experimentation that would otherwise be impractical.

The design of a simulator is inherently domain-specific. This is because every complex system, whether physical, social, or technological, has its own set of governing rules, scales, and interactions that must be captured accurately. For example, an aerodynamic simulator must model airflow physics with high precision, while a computer network simulator focuses on packet-level communication protocols. Attempting to build a single universal simulator would be impractical–not only due to the enormous computational demands, but also because of the difficulty of managing, maintaining, and learning such a system. Specialized simulators are therefore created, each optimized to represent a particular domain with the fidelity required by its users.

At the same time, modern technological systems are increasingly multidisciplinary, cutting across traditional domain boundaries. Transportation, energy, communication, and urban infrastructure, for instance, are deeply interconnected, and innovations in one area often depend on interactions with others. As the world becomes more cross-domain and multidisciplinary, the limitations of standalone simulators become evident. To address this, co-simulation techniques are employed, in which multiple specialized simulators are executed concurrently and exchange information in real time. This approach allows researchers to study complex systems-of-systems while leveraging the strengths of each simulator without overburdening any single tool.

In the context of ITS, simulation platforms are indispensable for design, testing, and validation. Modern ITS applications extend well beyond traditional traffic management, incorporating connected and autonomous vehicles, smart infrastructure, and real-time data exchange. At a minimum, practical ITS applications involve three complementary domains.

- **Vehicle dynamics and traffic flow modeling**, to capture both individual vehicle behavior and large-scale traffic interactions across roadway networks.
- **Sensor and environment modeling**, to reproduce how vehicles perceive their surroundings through cameras, LiDAR, radar, and other sensing modalities.

- **Connectivity and communications modeling**, to simulate vehicle-to-vehicle (V2V), vehicle-to-infrastructure (V2I), and broader C-V2X communication that enable cooperative and coordinated mobility.

No single simulator can natively represent all of these domains with sufficient fidelity. Consequently, researchers rely on specialized simulation platforms and, increasingly, on co-simulation platforms that integrate multiple simulators into a unified experimental environment. The following sections provide an overview of these specialized, standalone simulators and co-simulation platforms, and their roles in ITS research.

6.2.1 *Standalone Specialized Simulators*

Building on the three domains introduced earlier, standalone specialized simulators for ITS can be grouped into (i) vehicle dynamics and traffic flow simulators, (ii) sensor and environment simulators, and (iii) connectivity and communication simulators. Each group addresses a distinct aspect of modern ITS, offering depth within its scope but remaining limited outside of it.

Vehicle dynamics and traffic flow simulators such as SUMO, PTV Vissim (PTV Group, 2025), and Aimsun (2025) model large-scale roadway traffic. Well-established commercial tools such as PTV Vissim and Aimsun are widely used in professional traffic engineering practice. They offer advanced features, strong visualization capabilities, and extensive support for roadway design tools and operational analysis, making them valuable to consultants, practitioners, and government agencies. However, because they are closed-source, these platforms offer limited flexibility for experimental modifications and are less suited for exploratory research. In contrast, SUMO has emerged as the preferred choice in academic and research settings. As a highly scalable, open-source microscopic simulator, SUMO can model thousands of vehicles across extensive roadway networks. SUMO implements widely accepted car-following models such as the Intelligent Driver Model (IDM) and Krauss, as well as lane-changing models including DK2008, LC2013, and SL2015, enabling a detailed representation of vehicle interactions and lane-level dynamics (Blokpoel *et al.*, 2016; Erdmann, 2015; Lopez *et al.*, 2018). It also supports route assignment, multimodal traffic, and traffic signal control. Most importantly, SUMO's open-source nature and extensible architecture allow researchers to adapt models, extend functionalities, and integrate with external tools via interfaces such as TraCI (Wegener *et al.*, 2008). These features make it especially attractive for academia and cutting-edge ITS research, where flexibility and customizability are often more critical than polished user interfaces.

Sensor and environment modeling simulators focus on replicating how sensors perceive and interact with their surroundings, whether mounted on vehicles or deployed as part of smart infrastructure. Their distinctive contribution lies in providing physics-based simulations of sensing modalities, including LiDAR, cameras, radar, and GNSS, within photorealistic 3D virtual environments. By leveraging modern game engines like Unreal Engine and Unity, these platforms simulate not only the environment's geometry but also its physical properties, including lighting, weather, and material reflectivity. This enables the generation of realistic sensor data streams for testing perception algorithms in autonomous vehicles and roadside monitoring systems, supporting sensing applications across both vehicle autonomy and intelligent infrastructure. Prominent examples include CARLA, LGSVL (Rong *et al.*, 2020), and AirSim (Shah *et al.*, 2017). CARLA has become the de facto open-source standard among these simulators, offering configurable environmental conditions and synchronous, deterministic time-stepping that enable precise integration with other simulators. LGSVL was originally designed to support autonomous vehicle software, such as Apollo (ApolloAuto, 2025) and Autoware (Autoware Foundation, 2025). It provided similar features as CARLA, but its development was suspended in 2022. AirSim, created by Microsoft Research, was initially built for drones and later extended to ground vehicles; however, its support for vehicle-related applications remained limited, and development was discontinued in 2023. These simulators are also valuable for reinforcement learning, AI, and perception research, as they enable safe, repeatable experimentation with complex sensing setups. At the same time, their scope is narrower than that of traffic simulators: they do not model large-scale flows using microscopic traffic models, nor do they natively incorporate V2X communication stacks. This limits their applicability when evaluating connected transportation systems holistically.

Connectivity and communication simulators provide the third domain, focusing on networking and wireless communication. Tools such as OMNeT++ (OMNeT++ Community, 2025) and ns-3 (ns-3 Project, 2025) were originally developed as general-purpose computer network simulators and include comprehensive support for Internet Protocol (IP) stacks, from traditional IP and transport protocols to advanced wireless standards. Their flexibility and extensibility have made them the foundation for modeling vehicular communication technologies as well. Building on their core networking capabilities, researchers have introduced support for vehicular communication standards, including DSRC, LTE, and emerging C-V2X technologies. Extensions such as Veins (Veins Development Team, 2025) provide protocol-level implementations of V2X communication, while other frameworks adapt OMNeT++ and ns-3 to study wireless channel behavior and cooperative applications in transportation systems. Between the two, OMNeT++ is often favored in ITS contexts for its modular design, visual debugging tools, and ease of extension. ns-3, by contrast, provides lower-level and detailed protocol modeling, offering greater fidelity at the cost of accessibility and visualization. Both simulators remain central in ITS research, bringing the breadth of computer network simulation into the transportation domain and enabling exploration of how connectivity interacts with mobility and autonomy.

These specialized simulators illustrate how each of the three domains has dedicated tools that excel within their own boundaries. While indispensable, they cannot individually capture the cross-domain interactions that characterize modern ITS, motivating the use of co-simulation approaches discussed in the following subsection.

6.2.2 *Co-Simulation of Multiple Specialized Simulators*

As discussed earlier, standalone simulators excel within their domains-traffic flow, sensor and environment modeling, or networking- but remain limited for comprehensive ITS applications when applied in isolation. Traffic simulators such as SUMO or Vissim capture large-scale mobility patterns yet lack perception and connectivity. Simulators like CARLA provide photorealistic environments and sensor models, but are decoupled from standardized microscopic traffic models. Network simulators such as OMNeT++ or ns-3 support rich protocol stacks, but their mobility models are typically abstract and fail to reflect realistic vehicle dynamics.

These siloed approaches restrict the ability to evaluate closed-loop interactions among mobility, perception, decision-making, and connectivity. Such interactions are essential in scenarios involving mixed autonomy, cooperative driving, or cybersecurity in transportation applications, where the interplay between sensing, planning, and communication directly shapes outcomes. The lack of modular, extensible integration also hampers experimentation with new algorithms, protocols, or coordination strategies. Co-simulation addresses this gap by bridging specialized simulators into a unified framework, enabling multi domain studies that none of the tools can support individually.

One representative example is the native SUMO–CARLA co-simulation, introduced in CARLA v0.9.8, which synchronizes CARLA's high-fidelity vehicles and sensor models with SUMO's microscopic traffic simulation via an extended TraCI interface. This integration enables hybrid scenarios in which CARLA manages ego vehicles with detailed dynamics and perception, while SUMO controls surrounding traffic and infrastructure. Several works have extended this baseline: Li *et al.* developed a synchronized framework for multi-vehicle AV testing, Cantas and Guvenc designed a reinforcement learning–driven traffic co-simulation environment, and Roccotelli *et al.* (Cantas and Guvenc, 2023; Roccotelli *et al.*, 2024) proposed a modular CARLA–SUMO platform for studying mixed traffic autonomy. While these efforts bridge traffic modeling and perception, most lack C-V2X support and rely on simplified or abstracted representations for V2X communication.

In the networking domain, OMNeT++ and ns-3 have been extended to support vehicular communication, leading to frameworks such as Veins and OpenCV2X (McCarthy *et al.*, 2021). Among these, Veins has become the most widely used for ITS, offering a SUMO–OMNeT++ co-simulation pipeline. Veins enables bidirectional synchronization of traffic mobility and communication protocols, including DSRC, and even supports channel attenuation modeling with building polygons. However, its vehicles are limited to SUMO's node abstractions, and it lacks high-fidelity sensor modeling. For the development and testing of sensing and perception algorithms, platforms like CARLA remain necessary.

To overcome these limitations, newer co-simulation frameworks extend the scope further. Veins-Carla (Hardes *et al.*, 2023) replaces SUMO with CARLA to combine detailed vehicle dynamics with communication modeling. MS-VAN3T-CARLA (Risma Carletti *et al.*, 2024) connects CARLA and ns-3 using gRPC middleware, enabling photorealistic cooperative perception experiments with realistic C-V2X communication. CARLA–SUMO–Artery (Bouchouia *et al.*, 2022) integrates the Artery stack to support ETSI ITS-G5 protocols for security and cooperative automation studies. OpenCDA (Xu *et al.*, 2021) provides a unified platform for cooperative maneuvers such as platooning and coordinated merging. SimuTack (Finkenzeller *et al.*, 2023) integrates SUMO, CARLA, and OMNeT++ into a single co-simulation pipeline, while Eclipse MOSAIC acts as middleware, providing modular adapters to couple multiple domain-specific simulators. MOSAIC (Schrab *et al.*, 2022) functions as a modular co-simulation middleware connecting domain-specific simulators through dedicated adapters for each simulator. While offering a flexible and standardized interface between individual simulators, the CARLA adapter for MOSAIC is still under development. MOSAIC Extended, a commercial version of MOSAIC, provides support for 3D environments and vehicle physics simulation through a proprietary simulator, PHABMACS (DCAITI (Daimler Center for Automotive Information Technology Innovations), 2025). However, PHABMACS restricts environmental realism by procedurally generating minimal road furniture (no curbs, lamp posts, or benches) and targets small-scale ADAS prototyping, trading off high-detail physics for fewer vehicle support features.

Table 6.1 summarizes these co-simulation platforms against three essential capabilities for Connected and Automated Mobility (CAM): high-fidelity environment and sensor simulation, scalable microscopic traffic simulation, and flexible C-V2X communication support. A number of recent co-simulation platforms (e.g., Simutack, Finkenzeller *et al.*, 2023, ms-van3t-carla, Risma Carletti *et al.*, 2024) tightly couple CARLA with ns-3 or OMNeT++ and fixed deep-learning-based perception stacks, such as YOLO-based cooperative perception frameworks (Bouchouia *et al.*, 2022; Finkenzeller *et al.*, 2023). These approaches are well-suited for evaluating perception-driven

Table 6.1. Comparison of co-simulation platforms.

Platform	Environment & Perception Sensor	Traffic Simulation	Communication Simulation	Generalization
Veins (Sommer *et al.*, 2011)	○	●	●	○
Veins-Carla (Hardes *et al.*, 2023)	●	◐	●	○
ms-van3t-carla (Risma Carletti *et al.*, 2024)	●	◐	●	◐
Carla-SUMO-Artery (Bouchouia *et al.*, 2022)	●	●	●	○
PLEXE (Segata *et al.*, 2022)	○	●	●	◐
OpenCDA (Xu *et al.*, 2021)	●	●	◐	◐
Simutack (Finkenzeller *et al.*, 2023)	●	●	●	○
MOSAIC (Schrab *et al.*, 2022)	◐	●	●	●
OpenCAMS	●	●	●	●

● = full support ◐ = partial support ○ = no support

metrics under network constraints; however, they often embed a specific detector and perception pipeline into the simulation loop. In contrast, OpenCAMS intentionally remains perception-agnostic by exposing raw sensor outputs rather than embedding a predefined perception model. This design enables controlled substitution of perception modules (e.g., using RT-DETR, YOLO, or other detectors) and supports systematic studies in which perception models, communication models, and application-level logic can be modified independently. The comparison highlights remaining research gaps. Many frameworks support synchronous co-simulation but are narrowly tailored to specific applications. Communication stacks are often tightly coupled to particular protocol stacks, limiting generalizability. Moreover, synchronization mechanisms may not be modular, making it challenging to extend these platforms to diverse scenarios or additional simulation modules. These gaps underline the need for flexible, modular, and open co-simulation frameworks that can evolve alongside cutting-edge ITS research.

6.3 Co-Simulation Testbed for ITS Cybersecurity

To address these critical challenges, Ahmad *et. al.* (2025) present a generalized and time-synchronized co-simulation platform, OpenCAMS, that integrates three unique simulators for Connected and Automated Mobility (CAM) applications in ITS: (i) SUMO for scalable traffic modeling, (ii) CARLA for high-fidelity environmental perception and vehicle dynamics, and (iii) OMNeT++ for flexible C-V2X communication simulation. The co-simulation platform, OpenCAMS, enables researchers to explore the interplay between large-scale traffic dynamics, test vehicle perception stacks, and real-time C-V2X communications within a single, cohesive environment. Beyond cybersecurity research, by enabling the simulation of complex cyberattack scenarios and mitigation approaches such as signal spoofing, message tampering, and denial-of-service attacks across all layers of the transportation ecosystem, OpenCAMS also offers significant potential to improve safety and mobility. Its extensible design supports integrating emerging technologies and threat models, making it a valuable tool for developing and testing robust mitigation strategies.

Fig. 6.1. Simulator components of OpenCAMS.

Figure 6.1 presents the individual simulator components of the OpenCAMS co-simulation platform. OpenCAMS employs a multi-client TraCI loop with strict time step synchronization to ensure consistency across all three simulators. The unique feature of this co-simulation platform is that it avoids adding any abstraction layer on top of the simulators, preserving the flexibility and full potential of the underlying simulators. CARLA may simulate only a subset of vehicles that require advanced in-vehicle or roadside sensors, while SUMO handles the background roadway traffic. OMNeT++ dynamically maps communication nodes to vehicles and infrastructures, leveraging wireless propagation models and customizable computer networking stacks. This design supports reproducible, modular, and extensible research across various domains, including the development and evaluation of reactive and proactive safety, mobility, and cybersecurity integrations related to CAM applications. In essence, this platform provides a unified, scalable, and realistic testbed for transportation research that bridges the gap between what each simulator can do in isolation and what modern connected and automated transportation systems demand.

OpenCAMS operates on a deterministic, step-based execution model in which all simulators advance in lockstep using fixed temporal increments coordinated through a multi-client control loop. Simulation time is explicitly controlled by the co-simulation manager rather than wall-clock time, thereby avoiding clock drift and ensuring deterministic behavior across SUMO, CARLA, and OMNeT++. As the scale or complexity of a scenario increases, simulations may execute more slowly in wall-clock time due to higher computational load; however, temporal consistency and correctness of the simulated interactions are preserved since all components advance using the same simulation time steps. This design prioritizes reproducibility and controlled cyber-physical experimentation over real-time execution.

The development of OpenCAMS, as detailed in Ahmad *et al.* (2025), was driven by the need to capture the full complexity of modern ITS systems, prioritize open-source components, maintain minimal abstraction possible, and support generalized ITS applications. By fostering a collaborative, accessible environment, OpenCAMS can empower the research community to advance secure, efficient, and resilient next-generation transportation systems. For the latest details on its implementation, readers are encouraged to consult the project's GitHub repository at https://github.com/minhaj6/carla-sumo-omnetpp-cosim.

The OpenCAMS co-simulation platform architecture enables integrated development, evaluation, and validation, ensuring coherent evolution across cyber-physical layers. Diverse application domains include safety, mobility, and cybersecurity.

Safety: OpenCAMS allows rigorous testing of safety-critical applications such as Forward Collision Warning, Vulnerable Road User protection, and Blind Spot Warning/Lane Change Assist. CARLA models a suite of sensors, OMNeT++ transmits Basic Safety Messages (BSMs) and cooperative perception messages, and SUMO manages traffic flow in the road network. Researchers can test robustness under varying weather conditions, visibility, latency, and traffic density.

Mobility: The platform supports platooning, cooperative merging, adaptive following, and smart corridor management. SUMO manages background traffic and signals, CARLA handles perception sensors, and OMNeT++ simulates V2X communication. Applications like Green Light Optimal Speed Advisory (GLOSA), energy-aware routing, fleet electrification, and emergency vehicle preemption can be studied with end-to-end impacts on throughput, fuel economy, and safety.

Cybersecurity: OpenCAMS provides a controlled environment for evaluating attacks such as Sybil, replay, and infrastructure manipulation. OMNeT++'s application layer can model malicious messages, CARLA can model target vehicle responses, and SUMO can evaluate network-wide consequences. Researchers can test detection and mitigation mechanisms, including authentication, timestamp verification, and anomaly detection.

OpenCAMS enables comprehensive ITS cybersecurity research by simulating cyberattacks in a controlled environment. For example, researchers can simulate Sybil attacks, where multiple fake identities are injected into the network; replay attacks, where valid messages are retransmitted maliciously; or infrastructure compromise, where traffic signals are manipulated. This integrated approach allows for the evaluation of vulnerabilities across the entire system, from perception to communication, and the testing of mitigation strategies like intrusion detection systems or cryptographic enhancements. For example, a perception-level attack can be introduced in CARLA by degrading or spoofing roadside sensor detections, which affects conflict detection and safety logic running at the application level. The resulting safety messages are then generated, transmitted, delayed, or replayed in OMNeT++, where communication-layer attacks and security mechanisms can be evaluated. Finally, vehicle responses to these messages, such as braking, yielding, or failure to react, are reflected at the mobility level in SUMO, allowing researchers to observe roadway safety and operational impacts. Such scenarios illustrate why a tightly synchronized SUMO–CARLA–OMNeT++ co-simulation is required to evaluate cybersecurity effects that propagate across sensing, communication, and mobility domains. Overall, OpenCAMS provides a closed-loop, scenario-based benchmarking platform for emerging research in cooperative safety, mobility optimization, cybersecurity, and digital twin systems. Its extensibility supports frontiers such as human interaction, teleoperation, and city-scale integration. Beyond the specific case study presented in Section 6.4, the OpenCAMS architecture is open to implementation of a broad class of cybersecurity and cyber-physical attack scenarios that require tightly integrated simulation across mobility, sensing, and communication domains. These include perception manipulation (e.g., sensor degradation or spoofed detections), replayed or selectively delayed safety messages, infrastructure-side message tampering, and cross-layer attacks whose effects must be traced from sensor or communication disruptions to vehicle behavior and network-level safety outcomes. Such scenarios motivate the use of a tightly synchronized SUMO–CARLA–OMNeT++ stack in which causal relationships across layers can be explicitly observed.

The following section presents an example implementation of an infrastructure-enabled crash warning system that uses roadside sensors to detect vehicles approaching an urban intersection, calculates Time-to-Collision (TTC) between as they approach the intersection, and broadcasts warning messages verified by PQC signatures. This case study is intended as a feasibility and integration demonstration rather than a comprehensive stress test or performance characterization of post-quantum cryptography-enabled V2X systems. The primary objective is to validate end-to-end cryptographic integration within a synchronized cyber-physical co-simulation environment, rather than to exhaustively evaluate PQC performance under dense traffic or adversarial conditions.

6.4 Case Study: Demonstration of OpenCAMS for ITS Cybersecurity Research

Central to the operation and effectiveness of modern ITS architectures is connectivity. Connectivity through C-V2X enables a wide range of ITS applications that enhance security, safety, mobility, and operational efficiency. For instance, queue warning systems notify drivers of slow or stopped traffic ahead, helping them react in time and reducing the likelihood of rear-end collisions. Emergency vehicle alerts are another critical use case in which drivers and connected systems receive alerts about the approach and intended route of ambulances, fire trucks, or police vehicles, enabling surrounding traffic to clear lanes in advance (SAE International, 2020). Cooperative Adaptive Cruise Control (CACC) further demonstrates the potential of V2V communication by synchronizing acceleration and braking across multiple vehicles, which supports safe platooning, improves fuel efficiency, and helps ease congestion (5G Automotive Association, 2019; Liu *et al.*, 2021). Another prominent example of a safety application is the Intersection Collision Avoidance (ICA) warning system, in which vehicles approaching a connected intersection receive broadcasts from a roadside unit (RSU) when the smart intersection detects potential conflicts, improving intersection safety (SAE International, 2016, 2024).

C-V2X communication enables real-time information sharing between vehicles, infrastructure, and cloud services. C-V2X supports both PC5 sidelink for direct, low-latency communication and Uu interface over 4G/5G networks for broader coverage and cloud integration. These capabilities are indispensable for latency-sensitive applications such as platoon control, intersection coordination, cooperative perception, and dynamic traffic signal control. Today, connectivity is no longer optional; it is foundational. It enables not just functional capabilities but also determines the system's adaptability and resilience. However, this reliance on connectivity further enlarges the attack surface for adversaries, who may exploit vulnerabilities at the physical, communication, or application layers. Consequently, testing the cyber resilience of connected transportation systems has become an urgent research priority. Evaluating these systems under real-world cyber-physical attack scenarios is critical but fraught with logistical, financial, and regulatory challenges.

C-V2X is a communication technology standardized by 3GPP starting in Release 14 (3rd Generation Partnership Project (3GPP), 2017). It was designed to support direct communication between vehicles, infrastructure, and vulnerable road users as well as network-based communication through cellular operators (Sunuwar and Kim, 2024). These communication capabilities form the backbone for a wide range of ITS applications. C-V2X includes two complementary modes. The first is the PC5 sidelink mode, which enables direct V2V, V2I, and Vehicle-to-Pedestrian (V2P) communications over the 5.9 GHz ITS band without relying on cellular coverage (Molina-Masegosa and Gozalvez, 2017). The second is the Uu interface, which provides vehicle-to-network (V2N) communication using LTE or 5G infrastructure (5G Automotive Association, 2019). Together, these modes allow both low-latency safety message exchanges with surrounding vehicles and roadside infrastructure, as well as broader coverage through the cellular network, including communication with cloud services. Figure 6.2 presents the communication modes of C-V2X.

For safety-critical sidelink connectivity, 3GPP Release 14 introduced two new communication standards: Mode 3 (scheduled mode) and Mode 4 (autonomous mode). In Mode 3, the cellular infrastructure assists in scheduling communication resources for device-to-device communication, which requires devices to be within the cellular coverage area. However, in Mode 4, vehicles can select and manage their radio resources autonomously using sensing-based Semi-Persistent Scheduling (SPS). Mode 4 is especially important because safety applications must operate even when cellular coverage

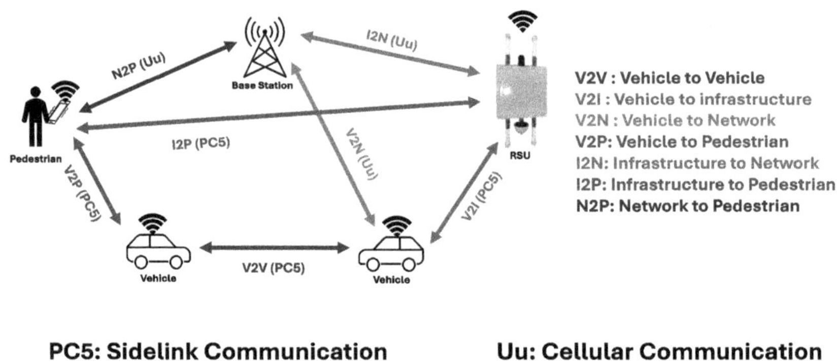

Fig. 6.2. Communication modes of C-V2X: direct PC5 sidelink for V2V, V2I, I2P, and V2P, and Uu interface for V2N, N2P, and I2N.

is unavailable. Subsequent releases, such as Release 16, extended C-V2X into 5G New Radio V2X (NR-V2X), improving reliability and adding support for advanced use cases, such as cooperative perception and high-density platooning (SAE International, 2024).

Intersection Collision Avoidance (ICA) is recognized as one of the most critical safety applications in ITS. Intersections are complex and high-risk environments where vehicles, pedestrians, and signals converge. A significant share of crashes and traffic fatalities occur at intersections, and many involve angle collisions that are particularly severe. ICA applications provide early warnings to approaching vehicles, helping drivers or automated systems to react in time and avoid dangerous conflicts. This ability to prevent side-impact collisions makes ICA a cornerstone of safety-critical ITS services (Intelligent Transportation Society of America, 2024).

While ICA requires reliable traffic modeling and communication, it also depends on the trustworthiness of the warning messages themselves. PQC provides digital signatures that remain secure and trustworthy against both traditional cyberattacks and in the presence of future quantum adversaries that could break classical algorithms (e.g., Rivest–Shamir–Adleman (RSA), Elliptic Curve Digital Signature Algorithm (ECDSA)) (Chen *et al.*, 2016; National Institute of Standards and Technology, 2022). By integrating PQC into C-V2X communication, ITS applications ensure that any C-V2X message exchange, including ICA warnings, cannot be forged or tampered with. Falcon, a lattice-based signature algorithm standardized by

NIST, is particularly well-suited for C-V2X applications. It produces compact signatures and offers very fast verification, both of which are essential for high-volume, low-latency C-V2X messaging (Ducas *et al.*, 2019; Giuliari *et al.*, 2023; National Institute of Standards and Technology, 2022). Evaluating PQC in a co-simulation environment is therefore required to confirm that stronger security does not undermine real-time safety. In this work, Falcon is adopted for ICA, providing both quantum-resistant security and practical performance for time-critical transportation safety applications. Conducting such evaluations in the real world is neither economical nor safe due to the risks of traffic disruption and potential hazards. This makes the OpenCAMS co-simulation testbed a vital tool for advancing secure and resilient ITS applications. Figure 6.3 presents the functional diagram of the PQC-enabled ICA application in the OpenCAMS platform.

6.4.1 *Implementation of PQC in ICA Application*

We selected a signalized intersection in CARLA Town 05 (Junction 829). We created a SUMO route file that generates 8 vehicles passing through the chosen intersection. We placed a RSU in OMNeT++ at the side of the intersection. OpenCAMS spawns and mirrors the

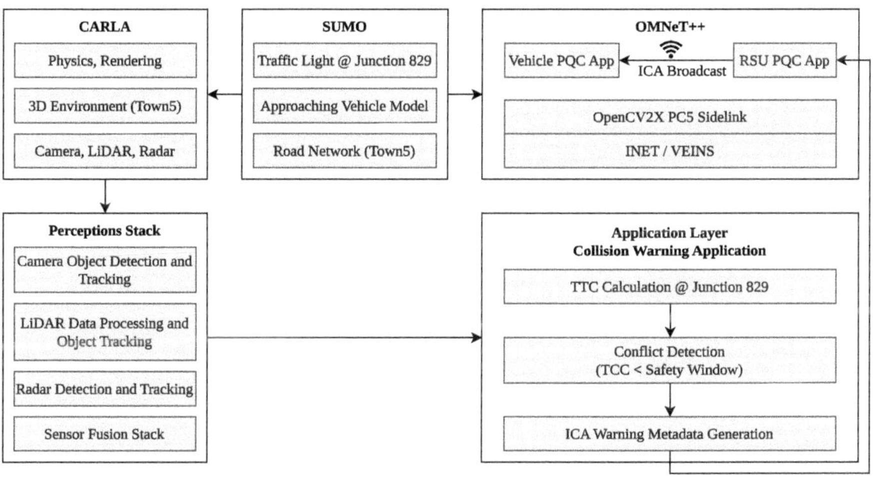

Fig. 6.3. Functional diagram of PQC evaluation in OpenCAMS: TTC at junction 829 and ICA broadcast.

same vehicles in SUMO, CARLA, and the corresponding mobile communication nodes in OMNeT++, ensuring that traffic, sensing, and communication are synchronized. We built a sensor view from the roadside unit's perspective. The RSU is equipped with LiDAR and radar. We defined the geometric center of the intersection as the reference point for calculating TTC. At each SUMO simulation step, we computed the time each vehicle would take to reach the reference point from its current location. If the time difference between any two vehicles was smaller than two seconds, the roadside unit generated an ICA message. For the message structure, we followed the SAE J2735 standard (SAE International, 2016), specifically the `MSG_IntersectionCollisionAvoidance` format, when defining and transmitting the ICA messages within OMNeT++. In addition, we adopted an X.509 certificate style to attach certificates directly to the ICA messages. The certificates contain the verification keys, enabling receiving vehicles to verify the authenticity of the broadcast message. Both vehicles and the roadside unit use C-V2X PC5 Mode 4 links for direct communication. As soon as an ICA message is created, the roadside unit signs the message with the Falcon-512 post-quantum digital signature. The roadside unit also builds a certificate that includes the public verification key and broadcasts the signed warning message.

Figure 6.4 shows the integrated view of the three simulators in the OpenCAMS co-simulation platform. The top row shows the (a) SUMO and (b) OMNeT++ interfaces. The bottom row shows the (c) CARLA bird's eye view and the (d) roadside perception view. This four-panel figure illustrates a time-synchronized co-simulation in which mobility, sensing, and communication are active at the same timestamp.

6.4.2 *Metrics and Evaluation of PQC-Enabled ICA*

We evaluated the warning pipeline end-to-end using metrics that capture both cryptographic performance and communication reliability. These metrics are critical for verifying that warning messages remain both trustworthy and effective in traffic.

Signature Generation and Verification Time: The cryptographic performance is measured in terms of signature generation

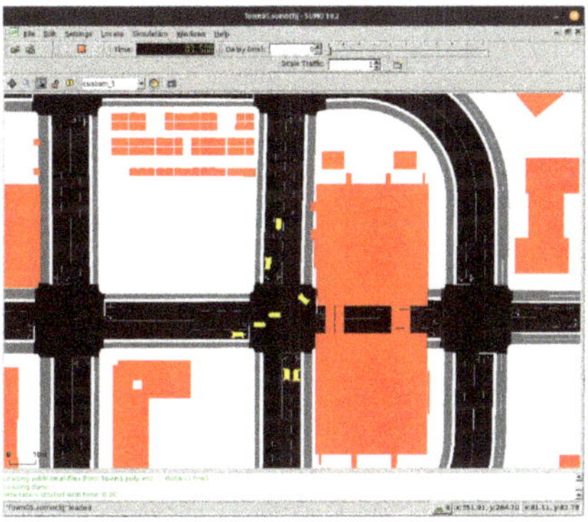

(a) SUMO interface in co-simulation

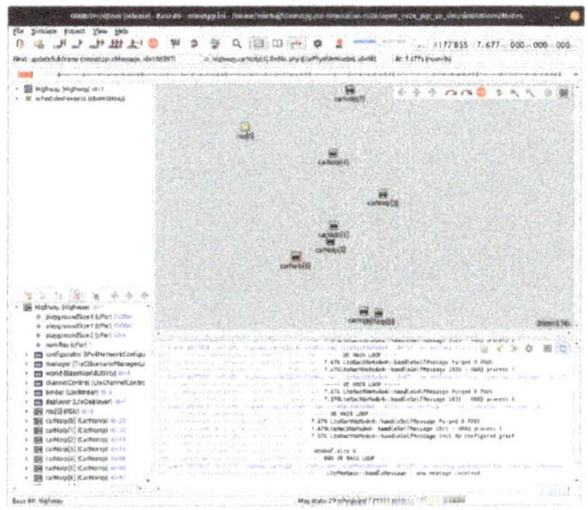

(b) OMNeT++ interface in co-simulation

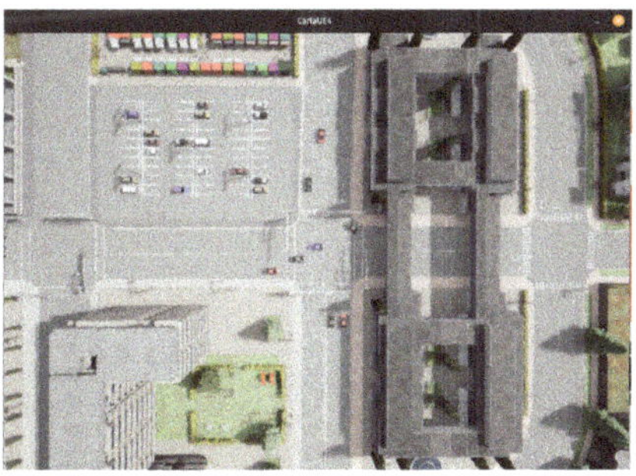

(c) CARLA bird's eye view

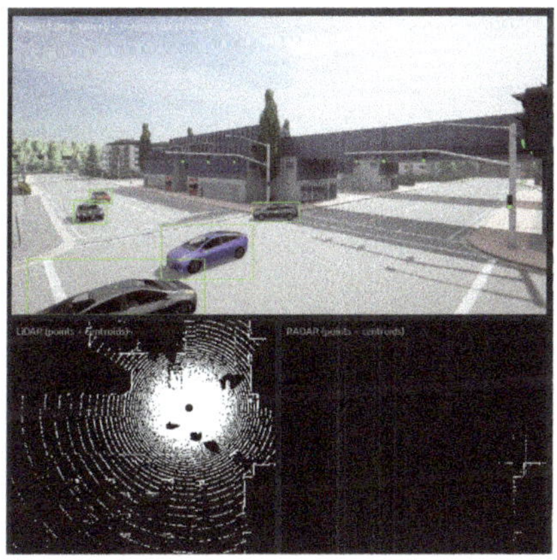

(d) Sensor view from RSU's perspective

Fig. 6.4. SUMO, OMNeT++, and CARLA running synchronously in the OpenCAMS co-simulation.

time at the roadside unit and verification time at the vehicles. If t_{gen} is the average signing time and t_{ver} is the average verification time, then

$$t_{\text{gen}} = \frac{1}{N}\sum_{i=1}^{N} t_{\text{gen}}^{(i)}, \quad t_{\text{ver}} = \frac{1}{N}\sum_{i=1}^{N} t_{\text{ver}}^{(i)}$$

where N is the number of messages signed or verified.

Let V be the number of vehicles. Let N_v be the number of signatures verified by vehicle v. Let $t_{\text{ver},v}^{(i)}$ be the ith verification time at vehicle v. The pooled mean verification time across all vehicles is

$$\bar{t}_{\mathrm{ver}} = \frac{\sum_{v=1}^{V} \sum_{i=1}^{N_v} t_{\mathrm{ver},v}^{(i)}}{\sum_{v=1}^{V} N_v}.$$

The pooled standard deviation is

$$s_{\mathrm{ver}} = \sqrt{\frac{\sum_{v=1}^{V} \sum_{i=1}^{N_v} \left(t_{\mathrm{ver},v}^{(i)} - \bar{t}_{\mathrm{ver}}\right)^2}{\sum_{v=1}^{V} N_v - 1}}.$$

For signing at the RSU, if N_{sig} warnings were signed, the mean signing time is

$$\bar{t}_{\mathrm{gen}} = \frac{1}{N_{\mathrm{sig}}} \sum_{i=1}^{N_{\mathrm{sig}}} t_{\mathrm{gen}}^{(i)}.$$

Packet Delivery Ratio (PDR): PDR represents the reliability of message delivery over the PC5 link. It is defined as

$$\mathrm{PDR} = \frac{N_{\mathrm{recv}}}{N_{\mathrm{exp}}}$$

where N_{recv} is the number of ICA messages successfully received by vehicles, and N_{exp} is the total number of messages expected.

Channel Busy Ratio (CBR): CBR measures the fraction of time the wireless channel is occupied. If T_{busy} is the total time the channel is sensed as busy and T_{total} is the total observation time, then

$$\mathrm{CBR} = \frac{T_{\mathrm{busy}}}{T_{\mathrm{total}}}.$$

Monitoring CBR is important because larger post-quantum signatures could increase network load.

We measured signature generation at the RSU and signature verification at the vehicle using the Falcon-512 post-quantum scheme. On average, the roadside unit required 0.30 ms to generate a signature, with a standard deviation of 0.09 ms. The vehicles required an average of 0.11 ms to verify each signature, with a standard deviation of 0.02 ms.

The measured PDR was 0.93 for all the simulated vehicles, as shown in Table 6.2, meaning that 93% of the broadcasted warning messages were successfully received. A high PDR ensures that vehicles consistently receive safety-critical messages, which is vital for preventing collisions even when cryptographic overhead is present.

The mean CBR was approximately 0.02 with a maximum standard deviation of 0.007, as shown in Table 6.3. This low value indicates that PQC signatures did not cause congestion and that sufficient capacity remained for other V2X messages.

Table 6.4 summarizes the main metrics used in our evaluation. Signature processing times quantify the computational cost of PQC,

Table 6.2. Summary of ICA message delivery reliability.

Vehicle ID	ICA Messages Expected	ICA Messages Received	PDR
1	61	57	0.93
2	61	57	0.93
3	61	57	0.93
4	61	57	0.93
5	61	57	0.93
6	61	57	0.93
7	61	57	0.93
8	61	57	0.93

Table 6.3. Channel busy ratio statistics across vehicles.

Vehicle ID	CBR Mean	CBR Standard Deviation
1	0.020	0.007
2	0.021	0.006
3	0.023	0.003
4	0.024	0.004
5	0.021	0.006
6	0.020	0.007
7	0.022	0.005
8	0.023	0.005

Table 6.4. Application-level metrics for ICA warnings.

Metric	Measured Value	Relevance for Security Assessment
Signature generation time	0.30 ms (mean), 0.09 ms (std. dev.)	Demonstrates whether the RSU can generate signatures for warning messages within acceptable time constraints.
Signature verification time	0.11 ms (mean), 0.02 ms (std. dev.)	Demonstrates whether vehicles can verify incoming warnings promptly to maintain real-time responsiveness.
Packet Delivery Ratio	0.93 (mean), 0 (std. dev.)	Reflects the reliability of ICA message dissemination in the presence of PQC operations.
Channel Busy Ratio	0.02 (mean), 0.007 (std. dev.)	Indicates the extent of wireless channel utilization and whether PQC signatures contribute to potential congestion.

PDR shows communication reliability, and CBR reflects channel utilization. Together, these metrics provide a balanced view of cryptographic and network performance, ensuring that stronger security does not undermine the usefulness of ICA safety applications.

6.5 Summary

We discussed ITS-specialized simulators and co-simulation platforms. In addition, we presented a case study using OpenCAMS, a co-simulation testbed that integrates SUMO, CARLA, and OMNeT++ to model CAM systems for transportation cybersecurity research. The utility of the co-simulation testbed was further demonstrated through the implementation of a PQC-enabled safety application. Looking ahead, co-simulation platforms with capabilities like those of OpenCAMS can support emerging technologies and next-generation ITS research. For instance, integrating 5G NR sidelink protocols, ROS2-based autonomy stacks, real-time GNSS spoofing models, and cyberattacks on cooperative driving. As CAM systems continue to advance, co-simulation remains a critical tool for exploring safe, intelligent, efficient, and secure mobility solutions under realistic and tightly integrated simulation conditions.

There are several limitations of this study. The presented evaluation focuses on a single intersection scenario with a limited number of vehicles and does not include dense traffic conditions, adversarial communication environments, or perception-level attack evaluations. As such, the results are not intended to characterize system performance under worst-case or large-scale conditions. Instead, the book chapter demonstrates the feasibility and integration of OpenCAMS while providing a foundation for future, more extensive adversarial and large-scale studies.

Bibliography

3rd Generation Partnership Project (3GPP) (2017). Release 14: LTE support for V2X, Technical Specification Release 14, 3GPP, https://www.3gpp.org, Technical Specification Group Radio Access Network.

5G Automotive Association (2019). C-V2X roadmap: Timeline for deployment, White Paper, 5GAA, Munich, Germany, https://5gaa.org/5gaa-releases-updated-white-paper-on-c-v2x-deployment-timeline/.

Ahmad, M. U., Abrar, A., Dasgupta, S., and Rahman, M. (2025). OpenCAMS: An open-source connected and automated mobility co-simulation platform for advancing next-generation intelligent transportation systems research, *arXiv preprint* arXiv:2507.09186.

Aimsun (2025). Aimsun: Mobility intelligence for decisions that count, https://www.aimsun.com/, accessed: May 29, 2025.

Ammar, M., Janjua, H., Thangarajan, A. S., Crispo, B., and Hughes, D. (2020). Securing the on-board diagnostics port (OBD-II) in vehicles, *SAE International Journal of Transportation Cybersecurity and Privacy*, 11-02-02-0009, pp. 83–106.

ApolloAuto (2025). Apollo: Open source autonomous driving platform, https://github.com/ApolloAuto/apollo, accessed: May 29, 2025.

Autoware Foundation (2025). Autoware: Open-source software for autonomous driving, https://autoware.org/, accessed: May 29, 2025.

Blokpoel, R., Joueiai, M., Barthauer, M., Friedrich, B., Richter, A., Friedl, H., Scholz, M., Krumnow, M., Flötteröd, Y.-P., Erdmann, J., *et al.* (2016). SUMO 2016–Traffic, mobility, and logistics, in *Proceedings of the SUMO2016*.

Bouchouia, M. L., Monteuuis, J.-P., Labiod, H., Jelassi, O., Jaballah, W. B., and Petit, J. (2022). A simulator for cooperative and automated driving security, in *Workshop on Automotive and Autonomous Vehicle Security (AutoSec), Ndss-Symposium*.

Bozdal, M., Samie, M., and Jennions, I. (2018). A survey on can bus protocol: Attacks, challenges, and potential solutions, in *2018 International Conference on Computing, Electronics & Communications Engineering (iCCECE)*, pp. 201–205, https://doi.org/10.1109/iCCECOME.2018.8658720.

Cantas, M. R. and Guvenc, L. (2023). Customized co-simulation environment for autonomous driving algorithm development and evaluation, *arXiv preprint* arXiv:2306.00223.

Chen, L., Jordan, S., Liu, Y.-K., Moody, D., Peralta, R., Perlner, R., and Smith-Tone, D. (2016). Report on post-quantum cryptography, Technical Report NISTIR 8105, NIST.

DCAITI (Daimler Center for Automotive Information Technology Innovations) (2025). PHABMACS: Physically realistic vehicle simulation tool for prototyping cooperative ADAS, https://www.dcaiti.tu-berlin.de/research/simulation/phabmacs/?lang=en, accessed: July 12, 2025.

Dosovitskiy, A., Ros, G., Codevilla, F., Lopez, A., and Koltun, V. (2017). CARLA: An open urban driving simulator, in *Proceedings of the 1st Annual Conference on Robot Learning*, pp. 1–16.

Ducas, L., Prest, T., *et al.* (2019). Falcon: Fast-Fourier lattice-based compact signatures over NTRU, https://falcon-sign.info/, Submission to the NIST PQC Standardization Project.

Erdmann, J. (2015). Sumo's lane-changing model, in *Modeling Mobility with Open Data: 2nd SUMO Conference 2014 Berlin, Germany, May 15–16, 2014* (Springer), pp. 105–123.

Finkenzeller, A., Mathur, A., Lauinger, J., Hamad, M., and Steinhorst, S. (2023). Simutack - An attack simulation framework for connected and autonomous vehicles, in *The 2023 IEEE 97th Vehicular Technology Conference: VTC2023-Spring (VTC2023-Spring)*, pp. 1–7, https://doi.org/10.1109/VTC2023-Spring57618.2023.10200555.

Giuliari, G. *et al.* (2023). Post-quantum cryptography for V2X communications: Performance evaluation and trade-offs, in *IEEE Vehicular Networking Conference (VNC)*, pp. 1–8.

Hardes, T., Turcanu, I., and Sommer, C. (2023). Poster: A case for heterogenous co-simulation of cooperative and autonomous driving, in *2023 IEEE Vehicular Networking Conference (VNC)* (IEEE), pp. 151–152.

Intelligent Transportation Society of America (2024). The future of V2X in the 5.9 GHz band, Technical Report, ITS America, https://itsa.org.

Islam, T., Sheakh, M. A., Jui, A. N., Sharif, O., and Hasan, M. Z. (2024). A review of Cyber attacks on sensors and perception systems in autonomous vehicle, *Journal of Economy and Technology* **1**, 1, pp. 242–258, https://doi.org/10.1016/j.ject.2024.01.002.

Liu, Y., Shi, Y., and Li, F. (2021). C-V2x communications in autonomous driving: A survey, *IEEE Communications Surveys & Tutorials* **23**, 4, pp. 2377–2402.

Lopez, P. A., Behrisch, M., Bieker-Walz, L., Erdmann, J., Flötteröd, Y.-P., Hilbrich, R., Lücken, L., Rummel, J., Wagner, P., and Wießner, E. (2018). Microscopic traffic simulation using sumo, in *The 21st IEEE International Conference on Intelligent Transportation Systems* (IEEE), https://elib.dlr.de/124092/.

McCarthy, B., Burbano-Abril, A., Licea, V. R., and O'Driscoll, A. (2021). OpenCV2X: Modelling of the V2X cellular sidelink and performance evaluation for aperiodic traffic, arXiv:2103.13212 [cs.NI].

Molina-Masegosa, R. and Gozalvez, J. (2017). System level evaluation of LTE-V2V mode 4 communications and its distributed scheduling, in *IEEE VTC-Fall* (IEEE).

National Institute of Standards and Technology (2022). NIST announces first four quantum-resistant cryptographic algorithms, https://www.nist.gov/news-events/news/2022/07/nist-announces-first-four-quantum-resistant-cryptographic-algorithms.

ns-3 Project (2025). ns-3: A discrete-event network simulator for internet systems, https://www.nsnam.org/, accessed: May 29, 2025.

OMNeT++ Community (2025). OMNeT++ discrete event simulator, https://omnetpp.org/, accessed: May 29, 2025.

PTV Group (2025). PTV Vissim – Multimodal traffic simulation software, https://www.ptvgroup.com/en-us/products/ptv-vissim, accessed: May 29, 2025.

Risma Carletti, C. M., Casetti, C., Härri, J., and Risso, F. (2024). MS-VAN3T-CARLA: An open-source co-simulation framework for cooperative perception evaluation, in *2024 19th Wireless On-Demand Network Systems and Services Conference (WONS)*, pp. 93–96, doi:10.23919/WONS60642.2024.10449533.

Roccotelli, M., Volpe, G., Fanti, M. P., and Mangini, A. M. (2024). A co-simulation framework for autonomous mobility in urban mixed traffic context, in *2024 IEEE 20th International Conference on Automation Science and Engineering (CASE)* (IEEE), pp. 812–817.

Rong, G., Shin, B. H., Tabatabaee, H., Lu, Q., Lemke, S., Možeiko, M., Boise, E., Uhm, G., Gerow, M., Mehta, S., *et al.* (2020). LGSVL simulator: A high fidelity simulator for autonomous driving, *arXiv preprint arXiv:2005.03778*.

SAE International (2016). Dedicated short range communications message set dictionary, Standard SAE J2735, SAE International, Warrendale, PA, https://www.sae.org.

SAE International (2020). On-board system requirements for LTE-V2X V2V safety communications, Standard SAE J3161/1, SAE International, Warrendale, PA, https://www.sae.org.

SAE International (2024). LTE vehicle-to-everything deployment profiles and radio parameters, Standard SAE J3161, SAE International, Warrendale, PA, https://www.sae.org.

Schrab, K., Neubauer, M., Protzmann, R., Radusch, I., Manganiaris, S., Lytrivis, P., and Amditis, A. J. (2022). Modeling an its management solution for mixed highway traffic with eclipse mosaic, *IEEE Transactions on Intelligent Transportation Systems* **24**, 6, pp. 6575–6585.

Segata, M., Cigno, R. L., Hardes, T., Heinovski, J., Schettler, M., Bloessl, B., Sommer, C., and Dressler, F. (2022). Multi-technology cooperative driving: An analysis based on PLEXE, *IEEE Transactions on Mobile Computing* **22**, 8, pp. 4792–4806.

Shah, S., Dey, D., Lovett, C., and Kapoor, A. (2017). AirSim: High-fidelity visual and physical simulation for autonomous vehicles, in *Field and Service Robotics*, arXiv:1705.05065, https://arxiv.org/abs/1705.05065.

Sommer, C., German, R., and Dressler, F. (2011). Bidirectionally coupled network and road traffic simulation for improved IVC analysis, *IEEE Transactions on Mobile Computing* **10**, 1, pp. 3–15, doi:10.1109/TMC.2010.133.

Sunuwar, D. and Kim, S. (2024). Cross-layer performance evaluation of C-V2X, in *IEEE SoutheastCon* (IEEE).

Veins Development Team (2025). Veins: The open source vehicular network simulation framework, https://veins.car2x.org/, accessed: May 29, 2025.

Wegener, A., Piórkowski, M., Raya, M., Hellbrück, H., Fischer, S., and Hubaux, J.-P. (2008). TraCI: An interface for coupling road traffic and network simulators, in *Proceedings of the 11th Communications and Networking Simulation Symposium*, pp. 155–163.

Xu, R., Guo, Y., Han, X., Xia, X., Xiang, H., and Ma, J. (2021). OpenCDA: An open cooperative driving automation framework integrated with co-simulation, in *2021 IEEE International Intelligent Transportation Systems Conference (ITSC)* (IEEE), pp. 1155–1162.

https://doi.org/10.1142/9789819831548_0007

Chapter 7

FLUTE: A Federated Q-Learning Zero-Trust Platform for Secure and Scalable Mobility Data Endorsement in Connected and Autonomous Vehicles

Junaid Khan*, **Fan Zuo**†, **and Kaan Ozbay**†

**Western Washington University, Bellingham, Washington, USA*
†New York University, New York, USA

Abstract

Connected and Autonomous Vehicles (CAVs) generate vast streams of mobility data critical to intelligent transportation safety applications, yet trusting these data sources remains an open challenge. This chapter presents FLUTE (Federated Learning for Urban Trajectory Endorsement), a zero-trust blockchain platform that combines federated Q-learning with a permissioned consensus mechanism to enable scalable, privacy-preserving mobility data endorsement. Through a Proof of Mobility (PoM) primitive, nearby nodes cryptographically validate each other's trajectories without exposing raw location data. A novel reputation metric, the Trajectory Trustworthiness Score (T^2-Score), dynamically weights each node's influence based on endorsement history, incentivizing honest participation and penalizing malicious behavior. Simulations on realistic New York City traffic scenarios with up to 3,200 nodes demonstrate that FLUTE effectively balances endorsement coverage against communication overhead and latency, with adaptive Q-learning continuously tuning consensus parameters to maintain efficiency across varying network conditions.

Keywords: Connected and autonomous vehicles, Federated learning, Q-learning, Blockchain, Zero-trust architecture, Trajectory endorsement, Reputation systems, Privacy preservation, Intelligent transportation systems.

7.1 Introduction

There is a tremendous increase in devices with location information generated typically as GPS trajectories, including Connected and Autonomous Vehicles (CAVs), pedestrians with smartphones/watches, e-bikes, e-scooters. Researchers can now employ state of the art machine learning and AI algorithms in different data driven safety-related location-based Intelligent Transportation System (ITS) applications towards improving the overall safety for the users of these devices and to avoid road injuries and fatalities. One such application is cooperative perception (Luo *et al.*, 2023) using the 5.9 GHz C-V2X communication band where lightweight trajectories are exchanged in real-time between nearby CAVs to enhance each other's collective perception in avoiding other road users and obstacles, by timely communicating safety-related information to each other as well as Vulnerable Road Users (VRUs). However, there is a challenge associated with trusting data generated by the devices due to the privacy and security concerns associated with both the data source and the intended users leveraging it for different safety-related mission critical applications in ITS such as real-time cooperative perception using C-V2X communication (Sullivan *et al.*, 2024).

The two key problems to address are: (i) trusting the data source to potentially avoid generating fake or malicious trajectories that are harmful to usage in different safety-related applications, where inaccurate or manipulated data could lead to deadly consequences such as collision warnings based on false vehicle positions or fraudulent emergency vehicle alerts causing dangerous evasive maneuvers, thus requiring a trust model to rely on the generated mobility data and (ii) the data could be misused by the intended user where using sophisticated machine learning and AI algorithms can result in potential privacy leakage, inferring sensitive location information revealing the source users' Points of Interest (PoI), such as places of work, worship, etc., leading to easily infer their political, religious, and social affiliations.

Numerous studies have adopted Blockchain technology to address the issue and proposes to validate and store mobility data endorsement information on a distributed ledger (Khan and Ozbay, 2023). In Blockchain, a smart contract can be defined where nodes in the blockchain network first validate the trajectories generated by the source nodes using a zero-trust based consensus process and store this information on the ledger as an immutable transaction. Additionally, smart contracts can be defined for data access enabling the storage of information regarding nodes accessing data on the distributed ledger, thereby keeping track of the intended data usage to preserve data integrity and privacy.

The successful and salable deployment of Blockchain based zero trust models for mobility data endorsement faces some challenges; first, the high number of messages exchanged between peers in different communication rounds to reach a consensus consumes substantial resources at the node. Second, the constant mobility of nodes makes is difficult to maintain wireless connectivity enabling multiple communication rounds for the consensus process, making it inefficient and costly for network providers and users employs C-V2X communication. The third challenge is associated with scalability, to decide how frequently should the distributed consensus process and data storage be performed, how many nodes can participate and for how long the process should last.

We address the above challenges and propose **FLUTE: F**ederated **L**earning for **U**rban **T**rajectory **E**ndorsement, leveraging reinforcement learning, specifically Q-learning in a federated approach for trajectory endorsements using blockchain to preserve user privacy. Individual nodes learn and adapt the frequency at which the consensus process and storage are performed, along with considering other configuration decisions such as the number of participating nodes and timeliness requirements for the data generation and storage.

The decisions are based on the number and the familiarity and freshness of updates from nodes surrounding each other, thereby balancing the trade-off between increasing trajectory endorsements and reducing resource consumption at the participating nodes.

Additionally, we employ a novel reputation scoring mechanism, namely, **Trajectory trustworthiness Score (T^2-Score)** for nodes to establish trust over time, where a node actively participating in a consensus process and having its mobility data endorsed frequently by a majority (>51%) of neighbors increases its T^2-Score over time. On the other hand, for a node participating in a consensus process and having its mobility data not successfully endorsed by majority neighbors decreases its score, while nodes not participating in the consensus process have no change to their T^2-Score. The proposed Q-learning model along T^2-Score is evaluated using mobility data from New York City intersections and has been shown to achieve lower overhead. The framework of the implementation is depicted in Fig. 7.1.

FLUTE integrates three core operational loops: (i) a consensus loop where nodes cryptographically validate trajectories through Proof of Mobility, determining whether predicted trajectories match observed movements within threshold λ; (ii) a federated learning

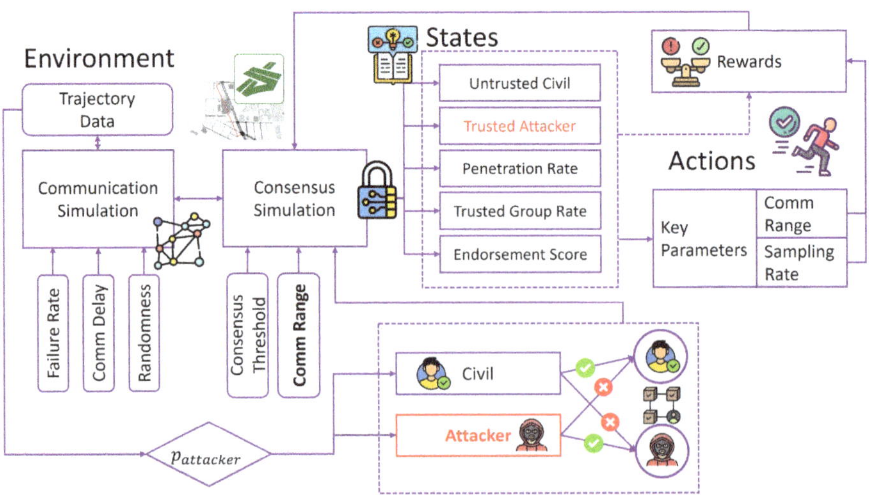

Fig. 7.1. The framework of this study.

loop where nodes train local LSTM models on trajectory data, share only model updates via FedAvg aggregation on the blockchain, and collectively improve prediction accuracy without exposing raw location data; (iii) a reputation loop where the T^2-Score dynamically weights each node's influence in consensus based on successful endorsement history, incentivizing honest participation. Data flows from local trajectory observations through federated prediction models to collaborative consensus validation, with only endorsement outcomes and aggregated model parameters recorded on the permissioned blockchain—preserving privacy while establishing cryptographic proof of mobility trustworthiness.

The contributions of this chapter are summarized as follows:

- We present FLUTE, a privacy-aware federated Q-learning-based adaptive zero-trust consensus process for mobility data endorsement in blockchain based on the number of participating nodes in each other's vicinity.
- We present T^2-Score, a reputation score for nodes endorsing each other's mobility data where actively endorsing a node's trajectory increases its score and vice-versa, incentivizing maximum nodes participation.
- We empirically evaluate the proposed zero-trust platform and reputation score on mobility data from New York city, with realistic trajectories from up to 3,200 nodes.

The remainder of the chapter is organized as follows. Section 7.2 discusses the background and related work. Section 7.3 describes the adaptive Q-learning model, FLUTE, followed by a discussion on the reputation score (T^2-Score). Performance evaluation and results are discussed in Section 7.4, and finally, conclusions and insights in Section 7.5.

7.2 Related Work: Secure and Trustworthy C-ITS via Blockchain, Federated Learning, and Zero-Trust

Recent work at the intersection of blockchain and federated learning (FL) aims to make cooperative perception, control, and data analytics in connected and automated transportation both privacy-preserving and verifiable. Permissioned/consortium-ledgers have been used to enforce provenance and auditability over mobility data sharing and vehicular messages, while FL prevents raw data exfiltration by keeping training local and aggregating only model updates (BelMannoubi *et al.*, 2023; Chai *et al.*, 2020; Fan *et al.*, 2023; Fernandes *et al.*, 2023; Lopez and Farooq, 2020; Sehar *et al.*, 2023; Wang *et al.*, 2023). Within this landscape, FLUTE advances the state of the art by tightly coupling a zero-trust architecture with *federated Q-learning* and a permissioned blockchain: (i) a *Proof of Mobility* (PoM) primitive binds model contributions to cryptographically verifiable trajectories; (ii) a T^2-Score fuses trajectory trust with contribution quality to gate aggregation and consensus; (iii) permissioned-chain governance logs, audits, and adjudicates updates and reputations. Rather than treating blockchain and FL as loosely connected components, FLUTE integrates trajectory-grounded attestation directly into the learning/control loop, aligning consensus weight with physical-world evidence and utility. For a system-level view of CAV/C-ITS threats and defenses that contextualizes these design choices, see Abdo *et al.* (2023).

Federated Learning for Privacy and Misbehavior Resilience: FL has been adopted in vehicular networks to respect data locality while enabling collaborative modeling across vehicles, RSUs, and edge/cloud. Asynchronous and hierarchical variants mitigate stragglers and topology churn common in IoV, improving efficiency and convergence under intermittent connectivity (Chai *et al.*, 2020; Lu *et al.*, 2020). Robust aggregation and decentralized/BFT designs address poisoning, free-riding, and Byzantine clients typical of adversarial road environments (Chen *et al.*, 2021). Privacy layers such as local differential privacy can further reduce leakage from gradients while maintaining task utility (Batool *et al.*, 2024). FLUTE complements these lines by *making trust endogenous to learning*: the trajectory-aware T^2-Score adjusts how (and whether) local Q-updates influence the global policy, thereby coupling robustness to misbehavior with the physical validity of each client's mobility claims.

Proof-of-Location/Mobility and Trajectory Trust: Location/trajectory attestation is critical to resist Sybil, ghost, and relay attacks and to ground claims about observations and actions in verifiable spatiotemporal evidence. Blockchain-supported vehicle location verification systems (e.g., V-Track) demonstrate practical, witness-based proofs for mobile assets (Peelam *et al.*, 2024), while recent decentralized PoL frameworks formalize scalability and privacy trade-offs for time–space witnessing under collusion and sensor spoofing (Brito *et al.*, 2025). Vehicular-tailored consensus and reputation mechanisms on permissioned/consortium chains provide low-latency integrity guarantees for safety-critical messages and model updates (Fan *et al.*, 2023; Fernandes *et al.*, 2023; Sehar *et al.*, 2023). FLUTE builds on these ideas by binding PoM to model update eligibility and weighting: only trajectories that pass cryptographic and consistency checks contribute to aggregation, and their influence is calibrated by the T^2-Score, ensuring that trust in mobility evidence meaningfully shapes learning and coordination outcomes across C-ITS.

7.3 The FLUTE Platform: System Architecture and Mechanisms

7.3.1 *System Model*

We define the blockchain network comprising of a set of nodes $V = \{v_1, v_2, \ldots, v_n\}$ as mobility data generating devices. These include vehicles, micro-mobility devices such as e-bikes and e-scooters, as well as pedestrians with smartphones connected through wireless communication technology such as Dedicated Short-Range Communications (DSCR) or Cellular (C-V2X) enabling nodes to exchange messages for the consensus process and information storage on the blockchain.

The mobility data for a node v is defined as the spatio-temporal trajectory $C_v(l, t) = \langle(l_v^1, t_v^1), (l_v^2, t_v^2), \ldots, (l_v^m, t_v^m)\rangle$ with sequence of locations, $l_v^1, l_v^2, \ldots$, at times $t_v^1, t_v^2, \ldots$, respectively. The node location is further composed of the geographical latitude and longitude $l_v = (\text{lat}_v,\ \text{lon}_v)$ as a simple GPS coordinate. We consider nodes sharing a common clock with the time as $T = \{t_1, t_2, \ldots\}$ representing time instants t_1, t_2 and so on, while time can be divided into regular slots $\Delta t_1 = t_2 - t_1, \Delta t_2 = t_3 - t_2$, and so on, defined at different granularity depending on the application.

7.3.1.1 *Notation Table*

Table 7.1 provides a comprehensive summary of the mathematical notation used throughout this chapter.

Parameter selection is discussed as follows: λ is chosen based on GPS accuracy (± 5 m typical) and application safety margins; α balances reputation history versus recent behavior (experiments use $\alpha = 0.7$); β and γ follow standard Q-learning practices ($\beta = 0.1$,

Table 7.1. Mathematical notation and parameter definitions.

Symbol	Description	Units	Range/Type
V	Set of nodes in blockchain network	—	$\{v_1, v_2, \ldots, v_n\}$
v	Individual node/device	—	$v \in V$
$C_v(l, t)$	Spatio-temporal trajectory of node v	—	Sequence
l_v	Location (latitude, longitude)	degrees	GPS coordinates
t	Time instant	seconds	$t \in T$
Δt	Time slot duration	seconds	>0
λ	Acceptable deviation margin	meters	$[0, \infty)$
$\Omega_v(t)$	T^2-Score (trajectory trustworthiness)	—	$[0, 1]$
α	Reputation update weight parameter	—	$[0, 1]$
β	Q-learning learning rate	—	$(0, 1]$
γ	Discount factor	—	$[0, 1)$
ϵ	Exploration rate (epsilon-greedy)	—	$[0, 1]$
θ_v	Local model parameters at node v	—	$\mathbb{R}^d$
θ_{global}	Global aggregated model parameters	—	$\mathbb{R}^d$
η	Learning rate for model training	—	$(0, 1]$
$\xi_v(t)$	Number of validated nodes at time t	count	$\mathbb{N}$
Γ	Number of participating nodes	count	$[\Gamma^{\text{th}}_{\text{min}}, \Gamma^{\text{th}}_{\text{max}}]$
$f_r^v(l, t)$	Resource consumption function	bytes	$[f^{\text{th}}_{r_{\text{min}}}, f^{\text{th}}_{r_{\text{max}}}]$
$t_d^v(l, t)$	Consensus delay	seconds	$[0, t_d^{\text{th}}]$
$S(l, t)$	State in MDP	—	$(v(l, t), g(l, t))$
$A(l, t)$	Action in MDP	—	$A \in G(l, t)$
$Q(S, A)$	Q-value function	—	$\mathbb{R}$
$r_S^A(l, t)$	Reward function	—	$\mathbb{R}$

$\gamma = 0.9$); ϵ decays from 0.3 to 0.05 for exploration–exploitation balance.

7.3.2 *Threat Model*

The device generating mobility data $C_v(l, t)$ could be compromised, and generate fake or malicious trajectories. We categorize potential adversaries into three classes: (i) **Compromised nodes** that deliberately report false trajectories to mislead other vehicles or disrupt consensus; (ii) **Sybil attackers** attempting to create multiple fake identities to manipulate voting outcomes; (iii) **Colluding nodes** that coordinate to endorse each other's false trajectories.

Therefore, we define a blockchain-based consensus for nearby nodes to verify and endorse each other's trajectories at regular time intervals. We do not require nodes to constantly share raw trajectories to preserve privacy and reduce frequent data exchanges between nodes. Instead, nearby nodes leverage machine learning to visually observe all neighboring nodes at time slot Δt_1 and predict their respective trajectories for the next time slot Δt_2. If the difference between the observed and predicted trajectories is within an acceptable margin of deviation λ, defined by the application, the nodes would endorse the corresponding node trajectories for the time slot Δt_2. The blockchain architecture we consider is permissioned, enabling only nodes authenticated through public/private key encryption to participate in the consensus process.

Mitigation Strategies: Our Proof of Mobility (PoM) mechanism and permissioned blockchain architecture provide several safeguards:

- *Against compromised nodes*: PoM requires physical presence verification through multiple independent witnesses. A node reporting false positions will fail validation by honest neighbors who observe actual movements.
- *Against Sybil attacks*: The permissioned blockchain requires cryptographic identity verification and limits one identity per physical device, preventing mass creation of fake identities.
- *Against collusion*: The T^2-Score mechanism penalizes nodes whose endorsements consistently deviate from the majority. Colluding

nodes endorsing invalid trajectories will see their reputation decline, reducing their influence in future consensus rounds.

Residual Threats and Detection: While PoM and permissioned identities significantly reduce attack surfaces, certain threats remain:

Relay attacks where an adversary physically relays messages between distant locations remain theoretically possible but are detectable through timing analysis and multi-hop verification. **Limited collusion** among small groups (<20% of local network) may succeed temporarily but will be detected over time as their endorsements diverge from honest majority patterns.

During consensus, we distinguish between **detectable failures** (nodes whose predicted trajectories deviate beyond λ are immediately rejected) and **tolerable failures** (occasional packet loss or temporary communication failures up to 5% are handled through retransmission). Byzantine behavior exceeding 33% of local participants will cause consensus to abort, triggering a flag for investigation while preventing corrupted block creation.

7.3.3 *"Proof of Mobility" Consensus*

The proof of mobility consensus process is defined as follows. Nodes scan the wireless communication channel for a potential consensus initiator node $v \in V$ as leader for the time slot Δt_2. In case no initiator node is available, the node with the highest T^2-Score $\Omega_v(t)$ (to be defined later) v_1 initiates the consensus process after a back-off wait timer inversely proportional to its T^2-Score, $t_b = e^{-\Omega t}$.

It broadcasts this information to all nodes in its wireless communication range announcing itself as the consensus initiator for the time slot Δt_2, shares the threshold for deviation λ, and the deadline time instants prior to which all nodes need to respond with their respective validation information to complete the consensus round to validate data for time slot Δt_2. Nodes receiving this message observe other nodes in their vicinity during time slot Δt_1 and predict their trajectories for the time slot Δt_2 before the deadline t_2. Nodes then observe each other for the duration of time slot Δt_2 to verify the truthfulness of the trajectories. For a neighboring node v_j, if the difference between the predicted and observed trajectories for time slot Δt_2 are within the margin of acceptable deviation λ, the node

v_j trajectory for the time slot Δt_2 is validated and this validation information is shared with the initiator node v by the deadline t_3, which then subsequently creates a mobility block on the distributed ledger. The mobility block records the list of node IDs participating in the consensus process and has their trajectories validated instead of raw trajectories to preserve privacy and resource consumption at the nodes.

7.3.3.1 *Consensus Toy Example*

Figure 7.2 depicts a toy example of the consensus process where the numbers (1–4) labelling the arrows correspond to: (1) the initiator node v_1 or leader announcing the process, (2) nearby nodes v_2 and v_3 predicting each others trajectories including v_1's, prior to time instant t_2. Nodes observe each other during the time slot Δt_2 and then, (3) share validation results followed by, (4) the initiator node v_1 creating the mobility block on IPFS (Benet, 2014). Here, IPFS is the distributed ledger platform used for block storage, storing node IDs only to ensure privacy preservation.

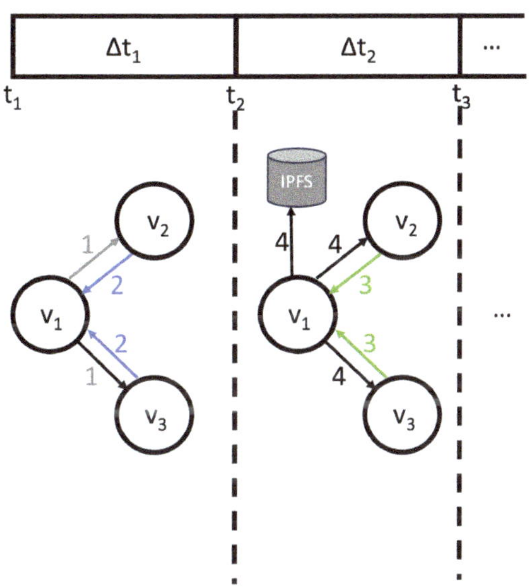

Fig. 7.2. Proof of mobility consensus.

7.3.4 *Federated Q-Learning for Adaptive Consensus*

The consensus is formulated as a Markov Decision Process performed at regular intervals. At each round, starting at time t_1, the initiator node decides what configuration to use for the consensus, this includes deciding the time during which it would perform the validation, Δt_2, the deadlines t_2 and t_3 for the nearby nodes to send their prediction and validation results, respectively, along deciding the frequency at which it creates the block on distributed ledger. Additionally, the configuration defines the threshold on the number of nodes (i.e., majority) it needs towards successful endorsement of a node's trajectory with the objective to maximize the number of nodes for which mobility data is endorsed over time. The optimization model therefore, is defined as

$$\text{maximize} \quad \sum_{V}\sum_{T}\frac{\xi_v(l,t)}{\Delta t(l,t)} \tag{7.1a}$$

subject to

$$f_{r_{\min}}^{\text{th}} \leq f_r^v(l,t) \leq f_{r_{\max}}^{\text{th}} \quad \text{for } v \in V, \tag{7.1b}$$

$$\Gamma_{\min}^{\text{th}} \leq \Gamma^v(l,t) \leq \Gamma_{\max}^{\text{th}} \quad \text{for } v \in V, \tag{7.1c}$$

$$t_d^v(l,t) \leq t_d^{\text{th}} \quad \text{for } v \in V \tag{7.1d}$$

The objective function here targets to maximize the number of nodes in the blockchain network for which the mobility data is validated, over time. Ideally, the aim is to have mobility data for each node validated for a maximum amount of time duration. Here, ξ_v represents nodes for which data is validated, and Δt represents the time it takes for one consensus round. It is defined by the time the initiator begins the process till the mobility block is created on the distributed ledger. The first constraint here refers to the individual resources at the nodes, the function $f_r^v(l,t)$, ensures the node has the minimum resources available to complete the process, represented as $f_{r_{\min}^{\text{th}}}$. Similarly, it ensures resource consumption does not exceed a maximum threshold value specified by $f_{r_{\max}^{\text{th}}}$. The second constraint defines the threshold to limit the number of nodes participating in the consensus process, where, $\Gamma_{\max}^{\text{th}}$ indicates the maximum and $\Gamma_{\min}^{\text{th}}$ represents the minimum number of nearby nodes allowed to participate in the consensus process. This constraint could be adapted in real-time by the initiator node with varying requirements on the number of nodes it needs to complete the consensus round, based on the delay and other application requirements. Lastly, $t_d^v(l,t)$ represents the time consumed by consensus process where this constraint ensures t_d^{th}, the threshold delay requirements are met.

The Q-learning parameters are defined as follows:

State: The state is defined by the tuple, $S(l,t) = \{v(l,t), g(l,t)\}$ with the initiator node $v(l,t)$ and the consensus configuration function $g(l,t)$ to decide the time requirements such as the deadlines t_2 and t_3 for the prediction and validation, as well as adapting the duration of time slots $\Delta t_2, \Delta t_3$ and so on, depending on the number of nodes in the neighborhood along other traffic conditions (average speed, density of vehicles, time of day, etc.) around the initiator node. Additionally, it configures λ, the acceptable margin of deviation of the predicted trajectory to the observed as well as t_d^{th}, the delay requirements for the consensus process at a given round.

Action: The action set is defined for the node responsible for the consensus round, where it learns the configuration function to be used for the next round based on the feedback from the current consensus round. The policy π that makes decision for a node v at a state $S(l,t)$, is a probability distribution on the action set $A(S(l,t)) \in G(l,t)$, representing the node v action to select a configuration function for the thresholds and deadlines along the number of nearby nodes to consider for the mobility data endorsement in state $S(l,t)$. The function $G(l,t)$ represents the configuration for the time delays, deadlines, and number of nodes to consider for the consensus while the initiator node is in state $S(l,t)$. The node's action enables it to learn the next state $S(l,t+1) = S'(l,t+1) = \{v(l,t+1), g(l,t+1)\} = A(S(l,t))$ for the configuration on deadlines and mobility block creation on the distributed ledger. Additionally, the function considers accepting or rejecting data from nearby nodes, where potential discrepancies in information from a given node could result in the action of not endorsing that particular node.

Reward: The reward of the action $A(S(l,t))$ is defined as

$$r_S^A(l,t) = f(S(l,t), A(l,t), \Omega_v(t), \forall v \in V)$$

It represents the node utility for time $t+1$ and is proportional to the T^2-Score of the node performing the consensus process when the current state is $S(l,t)$ and the selected action for the next time instant $t+1$ is A. Thus, an increase in T^2-Score positively contributes to an individual node reward.

The consensus initiator node v broadcasts the new configuration values for the threshold and deadlines for the next state/round $S(l,t+1)$ to the nodes participating in the consensus, and therefore, enabling each peer to use similar values in order to reduce its message exchanges and resource consumption. This privacy-by-design approach does not require nodes to share raw trajectories and therefore, all nodes learn their respective action for the new state in a distributed manner, keep their trajectories local to themselves, thereby avoid compromising their privacy.

Initializing with the state $S(0,0) = s_0$, the long term cumulative reward function for the node v can be represented as

$$R_\pi(0,0) = \sum_l \sum_t \gamma(l,t) r_S^{A_\pi}(l,t),$$

where γ is the discount factor. The objective function here for the consensus initiator node to select the deadlines and thresholds aiming to find a policy π that maximizes the expected cumulative reward, where the expected value of the cumulative reward R_π at the state S is represented by the state value function:

$$U_\pi(S) = E\left[\sum_l \sum_t \gamma(l,t) r_S^{A_\pi} | S(0,0) = S\right].$$

Thus, the optimal state-value function becomes

$$U^*(S_0) = \max_\pi U_\pi(S_0).$$

7.3.5 *Federated Learning Model*

To avoid individual node data sharing, we employ federated learning for the verification of trajectories enabling each node to locally run lightweight models to predict corresponding future trajectories of nearby nodes (i.e., at times $t_1, t_2 \cdots > t_0$) and endorse the node as trustworthy if the predicted trajectories match the shared one within an acceptable margin or error.

The local prediction models at a node v_i train to minimize the loss function as

$$\theta_{v_i} \leftarrow \theta_{v_i} - \eta \nabla_{\theta_{v_i}} L(\theta_{v_i}, D_{v_i})$$

where θ_{v_i} represents local model parameters for node v_i, η is the learning rate and $\nabla_{\theta_{v_i}} L(\theta_{v_i}, D_{v_i})$ represent the gradient of the loss function L for node v_i, computed using its local dataset D_{v_i}.

The global model aggregation (FedAvg) of the locally trained models from different nodes is regularly updated on the blockchain as

$$\theta_{\text{global}} \leftarrow \frac{\sum_{v_i}(n_{v_i}\theta_{v_i})}{\sum_{v_i} n_{v_i}}$$

where θ_{global} represents the global model parameters and n_{v_i} are the number of samples from node v_i.

FLUTE enables nodes to keep training data local and only share model updates or gradients on the distributed ledger for overall model improvements. FLUTE is not limited to a specific model and therefore, any model can be employed, however we suggest using lightweight models such as LSTM compared to more complex transformer based models for real-time predictions towards trajectory validation via consensus. The preference for LSTM over transformer-based models stems from three key factors: (i) lower computational overhead suitable for resource-constrained vehicular edge devices, (ii) reduced latency in real-time trajectory prediction critical for consensus deadlines, and (iii) smaller model size enabling faster federated aggregation and reduced communication overhead during model updates on the blockchain.

Algorithm 1 Preamble: The following algorithm describes how each participating node updates its local trajectory prediction model. It takes as input the node's local trajectory dataset and current model parameters, performs stochastic gradient descent over multiple epochs, and outputs updated model parameters that better predict neighboring vehicles' movements.

Algorithm 1 Local Model Update.

Require: Local dataset D_{v_i}, learning rate η, initial local model parameters θ_{v_i}

Ensure: Updated local model parameters θ_{v_i}

1: **for all** epoch e **do**
2: Sample a mini-batch of data from D_{v_i}
3: Compute the gradient of the loss function: $\nabla_{\theta_{v_i}} L(\theta_{v_i}, D_{v_i})$
4: Update the local model parameters: $\theta_{v_i} \leftarrow \theta_{v_i} - \eta \nabla_{\theta_{v_i}} L(\theta_{v_i}, D_{v_i})$
5: **end for**
6: **return** θ_{v_i}

Algorithm 2 Preamble: The following algorithm specifies the global model aggregation process performed by the consensus leader. It takes as input the locally trained model parameters from all participating nodes along with their respective dataset sizes, computes a weighted average (FedAvg) where weights are proportional to local data volume, and outputs the aggregated global model that is broadcast to all nodes and stored on the blockchain.

Algorithm 2 Global Model Aggregation (FedAvg).

Require: Local model parameters θ_{v_i} from all nodes $v_i \in V$, number of samples n_{v_i} from each node

Ensure: Updated global model parameters θ_{global}

1: Aggregate local model parameters: $\theta_{\text{global}} \leftarrow \frac{\sum_{v_i} (n_{v_i} \theta_{v_i})}{\sum_{v_i} n_{v_i}}$
2: Broadcast θ_{global} to all nodes $v_i \in V$
3: **return** θ_{global}

7.4 T^2-Score Computation

The Trajectory trustworthiness score as a reputation of the node is defined as follows:

$$\Omega_v(t+1) = (1-\alpha)\left(\frac{\xi_v(t+1)}{\Delta t(t+1)}\right) + \alpha \Omega_v(t)$$

For a node, the Trajectory trustworthiness score is proportional to ξ_v, the number of nodes for which data is validated over the time Δt. The $(t+1)$ in ξ_v and Δt represents the score computation for the next time slot, based on the current and past score.

The tuning parameter α decides the balance between freshness and staleness in score updates, where higher values of α would emphasize more on past scores.

Algorithm 3 Preamble: The following algorithm integrates Q-learning with the T^2-Score reputation mechanism to enable adaptive consensus configuration. Each node iteratively observes the current network state (including neighbor count and traffic conditions), selects actions using an ϵ-greedy policy to balance exploration and exploitation, receives rewards based on successful endorsements and updated T^2-Score, and updates its Q-table to learn optimal consensus parameters over time.

Algorithm 3 T^2-Score with Q-learning.

INPUT: $S(l,t), A(l,t)$ for $v \in V$,
OUTPUT: $U(S(l,t))$ for $v \in V$,
for each node $v \in V$ **do**
  repeat
    Initial State $S_0 = (v_0, g_0)$, $Q = 0$
    for $i = t_1, t_2, \ldots T$ **do**
      Scan for existing initiator node for next time slot
      if No initiator node detected **then**
        Broadcast initiator announcement after t_b
        Select action A based on ϵ-greedy policy
      end if
      Observe S, predict $\hat{r}$
      Send Threshold λ and deadline and other configuration
      $Q(S,A) = Q(S,A) + \beta[\hat{r} + \gamma \max_{A'} Q(S',A') - Q(S,A)]$
      $S = S'$
    end for
  until $Q(S,A)$ converge
  $g(l,t) = \arg\max_A Q(g(l,t-1),A)$
end for

Algorithm 3 shows the reputation mechanism with T^2-Score. Federated Q-learning is employed by nodes participating in the consensus to maximize their cumulative rewards, where an initiator node observes the feedback in terms of the number of nodes for which it successfully completed the consensus in previous rounds. It decides the threshold times and deadlines, along with other configurations based on the number of nodes in its vicinity. Nodes can learn long-term cumulative reward of the current times, thresholds, and other configurations by trial and error interactions with the constantly evolving topology (connectivity) of the mobile nodes in the surrounding. The algorithm employs Q-learning, where a node v takes action A at the current state S of the thresholds for the consensus process and mobility block creation to move to the next state S', by receiving an immediate reward $\hat{r}$ for updating the Q-table. The term $Q(S, A)$ represents the expected long-term reward for the node v for taking action A at state S. It selects the action for the state S based on an ϵ-greedy policy as

$$\pi(A|S) = \begin{cases} 1 - \epsilon + \frac{\epsilon}{|a(S)|}, & \text{if } A = \arg\max_A Q(S, A), \\ \frac{\epsilon}{|a(S)|}, & \text{otherwise,} \end{cases}$$

where $|a(S)|$ counts the number of alternative actions for the node at state S. The ϵ-greedy policy here serves as a balance between exploration and exploitation strategies, i.e., individual node takes local action that maximizes the Q-value as well as explore other actions.

7.5 Performance Evaluation and Results

The experiments were designed to evaluate the impact of communication range, attack penetration rate, and voting mechanisms in a CAV environment. The focus was on understanding how different settings influence the trustworthiness of groups of vehicles under various conditions, especially in the presence of malicious attackers. Modules are developed to emulate realistic communication conditions and the consensus process between nodes. These functions, developed in Python, consider communication range, time, delay, packet delivery failure, and randomness (C2SMART, 2024). These attributes are integrated into a calibrated traffic scenario created with the open-source microscopic traffic simulation software, SUMO (Krajzewicz *et al.*, 2012), and its internal extension API, Traffic Control Interface (TraCI) (Wegener *et al.*, 2008).

SUMO possesses the capability to simulate intricate and realistic vehicular driving behavior as well as traffic performance, while TraCI enables online interaction with the primary simulation by connecting external applications to SUMO via sockets. Below, we elucidate the details of the functions and the simulation settings.

7.5.1 *Simulation Setup*

The testing platform is a calibrated microscopic traffic simulation model built using Simulation of Urban MObility (SUMO) (Krajzewicz *et al.*, 2012). The simulation encompasses a 1.6-mile stretch on Flatbush Avenue between Tillary Street and Army Plaza, recognized as one of the USDOT Connected Vehicle Pilot Program test sites located in Brooklyn, New York.

The simulation parameters, such as acceleration rate (ranging from 2.0 to 3.5 m/s^2 for passenger vehicles), minimum gap (mean 2.5 m, standard deviation 0.5 m), reaction time (mean 1.0 s, standard deviation 0.2 s), distribution of different types of vehicles (70% passenger cars, 15% trucks, 10% buses, 5% e-bikes), and spatial and temporal features of local traffic flow, are regarded as random variables. They are calibrated as probability distributions founded on real trajectory data extracted from drone and CCTV camera videos collected during 2019–2020, as well as actual local traffic data from NYC DOT databases.

This study concentrates on an area between Fulton Street and Dekalb Avenue (Figure 7.3), characterized by high traffic volume and significant urban driving behavior, to economize computing resources. The simulation has 7200-second (2-hour) intervals and approximately 150 nodes (vehicles) in the network simultaneously at peak periods. Each segment's mobility data for the heterogeneous set of nodes (comprising passenger cars, trucks, buses, and e-bikes) are extracted from the simulation trajectory output and fed into the communication simulation as vital input.

Fig. 7.3. The network of study area in OSM map and SUMO environment.

The communication simulation incorporates parameters that govern the standard transmission of basic safety messages (BSMs) in connected vehicles. The experiments were conducted with three communication ranges: 300 m, 500 m, and 800 m. These ranges were chosen to reflect varying levels of vehicle connectivity, from moderate urban scenarios (300 m) to extended highway conditions (800 m). The communication range determines the distance within which vehicles can communicate with each other, influencing the formation of leader-follower relationships and the propagation of information (or misinformation).

The attack penetration rate (p_{attacker}) varied from 5% to 50% in increments of 5%. This parameter represents the proportion of malicious attackers among the total number of vehicles. By testing a range of penetration rates, we can observe the system's resilience and identify thresholds beyond which the system's trustworthiness significantly degrades. A voting frequency of 10 s was used to simulate the real-time decision-making process in a dynamic environment. This interval balances the need for timely updates with computational and communication overhead.

Multiple random seeds (five different seeds: 42, 123, 456, 789, 1024) were used to ensure the reproducibility and robustness of the experiments. Using multiple random seeds allows us to account for randomness in vehicle behavior and attacker placement, providing a more comprehensive analysis of the system's performance.

7.5.2 *Performance Metrics*

We evaluate FLUTE using the following key metrics:

Overhead (bytes): The total amount of exchanged data during a consensus round. Lower overhead is desirable as it correlates with reduced bandwidth and energy consumption.

Max Delay (s): The worst-case message round-trip time during the consensus process. Lower maximum delays indicate that the system can produce timely endorsements, which is critical for latency-sensitive applications.

To thoroughly analyze performance under uncertainty, we test each parameter combination (e.g., communication range, failure rate, and CV penetration) with multiple random seeds. This robust approach ensures that observed trends are statistically reliable and not artifacts of a particular run. Figure 7.4 presents the process of the PoM consensus.

7.5.3 *Results and Discussion*

This section presents and interprets the results obtained from our simulations, focusing on how communication range, CV penetration (MP), and packet failure rates impact overhead and delay for both leaders and followers.

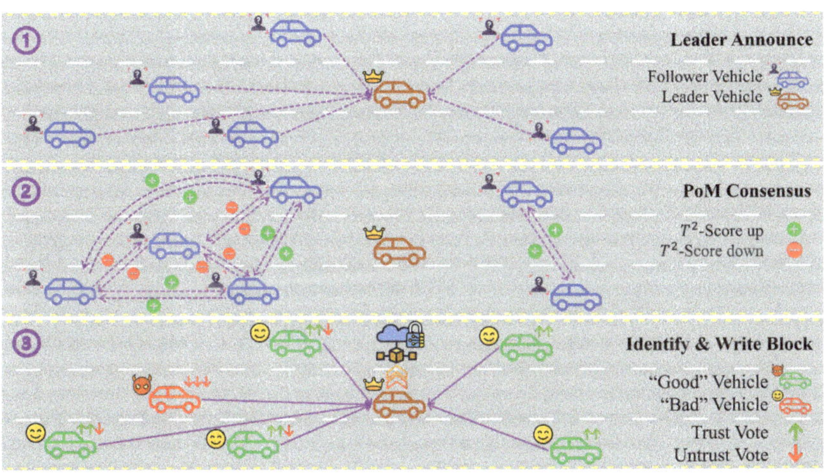

Fig. 7.4. The key step: PoM consensus.

7.5.3.1 *Overhead Analysis*

Figure 7.5 illustrates overhead distributions for different parameter configurations. Leaders and followers are analyzed separately.

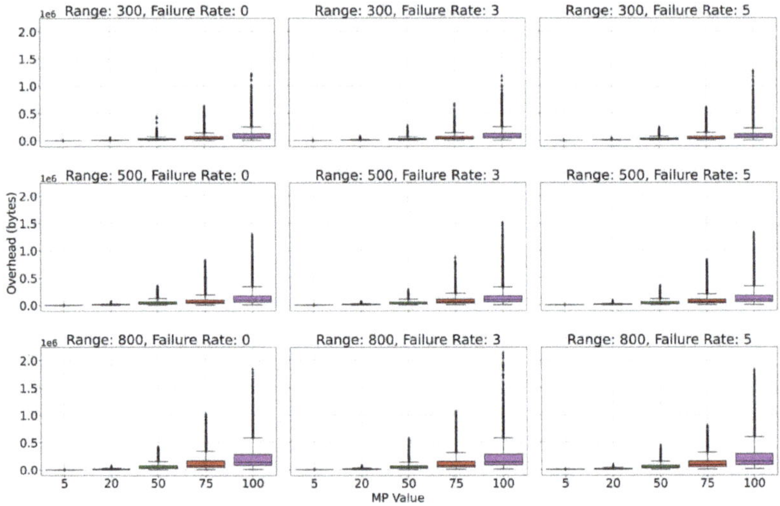

(a)

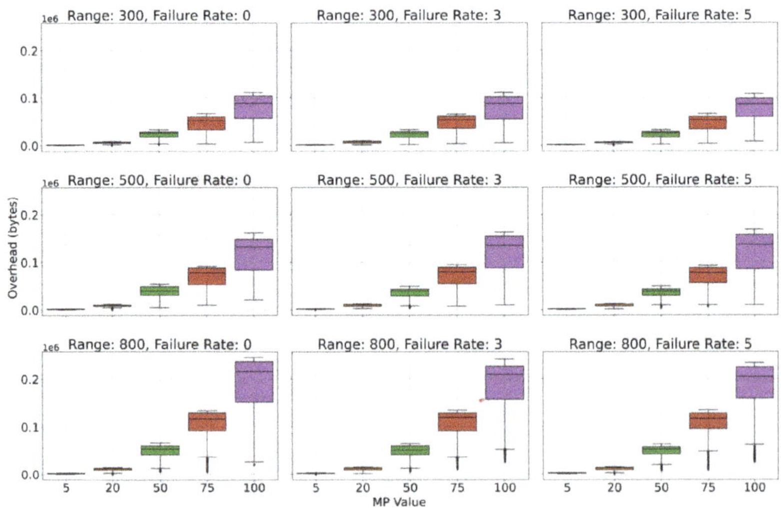

(b)

Fig. 7.5. The boxplots of overhead between leaders and followers over different communication ranges, failure rates, and market penetration rates. (a) Leaders' overhead performance. (b) Followers' overhead performance.

Leaders who initiate the consensus and coordinate multiple nodes exhibit increasing overhead with more extensive communication ranges and higher market penetrations. For instance, leader overhead is modest at a 300 m range and low MP, but at 800 m and high MP (100% CV), overhead can approach or exceed 200,000 bytes. Higher failure rates amplify this effect, as leaders must handle retransmissions and supplementary validation messages.

Followers generally have lower complexity roles, responding to leader queries and verifying fewer nodes. While their overhead remains lower than that of leaders under most conditions, it can still grow with MP and failure rate. This increase is logical: more connected vehicles in range means more endorsement-related data exchanges. However, follower overhead growth is less pronounced and shows less variability than for leaders, indicating that the leader's central coordination responsibility primarily drives overhead escalation.

These overhead results underscore the importance of choosing communication ranges and penetration targets carefully. A modest range (300 m) at mid-level MP (e.g., 50%) can strike a balance, achieving consensus with relatively lower overhead. Our Q-learning approach can exploit such observations to adjust parameters and maintain efficiency dynamically.

7.5.3.2 *Delay Analysis*

Figure 7.6 depicts the maximum observed delay for leaders and followers. Overall, delay increases with communication range and MP, mirroring trends in overhead. Message queues lengthen as more nodes join the consensus round, and packet collisions or retransmissions become more likely. This effect is exacerbated under higher failure rates, where error recovery steps add further latency.

Leaders are again more sensitive to these changes. Their position at the center of the consensus process—collecting endorsements, resolving disagreements, and finalizing blocks—means that any network instability or congestion affects them more heavily. Under challenging conditions (e.g., 800 m range, 100% MP, and 5% failure), leaders can experience delays approaching 0.25–0.30 s. Although this is still relatively low for many ITS applications, it indicates that future optimizations or hierarchical consensus schemes may be necessary as CV penetration and network complexity increase.

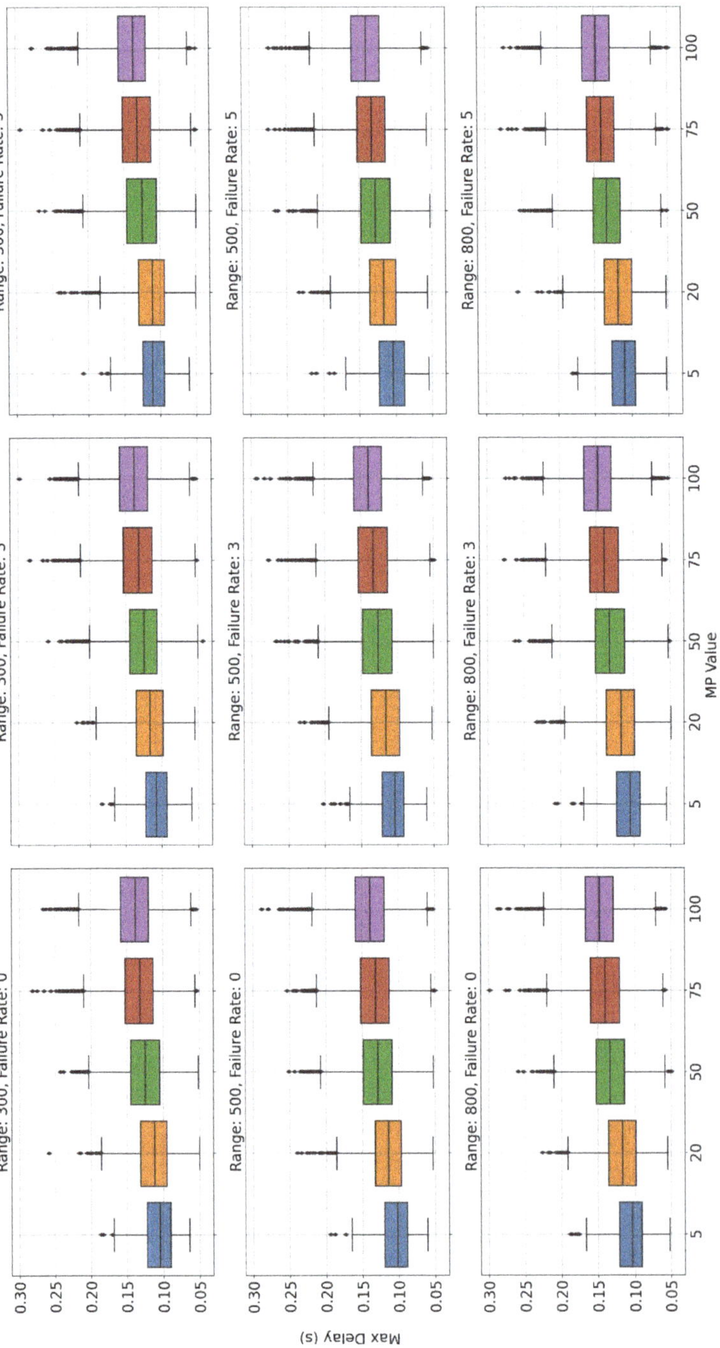

Fig. 7.6. The boxplots of max delays between leaders and followers over different communication ranges, failure rates, and market penetration rates. (a) Leaders' delay performance. (b) Followers' delay performance.

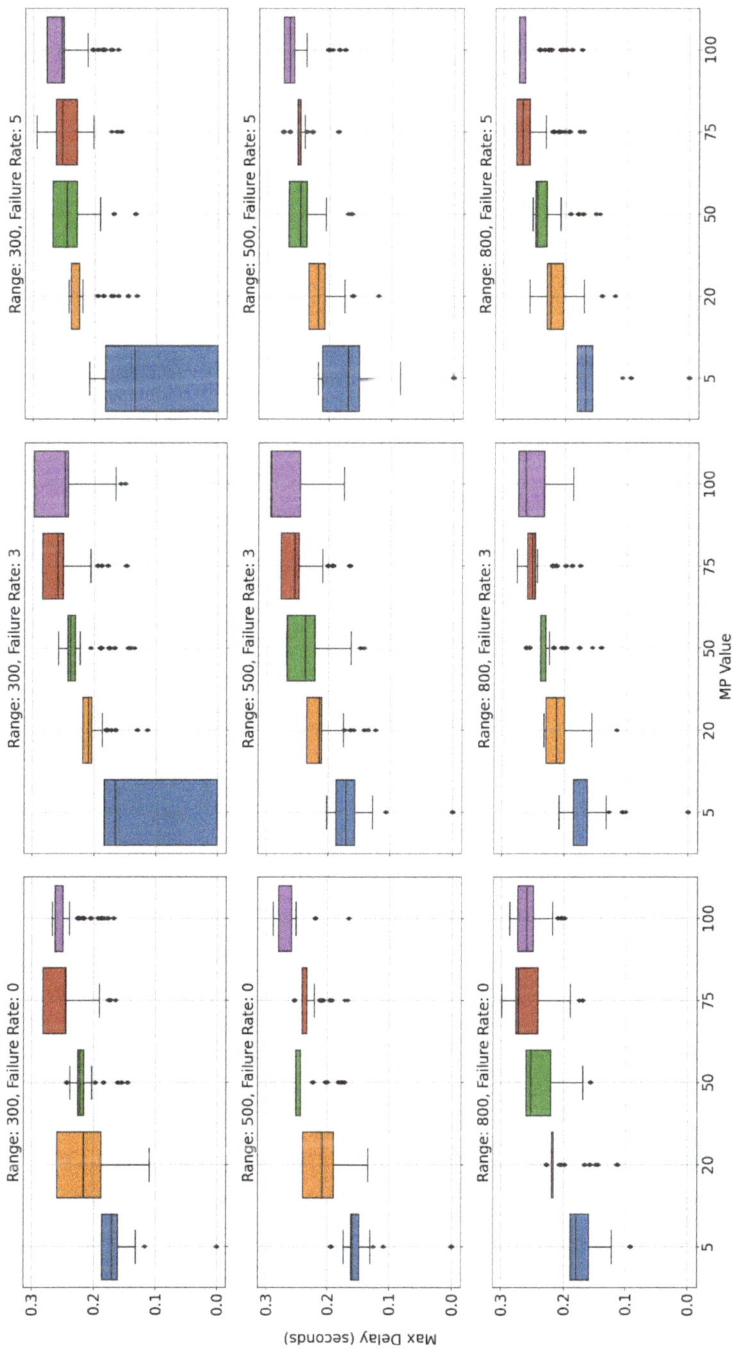

(b)

Fig. 7.6. *(Continued)*

Followers exhibit comparatively stable delay performance. They primarily receive leader instructions and benefit from a reduced decision-making load. Follower delays remain lower and less volatile even at higher MP or failure rates. This suggests that distributing responsibilities or reducing leader node burdens could improve overall responsiveness.

7.5.4 *Insights and Scalability Considerations*

The joint analysis of overhead and delays reveal a multifaceted relationship between network parameters, operating conditions, and system performance. High connected vehicle penetration (MP) rates and large communication ranges enhance the richness of endorsement data and potentially improve the overall trustworthiness of mobility information. However, these same factors inflate overhead for leader nodes and increase network delays, underscoring the challenge of achieving efficient scalability.

A key insight emerging from these results is the importance of adaptive parameter tuning. As network conditions, MP levels, and failure rates fluctuate over time and space, static configurations—such as fixed communication ranges or sampling frequencies—are unlikely to remain optimal. FLUTE's federated Q-learning framework addresses this by continuously adjusting parameters to balance reliability, latency, and resource efficiency. For instance, if the system detects persistent long delays, it can scale down the communication radius, reduce the endorsement frequency, or both. Conversely, if certain regions lack sufficient endorsement opportunities, FLUTE can increase the communication range or sampling frequency.

Hierarchical and clustered consensus architectures may further improve scalability. Instead of relying on a single leader to manage a large set of followers, a multi-tier structure can distribute responsibilities, reducing burdens on individual nodes. Similarly, incorporating contextual information, such as time-of-day effects, special events, or traffic incidents, can help the system anticipate resource bottlenecks and proactively adjust parameters to avoid congestion or excessive delay. This approach extends to accommodating edge devices with varying computational capabilities, as FLUTE can dynamically moderate overhead demands to suit less capable nodes.

Another noteworthy consideration is that integrating advanced privacy-preserving techniques or more robust cryptographic methods may introduce additional computational overhead. Balancing these cryptographic costs against communication overhead and delay calls for a holistic strategy that simultaneously accounts for privacy, security, and performance constraints. The insights gleaned from these experiments highlight FLUTE's potential for scalable and adaptable zero-trust consensus in large, complex urban settings. However, they also indicate that further refinement and research are needed to achieve optimal performance in ever-changing network environments.

7.6 Conclusion and Future Directions

FLUTE, as comprehensively presented in this chapter, marks a significant advancement in establishing trustworthy and scalable mobility data sharing for ITS applications. Understanding trajectory data from pedestrians with smartphones, CAVs, and micro-mobility devices, such as e-scooters and e-bikes, is critical for various safety-related ITS applications. However, trusting such mobility data shared by different sources is challenging. Therefore, we propose a blockchain-based consensus process enabling nearby devices to endorse each other's mobility data in a scalable manner. Nodes predict each other's trajectory and endorse corresponding neighbors' trajectory if the predicted trajectory is closer to the observed one for a given amount of time. To do so, we presented a Federated Q-learning-based endorsement protocol, FLUTE, for nodes to optimize the resource consumption for the distributed consensus in a scalable manner while preserving users' privacy.

We also presented trajectory trustworthiness (T^2-score) as a novel reputation score to incentivize nodes to participate in the endorsement process actively. By leveraging blockchain, federated learning, reinforcement learning, and an adaptive consensus mechanism, FLUTE fosters scalable and privacy-aware trajectory endorsements, encouraging nodes to participate through its trajectory trustworthiness scoring scheme. Evaluations using realistic urban traffic scenarios in New York City demonstrated that while increasing connectivity and participation enhances endorsement opportunities and trust, it simultaneously raises overhead and delay challenges. FLUTE's ability to dynamically tune parameters helps maintain efficiency and responsiveness, suggesting that adaptive strategies can mitigate the complexity of large-scale vehicular networks.

Future work will extend the geographical scope, deploying FLUTE in more extensive and topologically complex areas to assess scalability. Strengthening privacy and security will be crucial, leveraging AI integrated with techniques such as differential privacy, secure multiparty computation, or trusted execution environments potentially integrated to preserve confidentiality without imposing undue computational burdens. Comparative analysis against other frameworks such as fixed consensus intervals, systems without reputation mechanisms, or non-federated approaches will clarify FLUTE's unique advantages and limitations through controlled ablation studies. As mobility ecosystems evolve, FLUTE serves as a foundational step toward secure, trustworthy, and resource-efficient information sharing, supporting the growing demands of intelligent transportation systems and paving the way for safer, more reliable, and privacy-preserving urban mobility.

Bibliography

Abdo, A., Chen, H., Zhao, X., Wu, G., and Feng, Y. (2023). Cybersecurity on connected and automated transportation systems: A survey, *IEEE Transactions on Intelligent Vehicles* **9**, 1, pp. 1382–1401.

Batool, H., Anjum, A., Khan, A., Izzo, S., Mazzocca, C., and Jeon, G. (2024). A secure and privacy preserved infrastructure for VANETs based on federated learning with local differential privacy, *Information Sciences* **652**, p. 119717.

BelMannoubi, S., Touati, H., Hadded, M., Toumi, K., Shagdar, O., and Kamoun, F. (2023). A comprehensive survey on blockchain-based C-ITS applications: Classification, challenges, and open issues, *Vehicular Communications* **43**, p. 100607.

Benet, J. (2014). IPFS-content addressed, versioned, P2P file system, *arXiv preprint arXiv:1407.3561.*

Brito, E., Hadachi, A., Kamm, L., and Norbisrath, U. (2025). Decentralized proof-of-location systems for trust, scalability, and privacy in digital societies, *Scientific Reports* **15**, 1, p. 19808.

Chai, H., Leng, S., Chen, Y., and Zhang, K. (2020). A hierarchical blockchain-enabled federated learning algorithm for knowledge sharing in internet of vehicles, *IEEE Transactions on Intelligent Transportation Systems* **22**, 7, pp. 3975–3986.

Chen, J.-H., Chen, M.-R., Zeng, G.-Q., and Weng, J.-S. (2021). BDFL: A byzantine-fault-tolerance decentralized federated learning method for autonomous vehicle, *IEEE Transactions on Vehicular Technology* **70**, 9, pp. 8639–8652.

C2SMART (2024). T^2-score: A reputation score with an adaptive Q-learning based trust model to avoid malicious mobility data endorsements in blockchain consensus, https://c2smarter.engineering.nyu.edu/wp-content/uploads/2024/01/C8_Poster_TRB2024_T2Score_FanZuo.pdf.

Fan, Q., Xin, Y., Jia, B., Zhang, Y., and Wang, P. (2023). COBATS: A novel consortium blockchain-based trust model for data sharing in vehicular networks, *IEEE Transactions on Intelligent Transportation Systems* **24**, 11, pp. 12255–12271.

Fernandes, C. P., Montez, C., Adriano, D. D., Boukerche, A., and Wangham, M. S. (2023). A blockchain-based reputation system for trusted VANET nodes, *Ad Hoc Networks* **140**, p. 103071.

Khan, J. A. and Ozbay, K. (2023). AFFIRM: Privacy-by-design blockchain for mobility data in web3 using information centric fog networks with collaborative learning, in *2023 International Conference on Computing, Networking and Communications (ICNC)* (IEEE), pp. 456–462.

Krajzewicz, D., Erdmann, J., Behrisch, M., and Bieker, L. (2012). Recent development and applications of SUMO-simulation of urban mobility, *International Journal on Advances in Systems and Measurements* **5**, 3&4, pp. 128–138.

Lopez, D. and Farooq, B. (2020). A multi-layered blockchain framework for smart mobility data-markets, *Transportation Research Part C: Emerging Technologies* **111**, pp. 588–615.

Lu, Y., Huang, X., Zhang, K., Maharjan, S., and Zhang, Y. (2020). Blockchain empowered asynchronous federated learning for secure data sharing in internet of vehicles, *IEEE Transactions on Vehicular Technology* **69**, 4, pp. 4298–4311.

Luo, G., Shao, C., Cheng, N., Zhou, H., Zhang, H., Yuan, Q., and Li, J. (2023). EdgeCooper: Network-aware cooperative LiDAR perception for enhanced vehicular awareness, *IEEE Journal on Selected Areas in Communications*, **42**, 1, pp. 207–222.

Peelam, M. S., Shah, K., and Chamola, V. (2024). V-track: Blockchain-enabled IoT system for reliable vehicle location verification, *Digital Communications and Networks*, **12**, 1, pp. 119–133.

Sehar, N. U., Khalid, O., Khan, I. A., Rehman, F., Fayyaz, M. A., Ansari, A. R., and Nawaz, R. (2023). Blockchain enabled data security in vehicular networks, *Scientific Reports* **13**, 1, p. 4412.

Sullivan, B., Svendsen, S., and Khan, J. A. (2024). Federated learning for energy-efficient cooperative perception in connected and autonomous vehicles, in *IEEE INFOCOM 2024—IEEE Conference on Computer Communications Workshops (INFOCOM WKSHPS)*, pp. 1–2, doi:10.1109/INFOCOMWKSHPS61880.2024.10620757.

Wang, X., Zhu, H., Ning, Z., Guo, L., and Zhang, Y. (2023). Blockchain intelligence for internet of vehicles: Challenges and solutions, *IEEE Communications Surveys & Tutorials* **25**, 4, pp. 2325–2355.

Wegener, A., Piórkowski, M., Raya, M., Hellbrück, H., Fischer, S., and Hubaux, J.-P. (2008). TraCI: An interface for coupling road traffic and network simulators, in *Proceedings of the 11th Communications and Networking Simulation Symposium*, pp. 155–163.

Part III

Next Generation Autonomy and AI Native Security

https://doi.org/10.1142/9789819831548_0008

Chapter 8

Drones That Think on Their Feet: Sudden Landing Decisions with Embodied AI

Diego Ortiz*, Mohit Agrawal*, Yash Malegaonkar*, Luis Burbano*, Axel Andersson†, György Dán†, Henrik Sandberg†, and Alvaro A. Cardenas*

**University of California, Santa Cruz, USA*

†KTH Royal Institute of Technology, Stockholm, Sweden

Abstract

Autonomous drones must often respond to sudden events, such as alarms, faults, or unexpected changes in their environment, that require immediate and adaptive decision-making. Traditional approaches rely on safety engineers' hand-coding large sets of recovery rules, but this strategy cannot anticipate the vast range of real-world contingencies and quickly becomes incomplete. Recent advances in embodied AI, powered by large visual-language models, provide common-sense reasoning to assess context and generate appropriate actions in real time. We demonstrate this capability in a simulated urban benchmark in the Unreal Engine, where drones dynamically interpret their surroundings and decide on sudden maneuvers for safe landings. Our results show that embodied AI makes possible a new class of adaptive recovery and decision-making pipelines that were previously infeasible to design by hand, advancing resilience and safety in autonomous aerial systems.

Keywords: Embodied AI, Drones, Safety.

8.1 Introduction

Autonomous drones hold enormous promise for society, with potential applications in disaster response, infrastructure inspection, environmental monitoring, and transportation. Their ability to operate in dynamic environments and carry out missions without constant human supervision could transform how critical services are delivered. However, to fulfill this promise, drones must be trustworthy. In particular, they must be prepared to respond when alarms or unexpected events occur. These alarms may signal cyber attacks such as GPS spoofing, electromagnetic interference (EMI) signal injection (Jang *et al.*, 2023), device failures such as sensor degradation, or adverse conditions such as sudden weather changes or unexpected obstacles. Although rare, such events can have severe consequences, and autonomous agents must be able to assess the situation and determine an appropriate course of action in real time.

Previous work recognizes the need for automatic recovery when alarms or unexpected events occur, but existing approaches are tailored to *specific* failure models and lack general, context-aware reasoning (Barbosa *et al.*, 2025; Cardenas, 2025). **(A) Recovery to pre-specified targets.** A popular recovery method formalizes how to steer a system to a *pre-defined* safe set after an alert (Zhang *et al.*, 2024, 2020). These approaches assume that the target (safe) set is known in advance and valid throughout recovery; they do not revise that target if the environment becomes unsafe at runtime. **(B) Mitigation based on experience.** Complementary lines address particular attack or failure classes by learning from previous successful traces (Dash *et al.*, 2021); however, if the failure condition requires a completely new maneuver or destination not seen before, these approaches may struggle. In general, previous efforts are *scenario-bound*: they depend on pre-enumerated fault/attack models, static safe-target definitions, or fixed fallback behaviors, rather than dynamically *interpreting* complex scenes and choosing actions with common-sense reasoning in open-ended, evolving environments.

In this work, we introduce a new pipeline that leverages large visual-language models (LVLMs) to support real-time decision making when sudden events occur. Our approach focuses on the case of sudden landing maneuvers, where a drone must quickly assess its surroundings and select a safe course of action. The pipeline integrates traditional control modules with LVLM-based reasoning: perception modules identify candidate surfaces, the LVLM evaluates their suitability using common-sense reasoning, and a movement planner executes the maneuver. To evaluate this approach, we build a benchmark on top of open-source platforms, leveraging the Unreal Engine to propose realistic and dynamic-looking environments, with the Cosys-AirSim simulator, modeling the sensors and control stack of drones.

This benchmark provides diverse, realistic urban scenarios with dynamic obstacles and configurable conditions, enabling systematic testing of recovery pipelines. Our contributions are threefold:

(1) We propose an LVLM-driven pipeline for sudden landing decisions that combines perception, reasoning, and control.
(2) We develop a reproducible benchmark for evaluating such pipelines in realistic urban simulations. Our benchmark is openly accessible at https://github.com/RollingBeatle/Airsim-closeloop.
(3) We demonstrate through experiments that embodied AI enables adaptive recovery behaviors that were previously infeasible.
(4) We show the capabilities of out-of-the-box LVLMs to identify and reason over safety concerns and accomplish a safe landing with acceptable latency rates.

8.2 Related Work

Early work in mission recovery for autonomous systems often relies on the assumption that the region to recover to is known beforehand. Authors in Dash *et al.* (2021) implicitly assume that the original target location is safe to go to regardless of the detected anomalies, and propose a controller based on the simplex architecture, where the recovery controller is activated under adverse conditions. The recovery controller is a neural network trained to mimic the behavior of a PID controller. A similar idea is presented in Dash *et al.* (2024), where a reinforcement learning policy is learned to satisfy safety constraints specified using signal temporal logic under adverse conditions. Focusing on recovery from sensor attacks, the implicit assumption is that the originally intended target location is safe. In Roque *et al.* (2022), a safe model-predictive control problem is formulated by using control barrier functions, which is similar to

Dash *et al.* (2024) and Zhang *et al.* (2024) in that safe regions and constraints are formulated beforehand.

Safe landing zone (SLZ) detection is an active area of research. Current methods rely on geometric and semantic analysis of onboard perception from RGB-cameras and/or LiDAR. These methods leverage techniques from classical image analysis, machine learning, and deep learning. In Maturana and Scherer (2015), the authors train a 3D convolutional neural network (CNN) to detect SLZs in rural areas for autonomous helicopters from a LiDAR point cloud. Lee *et al.* (2020) uses a 2D CNN to find designated landing zones (helipads) from camera images with a feature matching algorithm. These deep learning based approaches rely on labeled data for supervised training of the models; hence, they would perform best in domains similar to those where the training data were collected. Complementary to these learning-based methods, Yang *et al.* (2018) proposes a GPS-denied monocular-vision SLAM pipeline that reconstructs a 3D point cloud and transforms it into a height-annotated grid to find flat landing zones. The paper Marcu *et al.* (2019) formulates safe landing zone (SLZ) detection as a semantic segmentation task. Using a semi-synthetic dataset of aerial imagery, the authors train a U-Net model to assign each pixel to one of three categories: horizontal, vertical, or other. Candidate landing areas are then selected based on the assumption that pixels classified as horizontal correspond to locally planar terrain suitable for landing. de la Torre-Vanegas *et al.* (2025) estimates SLZs by constructing an image-space risk map from the onboard camera stream. A segmentation module identifies object classes, each of which is assigned a risk score reflecting its presumed landing hazard. The resulting per-pixel risk map is then minimized to select a low-risk landing location in the image plane, which is used as the target landing position.

A survey on SLZ detection (Shah Alam and Oluoch, 2021) highlights several open directions. One of them is to improve terrain-awareness. The surface of water is often flat and might seem like a viable landing option to an algorithm that successfully detects flat areas, but is unaware of the terrain. Another direction suggested in the survey is to couple SLZ localization with path planning and obstacle avoidance.

Foundation Models (FMs) have recently been integrated into various robotic systems for their reasoning capabilities. End-to-end models, such as Gemini Robotics (Gemini Robotics Team *et al.*, 2025), use a Vision-Language-Action (VLA) model that takes instructions in natural language and outputs low-level actuator commands in order to complete the task. Compound models combine FMs and classical control algorithms such as PID or model-predictive controllers. In Sinha *et al.* (2024), the authors use FMs in a drone to detect anomalies during flight and to select an appropriate region from a pre-defined set of regions. A VLM is used to convert camera images from the drone to text that determines which recovery region is selected. The main shortcoming of such approaches is that training and inference using such models is computationally intensive. End-to-end solutions for autonomous cars are starting to appear in the literature. In Xiao *et al.* (2022) the authors use conditional imitation learning to predict steering angle, throttle and brake actions from raw LiDAR and camera data. They investigate different methods to fuse these modalities in neural networks. Prakash *et al.* (2021) leverages the transformer architecture to predict suitable way points from LiDAR and camera data, but offloads the task to low-level control to traditional PID controllers.

A different line of work provides formal guarantees of recovery to pre-defined regions using control-theoretic approaches (Garg *et al.*, 2022; Leudo *et al.*, 2023; Zhang *et al.*, 2024, 2021, 2023). These works primarily focus on recovering control systems, such as drones, from sensor faults or attacks. The issue is that it is not always clear how to define safe regions in practice, as in a dynamic environment, the safety of a region may change over time, e.g., due to moving obstacles.

In addition to computer-vision and control-oriented works, there exist decision support and situational awareness systems that reason over dynamic environments and safety constraints, even if they do not use visual reasoning for landing. For example, Insaurralde *et al.* (2022) builds an ontology-driven decision support system that evaluates hazards in airspace (e.g., weather, traffic, changing surveillance inputs) and helps determine whether operations (like takeoff or routing) remain safe in real time.

Contrary to previous work, in this work, we explore how to leverage the reasoning capabilities of FMs for planning emergency landings in diverse realistic scenarios. We adopt a compound model, i.e., low-level control is via a traditional controller (such as a PID controller or MPC), which keeps compute requirements moderate.

Conceptually, the FM assumes the role of a human operator who, in the event of an emergency, selects a safe landing spot based on visual depth information. Safety in this context means that the drone lands in a flat area without humans or obstacles nearby.

8.3 Assumptions and Problem Formulation

Assumptions. Modern drones are equipped with a variety of sensors that provide complementary views of the environment. Typical platforms carry RGB cameras (monocular or stereo), depth sensors or LiDAR, GPS units, and sometimes radar altimeters or thermal cameras. For the scope of this work, we assume the drone is equipped with a downward-facing RGB camera, a basic LiDAR, and state-estimation sensors (e.g., IMU). These assumptions are realistic for both commercial and research platforms, while keeping the sensing payload lightweight enough for small UAVs operating in urban environments. We also assume that the camera, LiDAR, and actuators (in addition to our pipeline) are trustworthy during the recovery maneuver.

Problem Formulation. We consider the situation in which a drone receives an alarm or experiences an unexpected event (e.g., an attack, a device failure, or a sudden environmental change). The drone must respond by identifying a feasible recovery maneuver, such as a sudden landing, that ensures safety. Formally, let the environment at time t be represented by sensor observations O_t (images, depth estimates, or state information). Given an alert event E_t, the recovery problem is to compute a control sequence that drives the drone from its current state to a safe area in a short time. We informally define a safe state as one in which the drone has landed on a valid surface that avoids imminent hazards. The challenge is that safe areas are not known a priori and may evolve dynamically due to moving obstacles or environmental changes. Thus, the agent must infer both (i) which regions of the environment are plausible candidates for recovery and (ii) which sequence of actions will reliably reach them.

Challenges. The problem formulation raises several challenges. (1) **Uncertainty in perception:** vision- and depth-based modules may misclassify surfaces due to clutter, reflections, or sensor noise. (2) **Dynamic environments:** a region judged safe at time t may become unsafe by $t + T$ due to pedestrians, traffic, or adverse weather. (3) **Reasoning under novelty:** alerts may correspond to scenarios never observed during training, requiring the system to extrapolate beyond predefined models. (4) **Real-time constraints:** all inference and planning must be completed quickly enough to be actionable on a resource-limited aerial platform.

While classical emergency recovery methods assume fixed safe sets or pre-enumerated failure modes, our objective is to enable drones to dynamically interpret their environment and compute recovery actions in real time, even when alarms reveal novel or unforeseen situations.

LVLMs and Drones: An additional consideration in our problem formulation is where LVLM-based reasoning can realistically be executed. One option is to run small models directly *onboard* the drone, which minimizes latency and dependence on communication links, but requires lightweight models that trade reasoning depth for efficiency. A second option is to leverage *edge devices* such as ground stations, nearby base stations, or, in the case of swarms, a more powerful UAV, which provides more compute capacity than onboard processors while still offering relatively low latency communication; this setting enables medium-sized LVLMs that balance accuracy and resource demands. Finally, *cloud deployment* would require our drone to talk to the cloud, allowing us to use the largest and most capable models, but introducing additional latency, network dependence, and potential reliability issues if connectivity is lost during an emergency. These options are illustrated in Fig. 8.1.

In practice, the choice of deployment setting reflects a tradeoff between the hardware available to the platform and the size and complexity of the LVLM that can be used for real-time decision making. Our focus in this chapter is on the algorithmic pipeline and its reasoning capabilities, but the deployment tier—onboard, edge, or cloud—will ultimately determine the operational envelope for embodied AI in safety-critical drones. In our experimental section, we will use three different-sized models, representing the capabilities of onboard, edge, or cloud LVLMs; in particular, our implementation will consider GPT-5 Nano, GPT-5 Mini, and GPT-5. For practical uses, however, the embedded or edge LVLM will likely be an open

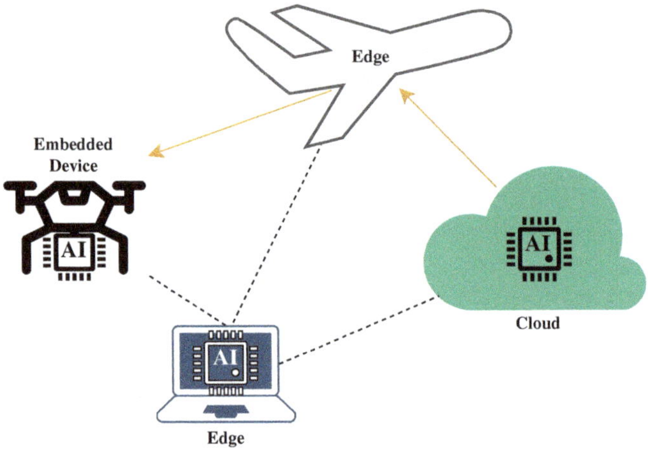

Fig. 8.1. LVLMs can be deployed on the device (drones), at the edge (ground control stations, other aerial vehicles), or in the cloud.

model, such as InternVL (Chen *et al.*, 2024), as black-box LVLMs accessible only via cloud APIs cannot be deployed directly on drones.

8.4 Design Tradeoffs and Pipeline Overview

Designing a recovery pipeline around LVLMs forced us to confront the tension between adaptability and reliability. We knew that a purely end-to-end approach—letting the LVLM decide directly from raw images where and how to land—would be maximally flexible, but too risky for safety-critical decisions: hallucinations, inconsistent outputs, or latency spikes could easily cascade into catastrophic failures. At the other extreme, a fully hand-engineered system would be predictable but brittle, unable to cope with novel or unexpected events. Our solution was to adopt a hybrid, modular design: give the LVLM responsibility only for the semantic judgments where its common-sense reasoning is indispensable, and anchor those judgments with conventional perception and control methods that are faster and more predictable.

This rationale led us to structure the pipeline into three interconnected modules (Fig. 8.2). The **Surface ID Module** handles raw

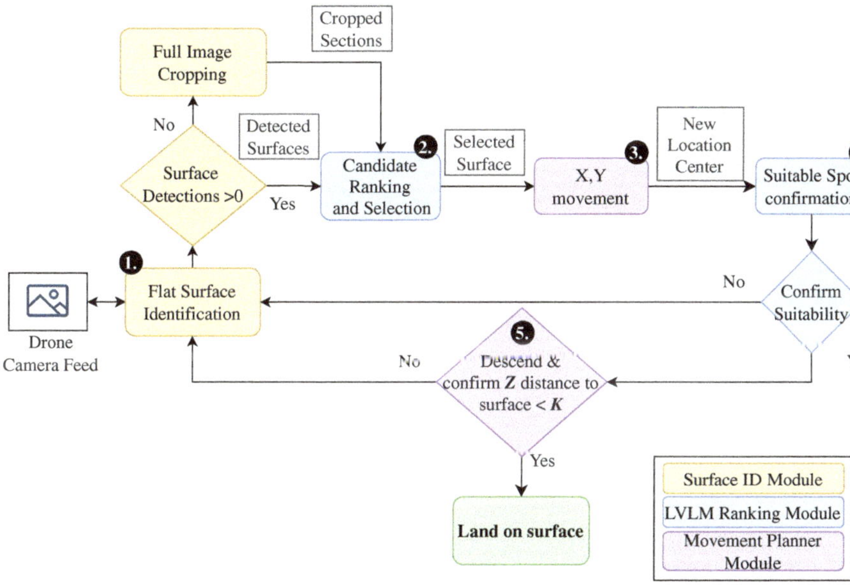

Fig. 8.2. Detailed pipeline.

perception and identifies plausible candidate surfaces. The **LVLM Ranking Module** interprets those candidates, applying contextual reasoning to select and later confirm a safe landing site. Finally, the **Movement Planner Module** ensures that high-level decisions translate into reliable motion by grounding LVLM outputs in precise world coordinates. This separation of roles keeps the LVLM "in the loop" where it adds unique value, while preventing it from exerting unchecked control over low-level behavior. The pipeline uses three sensor modalities: camera images, LiDAR point clouds, and range finder distances. Figure 8.3 shows how data flows between the described modules.

Together, these modules form a loop in which **(1) candidate surfaces are proposed, (2) semantically evaluated, and then (3) physically reached,** with multiple opportunities for validation before the final descent. The design is deliberately redundant—e.g., the LVLM evaluates surfaces twice, once before motion and once after repositioning—because redundancy is a feature, not a flaw, in safety-critical systems.

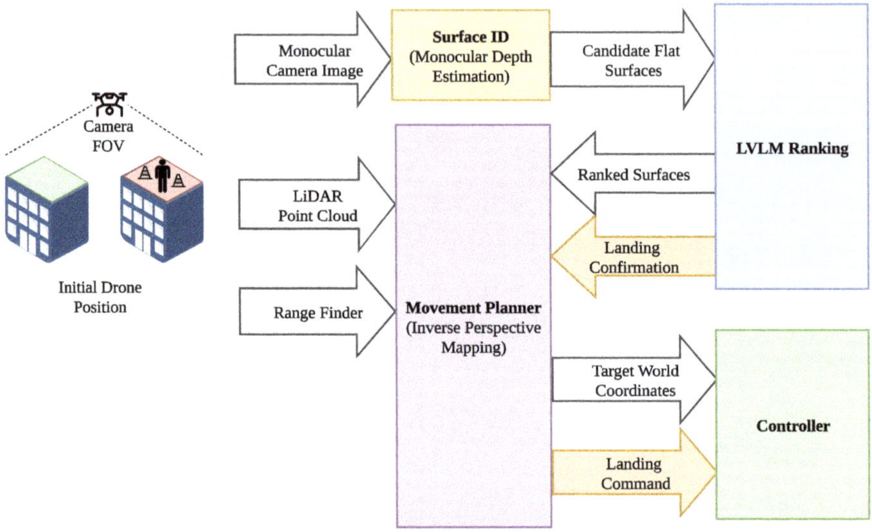

Fig. 8.3. Data flow diagram for the pipeline and its modules. Gray arrows denote transfer of data, orange arrows represent decisions.

8.4.1 *Surface ID Module*

The first stage of our pipeline is the Surface ID Module, whose task is deliberately narrow: rather than deciding where to land, it reduces uncertainty by pruning the search space and presenting the LVLM with only plausible candidates. This separation keeps the LVLM from being overwhelmed by raw imagery, limits the risk of hallucinations, and ensures that subsequent reasoning focuses on a small number of meaningful options. If no candidates are detected, the module partitions the input image into quadrants so that the LVLM can at least suggest a direction of motion, guaranteeing that the system always produces a next step instead of stalling. The tradeoff here is intentional: we sacrifice some end-to-end flexibility in order to gain robustness, interpretability, and consistent forward progress.

To identify candidate landing zones, we first compute a depth map from the drone's downward-facing monocular camera. Although a single RGB image does not directly provide depth, modern deep networks can infer it by learning geometric cues such as texture gradients, object sizes, and perspective lines from large training datasets (Eigen *et al.*, 2014; Yang *et al.*, 2024). The output is a dense depth map $D(u, v)$, where each pixel coordinate (u, v) is associated with an estimated distance from the camera.

The goal is to detect flat, planar regions in this depth map. Intuitively, a flat surface (like a rooftop or a road segment) will show a consistent change in depth across neighboring pixels, whereas cluttered or sloped areas will produce irregular variations.

We use the identified flat pixel regions to create a mask $M(u, v)$, which we refine through morphological filtering to remove noise from textures or depth artifacts, yielding a cleaned mask $M'(u, v)$. Connected components are then extracted, and any connected component that exceeds a minimum area threshold is marked as a valid flat surface. Finally, for each valid region, we compute a bounding box that localizes the surface within the original RGB image. These cropped patches are passed to the LVLM Ranking Module for semantic evaluation.

8.4.2 *LVLM Ranking Module*

The LVLM then evaluates the candidate surfaces with prompts designed to enforce safety and reduce hallucinations. We deliberately place the LVLM here—between perception and control—because its ability to interpret context is most useful at this semantic level. To address the challenge of dynamic environments, we invoke the module twice: once to rank the candidates before movement, and again to confirm the safety of the chosen site when the drone has repositioned. This double check introduces latency, but provides robustness by ensuring that conditions have not changed while moving.

In practice, the LVLM interaction is separated into two stages. In the **ranking stage**, the model receives up to five cropped candidate images from the Surface ID Module and orders them by suitability, considering both flatness and the presence of hazards. The top-ranked region is then passed to the Movement Planner. Once we reach the intended destination, we enter the **confirmation stage**, where the LVLM is queried again with a close-up view to verify that the site is still safe for descent. This two-pass process provides the system with an up to date evaluation of the target, to address the challenge of surfaces whose safety may change dynamically over time.

To maximize reliability, we carefully design prompts for both stages. A system prompt frames the LVLM as an emergency landing

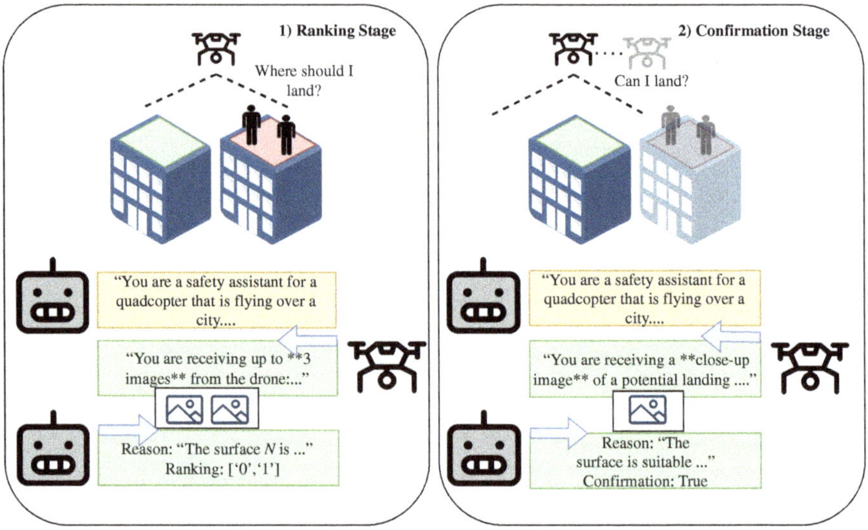

Fig. 8.4. Two-stage conversational prompting.

assistant, specifying its role, expected inputs, and output format. Conversational prompts (Figure 8.4) are structured using a Chain-of-Thought approach (Sahoo *et al.*, 2025; Wei *et al.*, 2023), encouraging the model to reason step by step rather than guessing directly. Safety constraints are reinforced through imperative formulations such as "MUST" (Amatriain, 2024), which obligate the LVLM to apply strict selection rules and always return a valid choice. While the prompts for ranking and confirmation share this structure, their emphasis differ: the ranking stage focuses on general surface suitability, while the confirmation stage focuses narrowly on detecting immediate threats that could compromise the final descent. The details of the prompts are in the Appendix.

A final threshold check on the vertical distance Z ensures that the drone is close enough for a conventional landing controller to take over, bridging LVLM-based reasoning with the reliability of low-level control.

8.4.3 *Movement Planner Module*

The LVLM's decision is expressed as an image crop, and we take the center of this crop as the target destination. To act on this decision, we must convert that central pixel in the camera image into a physical 3D location that the drone can navigate to. This translation from image space to world space is the purpose of inverse perspective mapping (IPM).

The process unfolds in three steps:

1. From LiDAR to Image Space. We first build a mapping between LiDAR points and image pixels using the pinhole camera model. A LiDAR point $p = (x_{\text{frd}}, y_{\text{frd}}, z_{\text{frd}})$ in the drone's body frame (Forward–Right–Down) projects into the image plane as

$$u = \frac{x_{\text{frd}} \cdot f}{z_{\text{frd}}} + c_x, \quad v = \frac{y_{\text{frd}} \cdot f}{z_{\text{frd}}} + c_y, \tag{8.1}$$

where (u, v) are image coordinates, f is the focal length, and (c_x, c_y) is the camera's principal point. Repeating this for all LiDAR points creates a lookup table linking pixels to 3D voxels.

2. From Image to 3D. When the LVLM outputs a candidate image, we take its center pixel (u, v) and look up the corresponding LiDAR point. If the exact pixel has no LiDAR match (due to sparsity), we select the nearest pixel with a valid correspondence. The result is a 3D point $(x_{\text{frd}}, y_{\text{frd}}, z_{\text{frd}})$ in the drone's body frame.

3. From Body to World. To express this location in the world frame, we rotate the FRD coordinates into the North–East–Down (NED) frame using the drone's yaw angle ψ:

$$\begin{bmatrix} x_{\text{ned}} \\ y_{\text{ned}} \end{bmatrix} = \begin{bmatrix} \cos\psi & -\sin\psi \\ \sin\psi & \cos\psi \end{bmatrix} \begin{bmatrix} x_{\text{frd}} \\ y_{\text{frd}} \end{bmatrix}. \tag{8.2}$$

We then provide $(x_{\text{ned}}, y_{\text{ned}})$ to the controller, which moves the drone horizontally above the chosen landing site. The vertical component is ignored at this stage; descent begins only after safety has been reconfirmed.

Once the drone has moved above the candidate site, the LVLM Ranking Module is invoked again for confirmation. If the site is still deemed safe, the drone uses a distance sensor to measure altitude. It first descends to a fraction k of this distance, giving the system another opportunity to reevaluate safety if new hazards appear. If the drone is already within a 5-meter threshold, it proceeds directly to final landing.

This design keeps the LVLM responsible only for high-level judgments—identifying which region looks safe—while the Movement Planner handles the geometric and control details needed to reach that region. In this way, the LVLM "suggests" and traditional control "guarantees," providing layered defenses against unsafe outcomes.

8.5 Implementation

Environment. To evaluate our pipeline, we build on the Unreal Engine 5 City Sample Project, a large-scale, photorealistic urban environment originally designed to showcase next-generation rendering. The City Sample is uniquely well suited to testing embodied AI for drones: it combines realistic rooftops, roads, vehicles, and pedestrians with full configurability, including weather and lighting. Unlike many static simulators, its scenarios are dynamic, with cars and pedestrians moving through the streets, creating the kinds of evolving conditions that challenge real-world decision-making. This realism is also practically valuable: recent work has demonstrated that synthetic data generated from the City Sample improves object detection models, thanks to its high fidelity and dynamic variety (Turkcan *et al.*, 2024).

To make scenarios more challenging, we added obstacles such as HVAC units to otherwise safe rooftops. The high degree of customization in City Sample allows future extensions. In the future, we plan to further stress-test our pipeline with rooftop fires, obstructing pedestrians, or adversarial markings (e.g., text or symbols) to probe system robustness under more challenging scenarios.

Drone. We integrated our pipeline with Unreal Engine 5 using Cosys-AirSim, a plugin developed by Cosys-Lab as an extension of Microsoft's AirSim. Cosys-AirSim lets us simulate drones with realistic physics and visuals, and it provides APIs that allow our pipeline to stream sensor data and issue motion commands in real time. It also handles the conversion between Unreal's coordinate system and the North–East–Down (NED) convention used in robotics.

During our experiments, we ran the built-in `SimpleFlight` controller. We configured the simulated drone with the sensors required for our pipeline: a downward-facing RGB camera for perception and

LVLM queries, a distance sensor for controlled descent, a LiDAR sensor for inverse perspective mapping, and IMU/GPS data for pose reference.

Surface Detection. We identify candidate landing surfaces using monocular depth estimation. Specifically, we employ a compact version of the Depth Anything V2 model (Yang *et al.*, 2024), which predicts dense depth maps from single RGB images captured by the drone's downward-facing camera. From these maps, we segment locally flat regions by thresholding gradient magnitudes and applying simple morphological filtering to remove noise. We then extract connected components that exceed a minimum area threshold and generate bounding boxes for each valid surface. These cropped image patches serve as candidate landing zones and are passed to the LVLM for semantic evaluation and ranking.

LVLM. We evaluated cloud deployed models, three OpenAI multimodal models of different scales—GPT-5, GPT-5-mini, and GPT-5-nano alongside Gemini Flash 2.5, Mistral 3 Large and a model deployed on an edge device, InternVL 2.5, in order to study the tradeoff between reasoning strength and computational efficiency. GPT-5 and Gemini Flash 2.5 provide the most powerful multimodal reasoning and serve as an upper bound for performance. GPT-5-mini delivers nearly comparable accuracy while requiring fewer resources, making it attractive for edge deployment. GPT-5-nano is the most compact and recent GPT model, and represents state of the art models that could be run on constrained hardware. Likewise, Mistral 3 and InternVL are small open-source models that could potentially run on an edge device and have faster response time. This range of models allows us to explore how the size and capacity of an LVLM affect decision-making in safety-critical drone scenarios.

8.6 Evaluation Metrics

To evaluate the pipeline, we first tested each module in isolation (20 iterations per module) before moving to end-to-end experiments. This modular evaluation allowed us to verify that every component operated as intended and to quantify its specific contribution to the overall system. We relied on three metrics to assess module performance.

Jaccard Index (JI): The Jaccard Index (JI) (Everingham *et al.*, 2010) quantifies how well a predicted bounding box overlaps with the ground-truth surface. A value of 1 indicates perfect agreement. For a ground-truth area B and a predicted area A, we define

$$\mathrm{JI} = \frac{A \cap B}{A \cup B}. \tag{8.3}$$

We use this metric to evaluate the accuracy of the Surface ID module.

Landing Distance. We evaluate the precision of the Movement Planner by measuring the Euclidean distance between the drone's final landing position and the ground-truth target in the UE5 coordinate space. Let $(x_{\mathrm{gt}}, y_{\mathrm{gt}})$ denote the ground-truth location and (x_r, y_r) the position reached by the drone. The distance is

$$d = \sqrt{(x_{\mathrm{gt}} - x_r)^2 + (y_{\mathrm{gt}} - y_r)^2}. \tag{8.4}$$

A smaller value indicates more accurate guidance toward the intended landing site.

LVLM Success Rate. To evaluate the LVLM, we designed two test scenarios, each with a pair of candidate rooftops: one clear of obstacles and one cluttered with debris such as rubble or HVAC units. In each trial, success is counted when the LVLM selects the clear rooftop. The success rate is the percentage of trials in which the model made the correct choice.

8.7 Benchmarking Modules and the Pipeline in Curated Scenarios

For our initial tests we use two fixed scenarios, as illustrated in Fig. 8.5. Both scenarios have two rooftops, one cluttered with obstacles and one clear of any obstacles (this should be the preferred landing option).

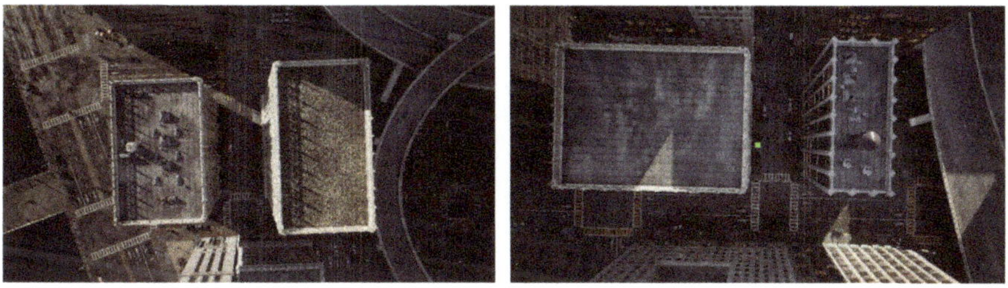

(a) (b)

Fig. 8.5. Testing scenarios, original views. (a) Scenario 1. (b) Scenario 2.

Table 8.1. Success rate for surface ID and movement planner modules.

	JI Score Avg. (%)	Avg. Distance (m)
Scenario 1	87.09	0.802
Scenario 2	59.55	1.145

8.7.1 *Results of Individual Modules*

Table 8.1 summarizes the performance of the Surface ID and Movement Planner modules. The Surface ID Module consistently detected the correct candidate surface, with a higher Jaccard Index in Scenario 1 (87%) and a lower but still acceptable score in Scenario 2 (60%). After selecting the center-point of these locations, the Movement Planner guided the drone to within roughly one meter of the ground-truth target in both scenarios, with slightly larger drift in Scenario 2. These results show that both modules provide accurate and reliable inputs for the overall pipeline.

Table 8.2 compares the performance of different LVLMs in the Ranking Module. Most model such as Gemini 2.5 F., GPT-5 and GPT-5-mini consistently selected the correct landing surface in both scenarios. However, differences start to appear on the confirmation step with GPT-5-mini and Mistral showing slightly lower accuracy on the confirmation step in Scenario 1, likely due to a more cautious bias, but overall demonstrated the potential of smaller models.

Table 8.2. Operation success rate in Scenarios 1 and 2 for LVLM operations.

	Stage	GPT-5 (%)	GPT-5 mini (%)	GPT-5 nano (%)	Gemini 2.5 F (%)	Mistral 3 L (%)	InternVL 2.5 (%)
Scenario 1	Rank.	100	100	20	100	100	100
	Conf.	100	95	100	100	90	100
Scenario 2	Rank.	100	100	100	100	100	100
	Conf.	100	75	20	100	20	100

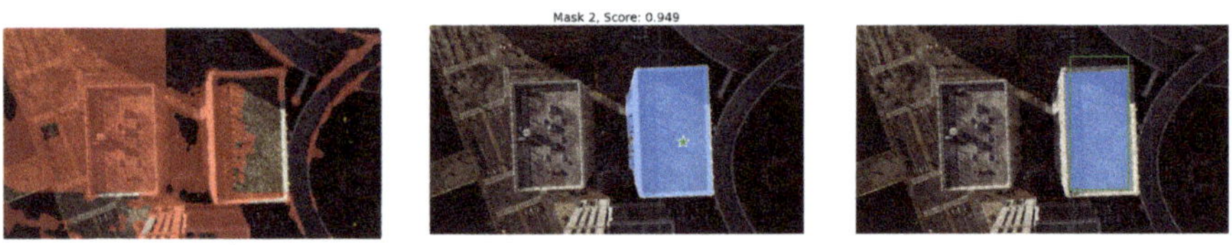

(a) (b) (c)

Fig. 8.6. Comparison of vision-based approaches for Scenario 1. (a) OpenLander segmentation. (b) SAM point segmentation. (c) SAM bounding box.

In contrast, GPT-5-nano performed much worse. It struggled with ranking in Scenario 1 (20%) and failed the confirmation step in Scenario 2 (20%), often rejecting valid rooftops by labeling tiles as "shingled," "irregular," or "uneven." Similarly, Mistral 3 L. starts showing its limitations by failing most of the iterations in the Scenario 2 confirmation step (20%), repeatedly mistaken the roofs as "brick" or "stone" walls.

8.7.2 *Qualitative Comparison with Vision-Based Baselines*

To further contextualize our results, we performed a qualitative and quantitative comparison using two established vision-based baselines on our ground-truth data: OpenLander (Sturges, 2022) and the Segment Anything Model (SAM) (Kirillov *et al.*, 2023).

As illustrated in Fig. 8.6, SAM demonstrates exceptional geometric precision. When evaluated against our ground-truth landing coordinates using Intersection over Union (IoU), SAM achieved a score of 0.951 in Scenario 1 and 0.787 in Scenario 2. These scores indicate that SAM can isolate potential landing surfaces with high spatial accuracy.

However, this geometric proficiency comes with significant trade-offs. First, SAM is purely objective; it provides high-fidelity segmentation but lacks the semantic capacity to distinguish between a safe landing zone and a restricted or dangerous area, such as an active highway or a private balcony. Second, while we could use SAM to refine our candidate proposals, its computational overhead is substantial. Given the hardware constraints of small UAVs, running a large model like SAM alongside our depth map estimation module would likely increase latency and potentially compromise the drone's ability to respond to sudden events.

Similarly, OpenLander's safety masks successfully identify "flatness" and provide basic risk-mitigation categories, such as obstacles or humans. Yet, as seen in our urban evaluations, it fails to incorporate the higher-level common-sense reasoning required to navigate evolving urban contexts, such as predicting the reappearance of traffic on a currently empty roadway, a capability that our LLM-driven pipeline addresses through contextual reasoning.

8.7.3 *LVLM Context Differences*

To probe how input context affects LVLM decisions, we varied the information provided to each model (Figure 8.7). In addition to cropped candidate surfaces from the Surface ID Module, we tested two variations: (i) adding 20–40% padding around the cropped region, and (ii) supplying the entire camera image alongside the cropped surface.

Figure 8.8 and Table 8.3 summarize the results. Most models were robust across all conditions, achieving perfect accuracy regardless of context in Scenario 2 and consistently correct choices in Scenario 1. In contrast, GPT-5-nano showed pronounced sensitivity. With partial padding, its accuracy fluctuated between 15–25%, and at times it misclassified safe rooftops as unsuitable due to surface textures such as gravel or shingles. Providing the full image greatly improved Nano's performance (up to 70%), yet it still fell short of the larger models. Similarly, InternVL struggled only with Scenario 1 being unable to select the correct surface when context of 40% margin was added, consistently describing the surface as "covered in a dense layer

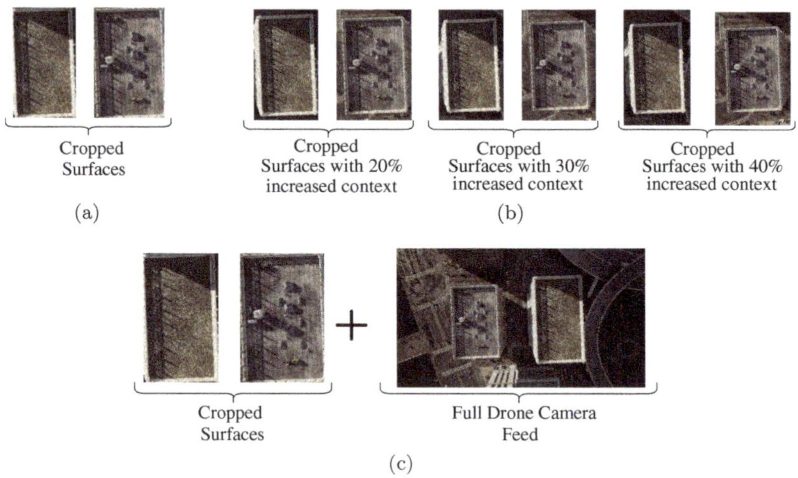

Fig. 8.7. Different context levels provided to the LVLM. (a) Cropped view. (b) Increased partial context views. (c) Full image view.

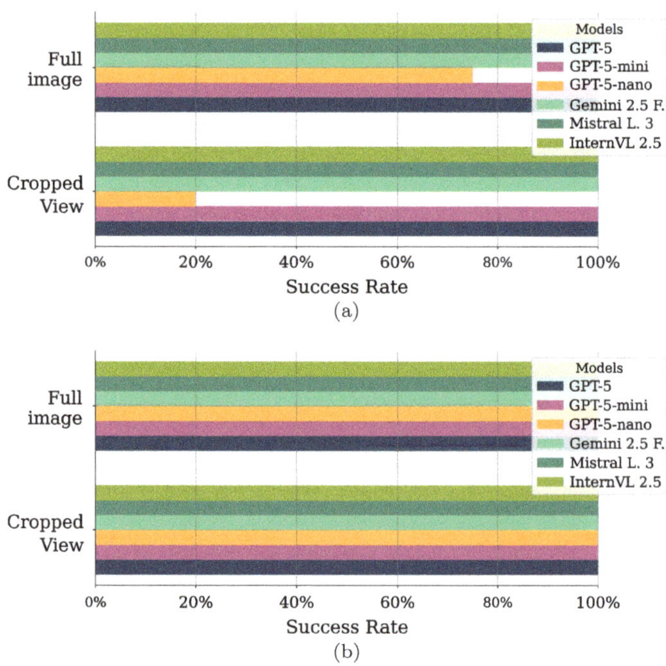

Fig. 8.8. Full image view and cropped view evaluation. (a) Scenario 1. (b) Scenario 2.

Table 8.3. Success per context level in Scenario 1.

Partial Context Level (%)	GPT-5 (%)	GPT-5 mini (%)	GPT-5 nano (%)	Gemini 2.5 F. (%)	Mistral L.3 (%)	InternVL 2.5 (%)
20	100	100	15	100	100	100
30	100	100	25	100	90	100
40	100	100	15	95	80	0

Table 8.4. Average response time per model.

	GPT-5	GPT-5 mini	GPT-5 nano	Gemini F. 2.5	Mistral L. 3	InternVL 2.5
Response Time (s)	25.783	15.152	16.938	3.755	7.184	8.579

of objects" while stating that the incorrect surface had less objects making it more suitable to land.

These findings indicate that while larger LVLMs reliably focus on relevant features even with minimal input, most lightweight models benefit substantially from additional context. Supplying the full scene helps smaller models better situate candidate surfaces in their surroundings, reducing—but not eliminating—their cautious bias.

8.7.3.1 *Model Latency*

To analyze the feasibility of our approach and how it would differ in different deployment types (cloud, edge). Cloud-deployed models were accessed via the OpenAI API and OpenRouter, while InternVL was deployed on a local device on the same network. We measured latency during the crop-view experiments and calculated the mean across all experiments. We show these results in Table 8.4.

Overall, all models demonstrated response times below 30 seconds, indicating they can handle the pipeline's input in a reasonable time. However, the response time of the largest models widely varies, with GPT-5 taking more than six times as long to respond as Gemini 2.5 F. with the same input. This could be due to GPT-5's larger scale and increased computational cost, while Gemini 2.5 F. is optimized for performance.

Within this context, our edge model, InternVL, achieves lower latency than all GPT models, demonstrating the benefits of reduced network overhead. However, both Mistral 3 L. and Gemini 2.5 F. outperform InternVL, with Gemini 2.5 F. having the lowest latency overall. These results show that although edge models can offer advantages and reduce latency, larger, optimized models can outperform them in both latency and output quality.

8.7.4 *Full Pipeline in the Two Scenarios*

With the individual modules validated, we next evaluated the full pipeline (running each module one after the other) in the two curated scenarios. Each run (20 iterations per scenario) started from the same initial position and tested different levels of context (cropped, partial 30% padding, and full image). Unlike earlier tests, the full simulation exposed the pipeline to a broader variety of surfaces and dynamic conditions. We defined a landing as successful if the chosen surface was flat, free of people and objects, away from traffic, edges, and water, and therefore safe under real-world constraints. Due to hardware constraints we did not run the full pipeline on InternVL.

Figures 8.9 and 8.10 categorize the types of surfaces selected by each model. GPT-5 consistently chose empty rooftops regardless of context, demonstrating robust reasoning. GPT-5-mini performed similarly with cropped inputs but showed more variability when additional context was provided, occasionally selecting less safe options such as roads or piers. GPT-5-nano displayed the widest variety of choices, performing best on rooftops with partial context but also producing errors, including hallucinations where it identified non-existent surfaces.

Similar to GPT-5 Mini, Gemini 2.5 F. performed best on cropped images, consistently landing on rooftops, with only a few exceptions. However, with partial context, performance diminishes, making the vehicle land on other types of surfaces and generating system errors. Mistral 3 L. had a diverse selection of surfaces across all levels of context, particularly favoring roads in Scenario 2. We can attribute this to its inability to distinguish between particular wall and roof textures.

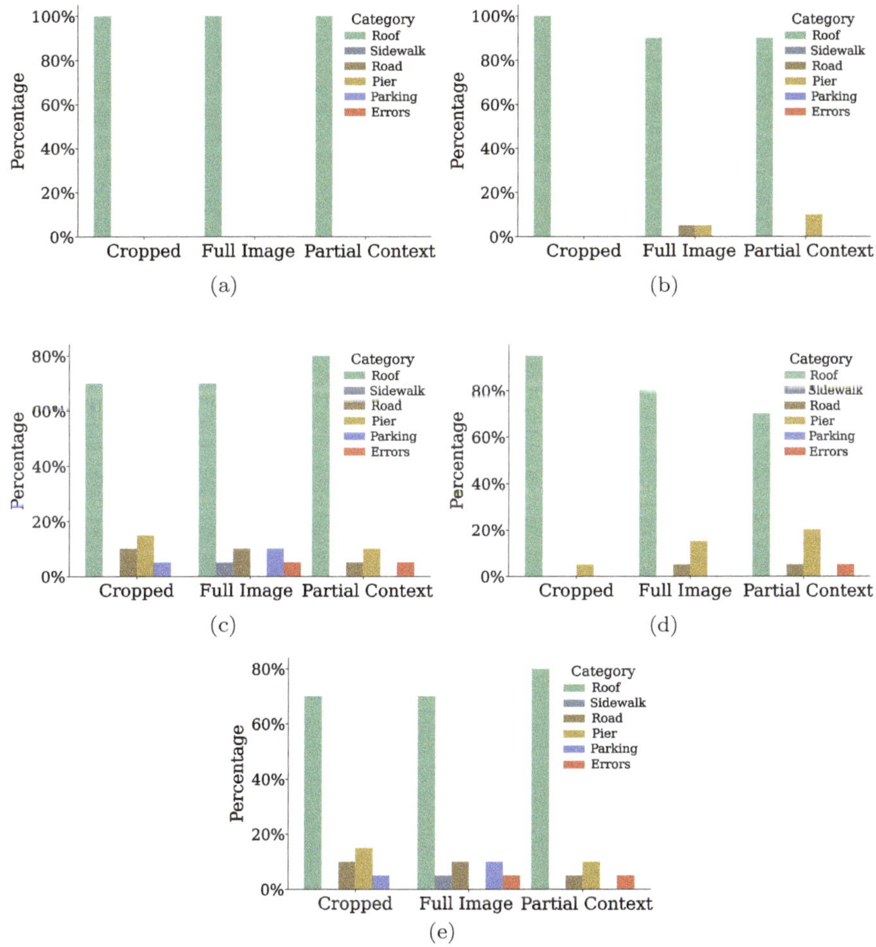

Fig. 8.9. Scenario 1: Type of surface landed on by each model by information level. (a) GPT-5. (b) GPT-5 mini. (c) GPT-5 nano. (d) Gemini F 2.5. (e) Mistral L 3.

Applying the safety criteria, we computed overall success rates (Figure 8.11). GPT-5 achieved perfect performance, while GPT-5-mini maintained success rates above 90%, with occasional failures near piers. Nearby, Gemini 2.5 Flash performed 80% or higher in all conditions across both scenarios, indicating an overall good performance with occasional landings in unsafe areas, such as roads, and system errors. Of the GPT models, GPT-5-nano performed worst,

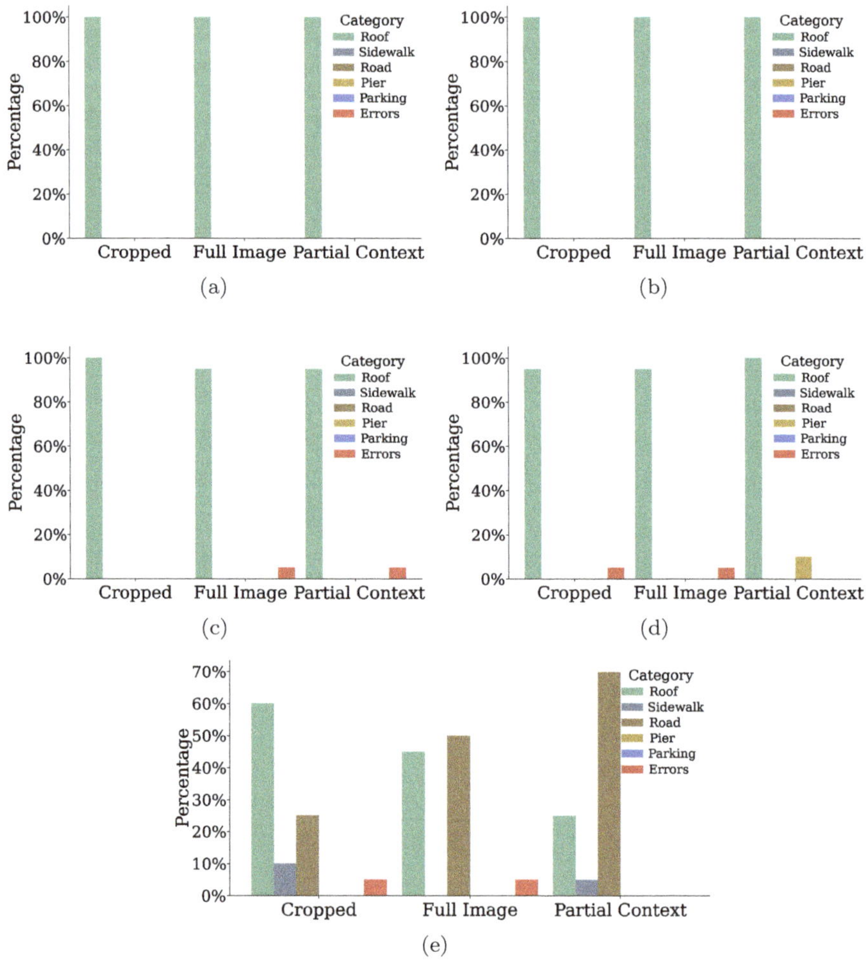

Fig. 8.10. Scenario 2: Type of surface landed on by each model by information level. (a) GPT-5. (b) GPT-5 mini. (c) GPT-5 nano. (d) Gemini F 2.5. (e) Mistral L 3.

especially in Scenario 1, where shadows, textures, and rooftop walls caused frequent misclassifications. Scenario 2, with more uniform surfaces, was easier for Nano to handle, though it still lagged behind larger models. Mistral L. 3 performed the worst of all models, especially on Scenario 2, where most of the unsafe landings occurred around roads or highways that could include live traffic. These results highlight a clear trade-off: larger LVLMs are consistently reliable,

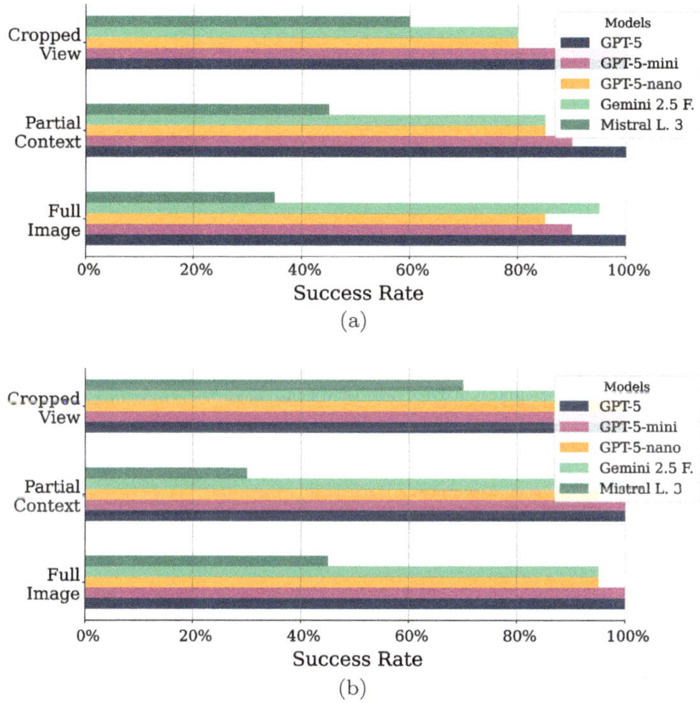

Fig. 8.11. Pipeline performance per scenario. (a) Scenario 1. (b) Scenario 2.

while smaller ones require additional context and remain more prone to errors in complex urban settings.

When calculating resource consumption for these experiments, it show that our base pipeline consumes, on average, 715.629 MB of system memory and uses 0.31GB of GPU memory, with a peak allocation 0.41 GB. These results demonstrate that, despite using complex and intensive processes such as monocular depth-estimation, our pipeline remains lightweight enough while using cloud-based LVLMs to run on commercially available laptops without specialized hardware.

The controlled experiments with two scenarios served as an important validation step. By creating situations with an obvious safe landing option, we confirmed that the modules integrate smoothly, the LVLM can reliably recognize the intended choice, the pipeline can execute an end-to-end maneuver when the decision space is constrained and resource consumption is lightweight enough to run on an edge device. These results establish a baseline of correctness and give confidence in the system's operation under simplified conditions, while also underscoring that such settings are only a first step toward the more complex realities drones encounter in practice.

8.7.4.1 *Pipeline Performance under Adverse Weather Conditions*

To evaluate the adaptability of our pipeline, we used Unreal Engine to extend our two previous scenarios and add adverse weather conditions, specifically fog and rain as shown in Fig. 8.12. We assess each scenario for 20 iterations using GPT-5 Gemini 2.5 F. and Mistral L. 3, following the same success criteria of the previous experiments. The results for Scenarios 1 and 2 are available in Tables 8.5 and 8.6 respectively.

Across both scenarios, GPT-5 and Gemini 2.5 F. performed similarly successfully landing on safe surfaces 95% of the time or more. In contrast, Mistral L. 3 showed inconsistent performance, improving during fog conditions but maintaining a poor record in the rain, with both scenarios having a 50% or lower success rate. These results show that larger models are more robust to dynamic environmental

(a)

(b)

Fig. 8.12. Adverse weather effects. (a) Scenario 1 under rain conditions. (b) Scenario 2 under fog conditions.

Table 8.5. Success rate under adverse weather conditions (Scenario 1).

Condition	GPT-5 (%)	Gemini 2.5 F (%)	Mistral L 3 (%)
Fog	100	90	30
Rain	95	95	50

Table 8.6. Success rate under adverse weather conditions (Scenario 2).

Condition	GPT-5 (%)	Gemini 2.5 F (%)	Mistral L 3 (%)
Fog	95	100	90
Rain	100	100	35

changes, while smaller models can exhibit inconsistent behavior due to perception degradation, which could significantly alter mission success.

8.8 Evaluation in a Realistic Urban Environment

The previous experiments validated our pipeline in curated scenarios where there was a clear and unambiguous safe landing choice. These settings were essential to confirm that the modules integrate correctly and that the LVLM can consistently select the intended option. Real cities, however, present a much harder problem: there is rarely a single "obvious" choice, and drones must instead reason about multiple possibilities that may all be partially safe or unsafe. Rooftops vary in slope and clutter, sidewalks and roads may appear flat but are risky due to traffic, and vehicles or pedestrians can suddenly appear in areas that initially seemed safe. The challenge in this section is therefore to evaluate how well our pipeline performs when it must navigate this ambiguity and make context-aware decisions without a predefined ground truth.

To keep the focus on system behavior rather than hyperparameter search, we use a set of parameters identified in the previous section with 100% full-pipeline success rate (GPT-5, cropped view), and carry them forward unchanged into this open evaluation.

To structure this evaluation, we design a set of city-wide experiments that stress-test different aspects of the pipeline. To avoid biasing the results toward a handful of locations, we seeded drone launch sites using a two-dimensional Halton sequence (Halton, 1960) (a quasi-random sequence with low discrepancy that has been applied in Monte Carlo simulations and falsification studies, Dreossi *et al.*, 2019) which distributes points evenly across the map while avoiding the clustering that can arise from purely random sampling.

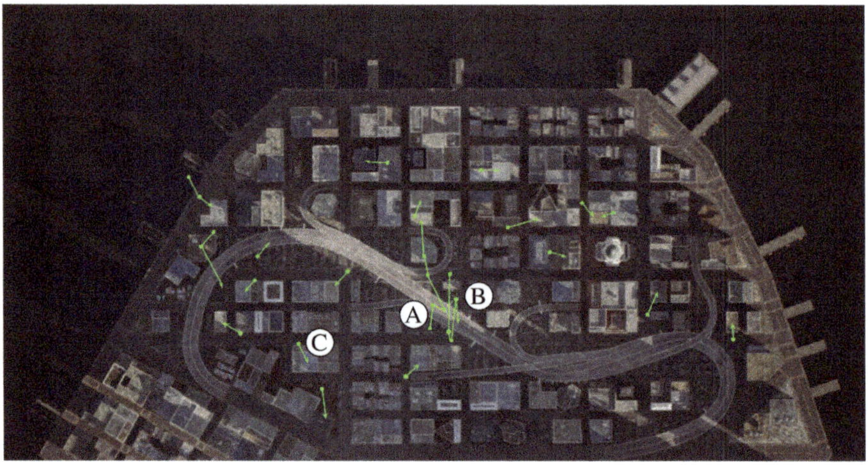

Fig. 8.13. Bird's eye view of city with all drone trajectories.

This ensures that the pipeline is exposed to a diverse range of urban contexts—dense blocks, open rooftops, highways, and mixed-use areas—rather than being confined to a few convenient spots.

We initialized the drone in 20 random locations selected by our Halton sampler. Figure 8.13 shows the drone's initial positions, trajectories, and final landing positions, with dots indicating stop positions and arrows the direction of the movement and its final position.

With this setup, we ask three key questions: (i) how safe are the landings that result, including both successes and failure cases such as obstructed rooftops or highways; (ii) how quickly can the system converge to a landing decision, measured by the number of rounds before descent; and (iii) what reasoning patterns does the LVLM exhibit, and how do they vary across different contexts? Together, these experiments provide a comprehensive picture of the pipeline's behavior in a dynamic city environment, highlighting both its strengths and its current limitations.

8.8.1 *Safety Analysis*

We study the 20 scenarios and qualitatively analyze whether the final landing area can be considered safe. Our safety assessment is a combination of areas without major obstacles, people, or other hazards such as vehicles.

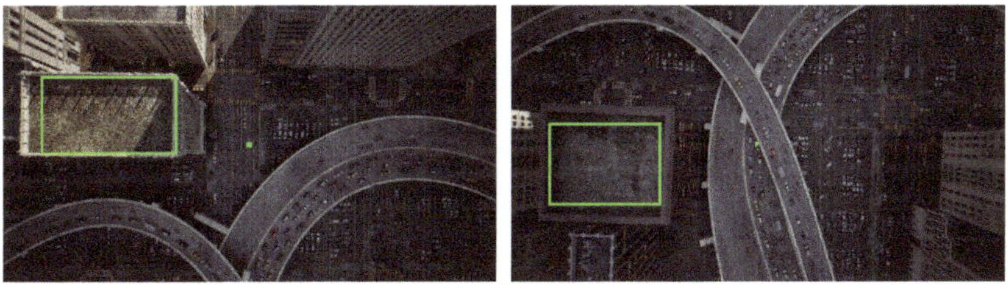

Fig. 8.14. Examples of successful landing scenarios.

During 20 iterations, the drone managed to land on a safe surface in 75% of the cases. All of these cases occurred on open rooftops, as illustrated in Fig. 8.14.

We observed five problematic outcomes:

Obstructed Rooftops (2 Cases). The LVLM cleared landings on rooftops with dense HVAC structures. Although these surfaces were geometrically flat, clutter increased collision risk and jeopardized relaunch, suggesting that our prompts should penalize rooftop clutter more aggressively.

Highways (2 Cases). The drone landed on a highway twice. In one instance, no traffic was visible at decision time, but the roadway remained unsafe due to potential traffic reappearance; in the other, a vehicle was already present and alternatives were heavily obstructed. This indicates we should encode stronger relative risk between "flat but risky" (roads) and "flat and benign" (rooftops, parking rooftops).

Timeout (1 Case). The pipeline exceeded the maximum rounds (10 in our case) and forced a landing after repeatedly rejecting otherwise safe sites; a misclassification near the end (the LVLM confused pipes for wildlife, and decided that in order to protect the birds, it was not going to land on an otherwise clear rooftop).

8.8.2 *Number of Rounds Needed to Complete a Landing*

A *round* is one full pipeline loop (Fig. 8.2): surface detection → LVLM ranking → movement → LVLM confirmation. Fewer rounds imply faster and more decisive behavior in a dynamic scene.

As shown in Fig. 8.15, more than 50% of the experiments concluded within two rounds, typically on nearby rooftops (e.g., marker C in Fig. 8.13).

Most of our safe landings take on average 2.2 rounds, indicating that our process can act quickly and correctly in dynamic environments. The most common example of our pipeline is illustrated in Fig. 8.16 where the drone identifies and navigates to a safe landing surface in the first run while confirming and arriving at the surface in the second round.

8.8.3 *Edge Cases*

Edge Case 1: We now explore some of the edge cases that required multiple rounds of the pipeline.

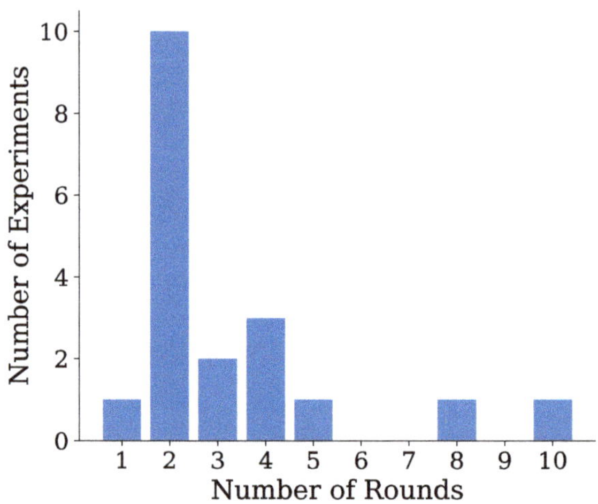

Fig. 8.15. Number of rounds needed for landing.

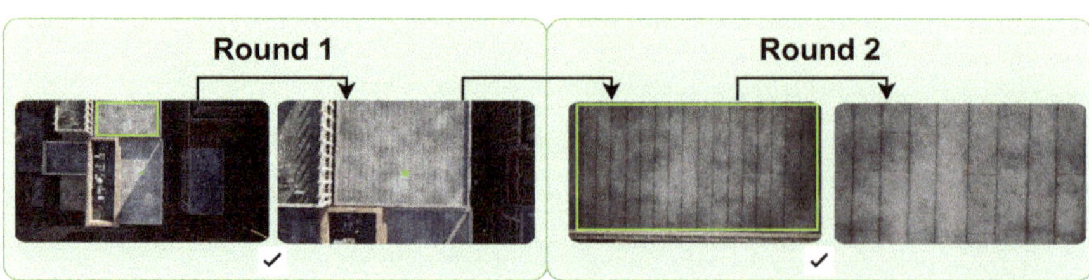

Fig. 8.16. Most common example.

Marker *B* in Fig. 8.13 highlights a launch over a sparse area with few suitable zones in view. Figure 8.17 shows all five rounds.

In this scenario, the drone experiences multiple failed attempts due to two primary factors: (1) dynamic changes in the landing environment between detection and approach phases, and (2) inconsistent safety assessments by the LVLM. This scenario demonstrates the challenges inherent in real-time autonomous decision-making systems operating in dynamic urban environments. Below we pair the LVLM's decisions with the key quotes from the LVLM.

Round 1: The LVLM initially identified a highway segment as the optimal landing zone, reasoning that the absence of traffic outweighed the explicit instruction to avoid highways, particularly given that alternative zones contained excessive obstacles.

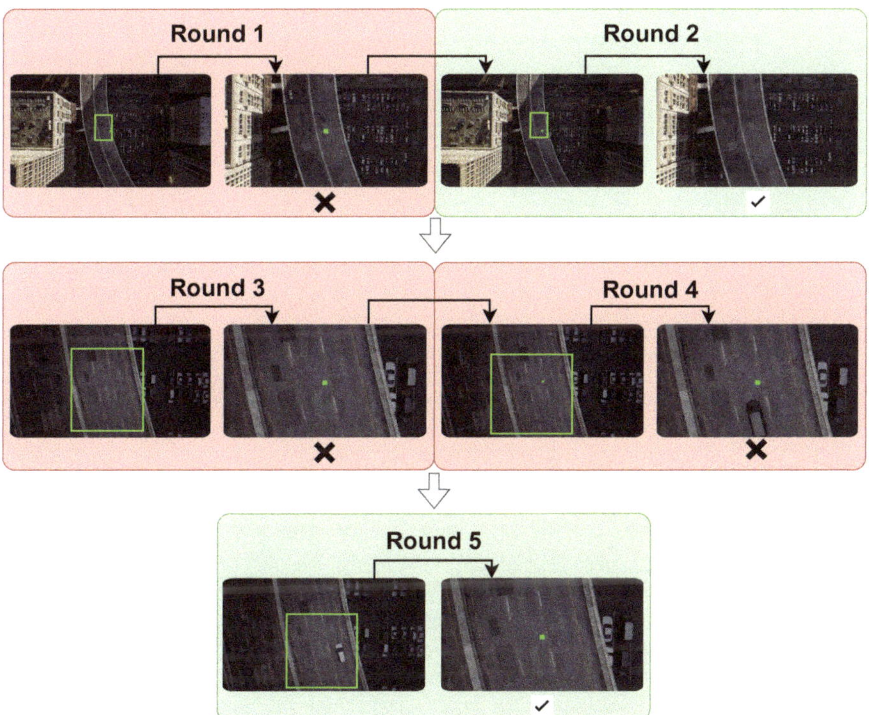

Fig. 8.17. Hesitation for landing on a highway.

```
Wide elevated roadway appears clear of vehicles and
pedestrians in the landing zone. Surface is flat and
unobstructed, offering ample space. Risk remains from
potential oncoming traffic and guardrails at the
edges, but within the frame it meets the constraints
best.
```

However, during the transit time to the proposed landing site, a vehicle entered the highway. Upon reaching the confirmation stage, where the drone reassesses landing zone safety from its new position directly above the target, the LVLM correctly identified the presence of the vehicle and aborted the landing attempt.

```
Not suitable for final approach. The marked spot is
in an active traffic lane on a roadway. There are
vehicles nearby, making the area not clear of
obstructions and posing collision risk. Although the
road appears generally flat, it is not clear of
traffic/obstacles and is near roadway
edges/structures, which further increases risk.
```

Round 2: As the LVLM does not confirm the landing, we run the pipeline again. The system again selected the same highway segment, which had cleared of traffic between rounds. During the confirmation stage, the highway remained clear, prompting the drone to initiate a controlled descent. However, as the drone had not yet reached sufficient proximity for final landing, another decision round was triggered.

Round 3: The LVLM identifies the highway once more as the safest option. Upon reaching the confirmation stage, the physical environment remained unchanged, with a clear highway.

```
Best option. It shows a wide, flat asphalt roadway
with clear lane markings. The central lanes appear
unobstructed and free of people and vehicles; parked
cars are confined to the right-side parking bay
behind a curb. Surface looks level and continuous,
with only minor patches/markings that don't pose
hazards. Risk remains from potential unseen traffic,
but among the options this is the widest, clearest
flat area.
```

Round 4: Similar to Round 1, the drone initially assessed the highway as safe, but vehicular traffic appeared during the transit period. The confirmation stage correctly identified this new obstacle, leading to another cycle repetition.

Round 5: Despite the presence of a vehicle on the highway, the LVLM determined this location remained the safest available option, as the alternative parking lot was deemed too cluttered with obstacles for safe landing. Fortuitously, the highway cleared during the drone's approach, enabling successful confirmation and completion of the landing sequence.

```
The highlighted area is on a wide, flat roadway.
The surface appears level with no visible debris or
protrusions near the target. There are no people in
the landing zone. Parked cars and the curb/guardrail
are off to the right but not within the immediate
touchdown area, leaving adequate clearance if you
stay centered between lane markings. Visually, this
meets the criteria for a clear, flat surface, though
remain vigilant for any approaching vehicles during
descent.
```

Takeaways. Safe choices can become unsafe during motion. Although the highway might be empty at some point in time, it becomes unsafe at others. As the LVLM correctly identified in the first round, "Risk remains from potential oncoming traffic", so this is something we should emphasize in future prompts; if the future risk is high, we should ignore that surface and search for better alternatives in other places, and not keep hesitating on whether or not to land on a highway.

Edge Case 2: We now analyze the case where we reached 10 rounds without landing. Marker A (Figure 8.13) shows this run. We illustrate the different rounds in Fig. 8.18.

We can attribute the failure of this experiment to (1) the limited availability of landing spots in the first rounds given to the LVLM by our surface detector and (2) the misclassification of an object as a bird in the final round.

Round 1–6: In these rounds, the LVLM detected different parts of the same highway moving slightly between lanes but failing to confirm due to the presence of vehicles moving beneath it and moving

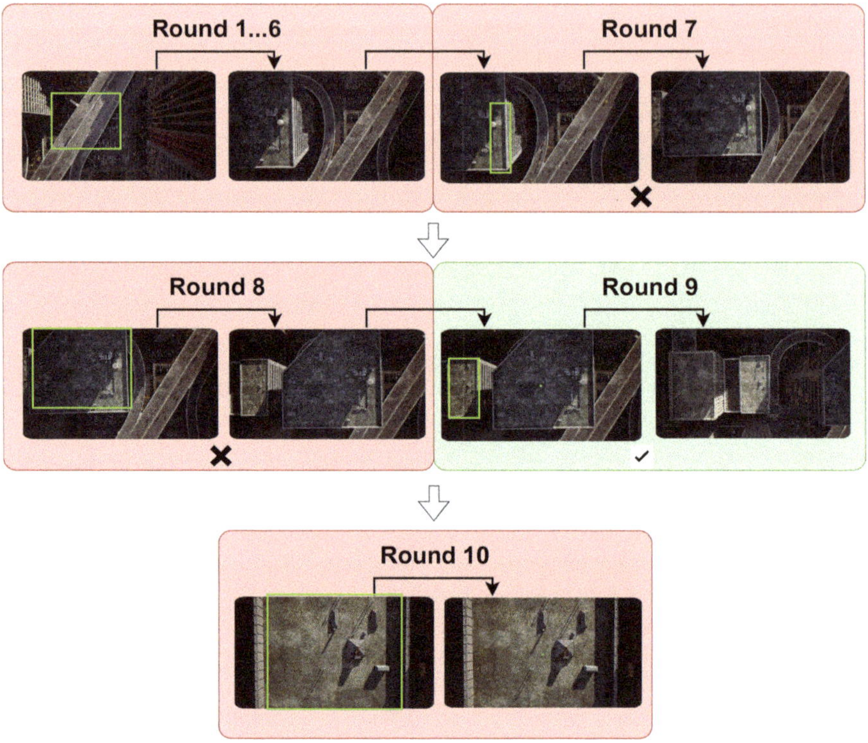

Fig. 8.18. Failure due to timeout.

at the end of Round 6 to an off-ramp from the highway, managing to change the camera view significantly.

```
Not suitable for final approach. The scene shows an
active roadway and a crowded parking lot. There are
many vehicles (moving on the street and parked in
rows), leaving no sufficiently large, obstacle-free
patch. Any gaps between cars are too tight for safe
rotor clearance. The elevated road surface is also a
traffic route with edge hazards and uncertain vehicle
presence. While surfaces appear flat, they are not
clear of obstructions and present collision risks.
```

Round 7: In this round, the LVLM selected a fraction of the adjacent rooftop as a suitable surface, finally leaving the highway area and positioning itself on top of a rooftop. However, the LVLM did not confirm the final approach due to the edge and obstacles on the roof.

```
The rooftop appears flat and there are no people
visible, but the selected spot is not safe. It lies
very close to multiple obstructions (HVAC
pipes/ducting and a vent) and is also near the roof
edge/parapet. These increase collision risk and
potential downdraft/edge turbulence during landing.
```

Round 8: In this round, the LVLM selected the complete rooftop as a possible landing spot, proceeding to its center, recognizing its possible obstacles and failing to confirm because of them.

```
The roof appears flat and there are no people or
vehicles. While there are many HVAC units, vents, and
pipes, there are multiple open gravel areas large
enough for a quadcopter to touch down, especially in
the lower-left and central-left zones away from
equipment and roof edges.
```

Round 9: With a better position, the system managed to detect an alternative landing surface at another roof further to the left, and unlike previous iterations, the LVLM cleared the drone for a descending operation.

```
Proceed with final approach to a clear section of
this roof, avoiding the central vent cluster and
maintaining margin from edges.
```

Round 10: In its final round, the drone identified the roof beneath it as the best option. However, it failed to confirm the landing, justifying its rationale on the presence of nonexistent wildlife on the roof.

```
However, there are two birds on the roof very close
to the proposed touchdown point. Moving wildlife is
an immediate obstruction risk during landing
(possible collision or sudden movement into the
rotors).
```

Takeaways. (i) When nearby safe surfaces are scarce, the agent can loop locally before exploring further; (ii) incorrect object identification can veto otherwise safe landings—motivating additional confirmations or sensor cross–checks.

Overall, the big takeaway is that when there are no viable candidates in the first round, the drone should explore different areas and not settle for what it was originally perceived. Of course this decision might depend on the criticality of the alert that requires the drone to land. If it has to land immediately perhaps a risky situation is more suitable than attempting to move somewhere else.

8.8.4 *Reasoning Analysis*

Finally, we study the reasons the LVLM provides to select a safe place to land, and understand the relevant elements in the decision.

We measure the semantic similarity of the LVLM final ranking responses, and then we embedded each response with OpenAI's `text-embedding-3-small`. This model allows us to convert the natural text output of the LVLM to a numerical representation (a vector). We then perform clustering of these vectors to identify patterns of decisions.

A visualization of these vectors using Principal Component Analysis (PCA) (Jolliffe and Cadima, 2016) is shown in Fig. 8.19. This figure suggests three clusters.

By inspecting the output of the LVLM in each cluster, we suggest the following three categories of LVLM reasoning: **Similar Candidates:** The largest cluster includes cases where the LVLM described multiple candidate surfaces as broadly alike, often using more generic descriptions of the terrain without explicitly identifying the specific type of surface. **Diverse Candidates:** This cluster

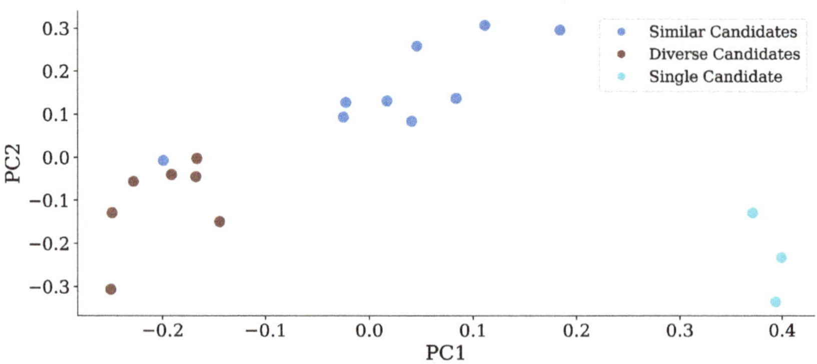

Fig. 8.19. Final decision clusters.

covers situations in which the LVLM has to make a selection over a broader variety of surfaces (e.g., roads, sidewalks, roofs), making these scenarios more complex than the previous cluster. **Single Candidate:** This cluster contains instances that only deal with a single detected surface, where reasoning is more straightforward and focused on acknowledging minor risks while ultimately following the prompt to select the sole option available.

These clusters illustrate that the LVLM adapts its reasoning style to the complexity of the decision space: when candidates look similar, it falls back on generic comparisons; when options differ substantially, it weighs trade-offs more carefully; and when only one option is present, it performs a simple risk check before proceeding. This variability suggests that the model is not applying a fixed heuristic but is instead modulating its rationale based on context, an encouraging sign for embodied AI operating in open environments. At the same time, the analysis highlights opportunities for improvement: more explicit guidance could help the model be less generic when faced with similar candidates, and better calibration is needed when evaluating diverse options with unequal risk profiles. In future work, we will try to focus on identifying the safety arguments the LVLM makes.

8.8.5 *Lessons Learned*

Our experiments taught us several valuable lessons about creating effective LVLM prompts and input strategies. For one, we discovered that when we explicitly told the model to both rank candidate surfaces and explain its reasoning, it became more reliable. Prompts that focused on safety constraints—like requiring surfaces to be clear of obstacles, people, and non-flat areas—led to more structured outputs and reduced the chances of missing hazards.

Second, we found that adding more context to the input did not always boost performance. Although we initially thought that including the entire scene would enhance reasoning, the most accurate results came from using only the cropped candidate surfaces. A key takeaway is that carefully curating the list of potential safe landing spaces can prevent the LVLM from getting overloaded with options and information.

Lastly, we need to account for cases where no suitable surfaces are found. Although our implementation reran the pipeline in these cases, alternative designs could use a set of predefined locations to evaluate or heuristics to enhance robustness and address uncertainty. We also observed that compact models like GPT-5-nano show varying sensitivities to context across different model variants, underscoring the broader point that efficiency-focused LVLMs may sacrifice accuracy for speed or easier deployment.

8.9 Conclusions and Future Work

We introduced a hybrid recovery pipeline that leverages LVLMs for high-level reasoning while grounding decisions in conventional perception and control modules. This modular design proved effective: the system can identify and execute safe landings in realistic city environments, demonstrating the promise of embodied AI for adaptive recovery. At the same time, our experiments revealed a key limitation—reliable performance currently requires large models, which are difficult to host directly onboard resource-constrained drones.

This tension between capability and deployability motivates future work on hierarchical inference architectures. A promising approach is to combine lightweight, onboard geometric checks with mid-sized distilled models at the edge and large LVLMs in the cloud. Such a design could balance efficiency, reliability, and resilience to connectivity constraints. Another direction is to explicitly model and quantify uncertainty in LVLM decisions (Bhatt *et al.*, 2024). Being able to detect when the model is unsure would enable active sensing or fallback heuristics, providing additional safeguards in dynamic environments.

Finally, our study highlights the need to evaluate robustness against adversarial and deceptive inputs (Byrd Victorica *et al.*, 2025). LVLM-based recovery systems remain vulnerable to visual perturbations such as phantom or patch attacks, and integrating adversarial resilience into the design will be essential for safety-critical deployments. Together, these directions chart a path toward recovery pipelines that are not only effective in simulation but also trustworthy and practical in real-world operations.

Acknowledgments

This material is based on work supported in part by the Air Force Office of Scientific Research (AFOSR) under award number FA9550-24-1-0015, and by the National Center for Transportation Cybersecurity and Resiliency (TraCR) (a US Department of Transportation National University Transportation Center) under grant numbers 69A3552344812 and 69A3552348317. Any opinions, findings, conclusions, and recommendations expressed in this material are those of the authors and do not necessarily reflect the views of TraCR, and the US Government assumes no liability for the contents or use thereof.

Prompts

System Prompt:

```
##Safe Drone Landing Assistant

You are a safety assistant for a quadcopter that is
flying over a city and needs to perform an emergency
landing on a surface. The quadcopter will provide you
with photos of several possible surfaces taken by its
camera bellow, and ask questions of the suitability
of the surfaces to perform the safest emergency
landing possible. To classify as safe a potential
landing site MUST take into consideration the
following factors:

## Constraints

- **MUST** be clear of any obstructions such as air
ducts, cars, rubble etc. - **MUST** be clear of
people - **MUST** be a flat surface

## Clarifications

- Ignore non-critical visual features such as
shadows, surface textures, tile patterns, or color
variations unless they **CLEARLY** indicate an actual
obstruction. - Do not assume a surface is unsafe just
because it might be something like a shingled roof
unless there is strong visual evidence of danger
(e.g., visible slope, fragile material, obvious
gaps). - If no surface is perfectly safe, select the
one with the lowest risk. - You **MUST** always
return a clear selection (never refuse).
```

Ranking Prompt:

```
You are receiving up to **5 images** from the drone:

- **Images 0 - 4** (maximum 5 images): Candidate
surfaces for landing.

### Task 1. Analyze each candidate surface (Images
0-4). 2. Rank the candidate surfaces from **most
suitable** to **least suitable** for landing. 3.
Provide detailed reasoning for each candidate,
explaining why it is more or less suitable. 4. Output
the ranking as a list of indices (corresponding to
the candidate images only).

### Output Format 1. **Detailed Explanation**
(paragraphs describing reasoning for each candidate).
2. **Final Ranking**: output the indices
corresponding to each photo, representing the ranking
from the most suitable place to land to the least
one. The index starts at 0.
```

Confirmation Prompt:

```
You are receiving a **close-up image** of a potential
landing surface that was previously selected by you.
Your task is to confirm whether this surface is safe
for a **final landing approach**.

### Task 1. Analyze the surface in the close-up
image. 2. Decide if the surface is **safe to land**
or **unsafe**. 3. Provide a detailed explanation of
your reasoning. 4. Output your decision as an index
array with a single value: - 1 in the indices array
to confirm or 0 to cancel the final approach.

### Notes - Ignore **non-critical details** such as
shadows, lighting variations, or surface patterns
unless they clearly indicate a real hazard. - Confirm
landing if the surface appears **flat, unobstructed,
and free of people**. - Deny landing only if there is
clear evidence of danger (obstructions, slope,
rubble, fragile/unsafe material). - Always return a
decision (never leave it blank).

### Output Format 1. **Explanation** (paragraph
justifying your decision). 2. **Final Decision** in
the form: 1 in the indices array to confirm or 0 to
cancel the final approach.
```

Bibliography

Amatriain, X. (2024). Prompt design and engineering: Introduction and advanced methods, arXiv:2401.14423 [cs.SE], https://arxiv.org/abs/2401.14423.

Barbosa, D. O., Burbano, L., Yang, S., Wang, Z., Cardenas, A. A., Xie, C., and Cao, Y. (2025). Robust and efficient AI-based attack recovery in autonomous drones, *arXiv preprint* arXiv:2505.14835.

Bhatt, N. P., Yang, Y., Siva, R., Milan, D., Topcu, U., and Wang, Z. (2024). Know where you're uncertain when planning with multimodal foundation models: A formal framework, *arXiv preprint* arXiv:2411.01639.

Byrd Victorica, M., Dán, G., and Sandberg, H. (2025). Saliuitl: Ensemble salience guided recovery of adversarial patches against CNNs, in *IEEE/CVF CVPR*.

Cardenas, A. A. (2025). Response and reconfiguration under attacks in CPS, in *Encyclopedia of Cryptography, Security and Privacy* (Springer Nature Switzerland AG), pp. 2100–2105.

Chen, Z., Wu, J., Wang, W., Su, W., Chen, G., Xing, S., Zhong, M., Zhang, Q., Zhu, X., Lu, L., *et al.* (2024). InternVL: Scaling up vision foundation models and aligning for generic visual-linguistic tasks, in *Proceedings of the IEEE/CVF Conference on Computer Vision and Pattern Recognition*, pp. 24185–24198.

Dash, P., Chan, E., and Pattabiraman, K. (2024). SpecGuard: Specification aware recovery for robotic autonomous vehicles from physical attacks, in *Proceedings of the 2024 on ACM SIGSAC Conference on Computer and Communications Security, CCS'24* (Association for Computing Machinery, New York, NY, USA), pp. 1849–1863, doi:10.1145/3658644.3690210, https://doi.org/10.1145/3658644.3690210.

Dash, P., Li, G., Chen, Z., Karimibiuki, M., and Pattabiraman, K. (2021). PID-Piper: Recovering robotic vehicles from physical attacks, in *2021 IEEE/IFIP International Conference on Dependable Systems and Networks (DSN)* (IEEE), pp. 26–38, https://people.ece.ubc.ca/zitaoc/files/Pid-Piper-DSN21.pdf.

de la Torre-Vanegas, J., Soriano-Garcia, M., Becerra, I., and Mercado-Ravell, D. (2025). Vision-based risk aware emergency landing for UAVs in complex urban environments, arXiv:2505.20423 [cs.RO], https://arxiv.org/abs/2505.20423.

Dreossi, T., Donzé, A., and Seshia, S. A. (2019). Compositional falsification of cyber-physical systems with machine learning components, *Journal of Automated Reasoning* **63**, 4, pp. 1031–1053.

Eigen, D., Puhrsch, C., and Fergus, R. (2014). Depth map prediction from a single image using a multi-scale deep network, arXiv:1406.2283 [cs.CV], https://arxiv.org/abs/1406.2283.

Everingham, M., Gool, L., Williams, C. K., Winn, J., and Zisserman, A. (2010). The pascal visual object classes (VOC) challenge, *International Journal of Computer Vision* **88**, 2, pp. 303–338, doi:10.1007/s11263-009-0275-4, https://doi.org/10.1007/s11263-009-0275-4.

Garg, K., Sanfelice, R. G., and Cardenas, A. A. (2022). Control barrier function-based attack-recovery with provable guarantees, in *2022 IEEE 61st Conference on Decision and Control (CDC)* (IEEE), pp. 4808–4813.

Gemini Robotics Team, A., Abeyruwan, S., Ainslie, J., Alayrac, J.-B., Arenas, M. G., Armstrong, T., Balakrishna, A., Baruch, R., Bauza, M., Blokzijl, M., *et al.* (2025). Gemini robotics: Bringing ai into the physical world, *arXiv preprint* arXiv:2503.20020.

Halton, J. H. (1960). On the efficiency of certain quasi-random sequences of points in evaluating multi-dimensional integrals, *Numerische Mathematik* **2**, 1, pp. 84–90.

Insaurralde, C. C., Blasch, E., and Sabatini, R. (2022). Ontology-based situation awareness for air and space traffic management, in *2022 IEEE/AIAA 41st Digital Avionics Systems Conference (DASC)* (IEEE), pp. 1–8.

Jang, J., Cho, M., Kim, J., Kim, D., and Kim, Y. (2023). Paralyzing drones via EMI signal injection on sensory communication channels, in *Network and Distributed System Security Symposium (NDSS)* (Internet Society), doi:10.14722/ndss.2023.24616, https://cybershafarat.com/wp-content/uploads/2023/06/ParalyzingDronesviaEMISignalInjection.pdf.

Jolliffe, I. T. and Cadima, J. (2016). Principal component analysis: A review and recent developments, *Philosophical Transactions of the Royal Society A: Mathematical, Physical and Engineering Sciences* **374**, https://api.semanticscholar.org/CorpusID:20101754.

Kirillov, A., Mintun, E., Ravi, N., Mao, H., Rolland, C., Gustafson, L., Xiao, T., Whitehead, S., Berg, A. C., Lo, W.-Y., Dollár, P., and Girshick, R. (2023). Segment anything, arXiv:2304.02643.

Lee, M.-F. R., Nugroho, A., Le, T.-T., Bahrudin, and Bastida, S. N. (2020). Landing area recognition using deep learning for unammaned aerial vehicles, in *2020 International Conference on Advanced Robotics and Intelligent Systems (ARIS)*, pp. 1–6, doi:10.1109/ARIS50834.2020.9205793.

Leudo, S. J., Garg, K., Sanfelice, R. G., and Cardenas, A. A. (2023). An observer-based switching algorithm for safety under sensor denial-of-service attacks, in *2023 American Control Conference (ACC)* (IEEE), pp. 2469–2474.

Marcu, A., Costea, D., LicăreŢ, V., Pîrvu, M., Sluşanschi, E., and Leordeanu, M. (2019). SafeUAV: Learning to estimate depth and safe landing areas for UAVs from synthetic data, in Leal-Taixé, L. and Roth, S. (eds.), *Computer Vision – ECCV 2018 Workshops* (Springer International Publishing, Cham), pp. 43–58.

Maturana, D. and Scherer, S. (2015). 3D convolutional neural networks for landing zone detection from LiDAR, in *2015 IEEE International Conference on Robotics and Automation (ICRA)*, pp. 3471–3478, doi:10.1109/ICRA.2015.7139679.

Prakash, A., Chitta, K., and Geiger, A. (2021). Multi-modal fusion transformer for end-to-end autonomous driving, in *Proceedings of the IEEE/CVF Conference on Computer Vision and Pattern Recognition (CVPR)*, pp. 7077–7087.

Roque, P., Cortez, W. S., Lindemann, L., and Dimarogonas, D. V. (2022). Corridor MPC: Towards optimal and safe trajectory tracking, in *2022 American Control Conference (ACC)*, pp. 2025–2032, https://doi.org/10.23919/ACC53348.2022.9867764.

Sahoo, P., Singh, A. K., Saha, S., Jain, V., Mondal, S., and Chadha, A. (2025). A systematic survey of prompt engineering in large language models: Techniques and applications, arXiv:2402.07927 [cs.AI], https://arxiv.org/abs/2402.07927.

Shah Alam, M. and Oluoch, J. (2021). A survey of safe landing zone detection techniques for autonomous unmanned aerial vehicles (UAVs), *Expert Systems with Applications* **179**, p. 115091, https://doi.org/10.1016/j.eswa.2021.115091, https://www.sciencedirect.com/science/article/pii/S0957417421005327.

Sinha, R., Elhafsi, A., Agia, C., Foutter, M., Schmerling, E., and Pavone, M. (2024). Real-time anomaly detection and reactive planning with large language models, arXiv:2407.08735 [cs.RO], https://arxiv.org/abs/2407.08735.

Sturges, S. (2022). OpenLander: Free-to-use landing-zone detection for UAVs, https://github.com/stephansturges/OpenLander.

Turkcan, M. K., Li, Y., Zang, C., Ghaderi, J., Zussman, G., and Kostic, Z. (2024). Boundless: Generating photorealistic synthetic data for object detection in urban streetscapes, *arXiv preprint* arXiv:2409.03022.

Wei, J., Wang, X., Schuurmans, D., Bosma, M., Ichter, B., Xia, F., Chi, E., Le, Q., and Zhou, D. (2023). Chain-of-thought prompting elicits reasoning in large language models, arXiv:2201.11903 [cs.CL], https://arxiv.org/abs/2201.11903.

Xiao, Y., Codevilla, F., Gurram, A., Urfalioglu, O., and López, A. M. (2022). Multimodal end-to-end autonomous driving, *IEEE Transactions on Intelligent Transportation Systems* **23**, 1, pp. 537–547, https://doi.org/10.1109/TITS.2020. 3013234.

Yang, L., Kang, B., Huang, Z., Zhao, Z., Xu, X., Feng, J., and Zhao, H. (2024). Depth anything V2, arXiv:2406.09414 [cs.CV], https://arxiv.org/abs/2406.09414.

Yang, T., Li, P., Zhang, H., Li, J., and Li, Z. (2018). Monocular vision slam-based UAV autonomous landing in emergencies and unknown environments, *Electronics* **7**, 5, doi:10.3390/electronics7050073, https://www.mdpi.com/2079-9292/7/5/73.

Zhang, L., Burbano, L., Chen, X., Cardenas, A. A., Drager, S., Anderson, M., and Kong, F. (2024). Fast attack recovery for stochastic cyber-physical systems, in *2024 IEEE 30th Real-Time and Embedded Technology and Applications Symposium (RTAS)* (IEEE), pp. 280–293, https://doi.org/10.1109/RTAS61025. 2024.00030.

Zhang, L., Chen, X., Kong, F., and Cardenas, A. A. (2020). Real-time attack-recovery for cyber-physical systems using linear approximations, in *Proceedings of the 41st IEEE Real-Time Systems Symposium (RTSS)* (IEEE), pp. 205–217, https://doi.org/10.1109/RTSS49844.2020.00028, https://par.nsf.gov/servlets/purl/10294493.

Zhang, L., Lu, P., Kong, F., Chen, X., Sokolsky, O., and Lee, I. (2021). Real-time attack-recovery for cyber-physical systems using linear-quadratic regulator, *ACM Transactions on Embedded Computing Systems* **20**, 5s, https://doi.org/10.1145/ 3477010, https://doi.org/10.1145/3477010.

Zhang, L., Sridhar, K., Liu, M., Lu, P., Chen, X., Kong, F., Sokolsky, O., and Lee, I. (2023). Real-time data-predictive attack-recovery for complex cyber-physical systems, in *2023 IEEE 29th Real-Time and Embedded Technology and Applications Symposium (RTAS)*, pp. 209–222, https://doi.org/10.1109/RTAS58335.2023.00024.

https://doi.org/10.1142/9789819831548_0009

Chapter 9

Leveraging Generative AI for Cybersecurity and Resiliency in Transportation Cyber-Physical Systems

M. Sabbir Salek*, Abdullah Al Mamun†, and Mashrur Chowdhury‡

** USDOT National Center for Transportation Cybersecurity and Resiliency (TraCR), Greenville, SC*
† Glenn Department of Civil Engineering Clemson University, Clemson, SC
‡ Glenn Department of Civil Engineering Clemson University, Clemson, SC

Abstract
Generative Artificial Intelligence (GenAI) is rapidly evolving as a transformative technology with the potential to substantially strengthen the cybersecurity and resiliency of transportation systems and critical infrastructure. As intelligent transportation systems (ITS) grow increasingly interconnected, they also become more vulnerable to cyberattacks, adversarial manipulation, and cascading failures. GenAI's unique ability to model complex data distributions, synthesize realistic scenarios, and adapt to evolving threats positions it as an enabler of proactive defense, anomaly detection, and resilient system design, as observed from related studies. Leveraging this capability is vital to ensure that next-generation transportation ecosystems remain safe, secure, and trustworthy in the face of emerging and evolving cyber challenges.

This chapter first briefly introduces GenAI and then explores its opportunities for enhancing transportation cybersecurity across multiple dimensions, including data-centric intrusion detection, AI-driven threat simulation and response, edge and physical layer protection, and system-level resilience within ITS ecosystems. It also critically examines the challenges of applying GenAI in these safety-critical domains. These challenges include dual-use risks, model and pipeline integrity, the scarcity of realistic cyber-physical system (CPS) datasets, and constraints associated with deploying GenAI models at the network edge. To demonstrate these opportunities in practice, the chapter presents a case study on a generative adversarial network (GAN)-based attack-resilient strategy for autonomous vehicle traffic sign classification. The proposed attack-resilient GAN (AR-GAN) demonstrates superior robustness against a variety of black-box and white-box adversarial attacks, underscoring the potential of generative approaches to safeguard critical transportation functions.

Finally, the chapter concludes by outlining promising future directions, including the integration of GenAI with digital twins for predictive threat and risk assessments, synthetic data generation to overcome data scarcity in rare and safety-critical transportation-related incidents, large language model-based copilots for cybersecurity operations, and multimodal approaches for threat detection and supply chain assurance. These discussions highlight how GenAI can be strategically leveraged to not only defend against emerging cyber threats but also to build resilient, adaptive, and future-ready transportation infrastructures.

9.1 Introduction to Generative AI

Generative Artificial Intelligence (GenAI) comprises models that learn a probability distribution over data and then synthesize new, plausible samples from it. In probabilistic notation, consider:

- $p(x)$ denotes probability distribution of the data x. In the context of intelligent transportation systems (ITS), x could be traffic sign images, traffic flow, density, etc. A model that learns $p(x)$ can directly generate new samples resembling those in the training set.
- $p(y \mid x)$ represents the conditional probability of a label or outcome y given an input x. This is the target of discriminative systems, similar to image or object classifiers. In the context of ITS, y could be a traffic sign label or any application-specific predictive outcome.
- $p(x \mid c)$ describes the probability of data x conditioned on some context or condition c. In practice, c can be a text prompt or a

class label. This conditional modeling allows controlled generation, producing outputs that align with the specified condition.

Whereas discriminative systems primarily model $p(y \mid x)$ to make predictions, a GenAI system models $p(x)$ or $p(x \mid c)$ to create new data samples. Modern GenAI systems are typically deep neural networks trained at scale, and they are deployed for different input data types, such as text, images, audio, video, and code, often within encoder–decoder pipelines that build a latent representation and then decode it into the target data types. Table 9.1 summarizes popular GenAI model families, along with their representative capabilities, strengths, and examples.

GenAI models have been explored for different functionalities in transportation cyber-physical systems (CPS). In addition to ensuring cybersecurity and cyber resilience in transportation CPS, which we discuss in detail in Sections 9.1 and 9.2, GenAI has shown promise in other transportation CPS use cases. Examples include GenAI's applicability in enhancing the robustness and coverage of autonomous perception by supplementing AI training with synthetic data (Tremblay *et al.*, 2018), supporting autonomous decision-making and planning with human-like reasoning and scenario foresight (Nie *et al.*, 2025; Zou *et al.*, 2025), and supporting traffic operations and network-level control (Bilal, H. *et al.*, 2025; Somvanshi *et al.*, 2025).

Sections 9.1 and 9.2 present the opportunities and challenges of GenAI, respectively, for enhancing the cybersecurity of transportation systems and infrastructure. Section 9.3 presents a case study of GenAI-enhanced cybersecurity, where we develop a generative adversarial network (GAN)-based attack-resilient traffic sign classification system for an autonomous vehicle. Section 9.4 presents a summary of the chapter, and Section 9.5 presents several promising directions for future research in GenAI-enhanced cybersecurity of transportation systems and infrastructure.

9.2 Opportunities of GenAI for Cybersecurity of Transportation Systems and Infrastructure

As GenAI techniques advance, they create practical opportunities to strengthen the cybersecurity posture of transportation CPS. The following subsections examine how GenAI supports the detection

Table 9.1. Overview of popular families of GenAI models.

Model Family	Typical Capabilities	Strengths	Examples
Variational Autoencoders (VAEs)	• Reconstruction and synthesis across images, audio, and text • Interpretable latent manipulation/editing • Foundational backbone for latent diffusion models	• Stable, likelihood-based training • Structured and controllable latent spaces • Effective for representation learning	Autoencoders with probabilistic latents, LSGM, etc.
Generative Adversarial Networks (GANs)	• High-fidelity image and video synthesis • Image-to-image translation, e.g., style transfer, domain adaptation • Super-resolution • Conditional and text-guided generation	• High perceptual sharpness and detail • Strong conditional control • Flexible architectures for translation and style transfer	cGAN, AC-GAN, StyleGAN, WGAN-GP, etc.
Diffusion/ Score-based Models	• Text-to-image generation • Inpainting and outpainting • Super-resolution • Text-to-speech (TTS) and audio synthesis • Text-to-video and motion generation	• State-of-the-art sample fidelity and diversity • Stable and scalable training procedures • Flexible conditioning with text, images, and audio	DDPM, DDIM, GLIDE, Imagen, etc.

(Continued)

Table 9.1. (*Continued*)

Model Family	Typical Capabilities	Strengths	Examples
Normalizing Flows	• Exact likelihood-based density modeling • Tractable sampling and inference with invertible transforms • Uncertainty estimation and out-of-distribution detections	• Tractable density estimation with exact log-likelihood • Calibrated uncertainty estimates • Efficient sampling and invertibility for analysis tasks	NICE, RealNVP, Glow, etc.
Autoregressive Transformers	• Text and code generation • Summarization and question answering • Reasoning, planning, and tool use • Act as decoders in multimodal pipelines	• Long-range dependency modeling in sequences • Strong transfer from large-scale pretraining • Versatile few-shot and zero-shot generalization	GPT-style decoders, T5, BART, etc.
Hybrids and Multimodal Encoder–Decoder Models	• Cross-modal generation • Controllable and conditional editing • Retrieval-augmented and tool-augmented generation	• Integration with retrieval and external tools for augmentation	Vision–language encoders with transformers, LDMs with text embeddings, etc.

Notes: LSGM: Latent Score-Based Generative Model; cGAN: Conditional GAN; AC-GAN: Auxiliary Classifier GAN; StyleGAN: Style-based GAN; WGAN-GP: Wasserstein GAN with Gradient Penalty; DDPM: Denoising Diffusion Probabilistic Model; DDIM: Denoising Diffusion Implicit Model; GLIDE: Guided Language-to-Image Diffusion for Generation and Editing; Imagen: Google's text-to-image diffusion model; NICE: Non-linear Independent Components Estimation; RealNVP: Real-valued Non-Volume Preserving transformation; Glow: a flow-based generative model with invertible 1×1 convolutions; GPT-style decoders: Generative Pretrained Transformer; T5: Text-to-Text Transfer Transformer; BART: Bidirectional and Auto-Regressive Transformers; and LDM: Latent Diffusion Model.

and diagnosis of cyber intrusions, anomalous network behaviors, and safety-relevant deviations in control and physical processes, enables proactive simulation and response to attacks and failures, strengthens edge and physical-layer robustness, and builds system-level resilience in ITS.

9.2.1 *Data-Centric Intrusion and Anomaly Detection*

In vehicle-to-everything (V2X) communication networks, where data exchange occurs among vehicles, roadside units (RSUs), and back-end services, GenAI can synthesize realistic and diverse traffic and attack traces. This helps mitigate class imbalance and improves generalization to zero-day behaviors by exposing the detector to attack patterns that have not appeared in training data and therefore lack labels. Figure 9.1 illustrates a representative workflow in which real-world vehicular telemetry and limited observed attack samples are augmented using a GAN to produce a richer training dataset for intrusion detection. By sampling across mobility conditions (e.g., traffic density and routes), wireless channel conditions (e.g., signal fading and interference), and protocol behavior (e.g., message timings and formats), the generative stage exposes the detector to a broader portion of the underlying data distribution. In practice, this helps the GenAI model learn decision boundaries that capture rare and emergent threats rather than overfitting to known signatures. One study in simulated V2X environments reported high overall detection accuracy with low false positives when a generative stage augmented training data before classification, illustrating the benefit of model-driven augmentation in dynamic, latency-constrained settings (Sun and Ortiz, 2024).

Railway operations face analogous high-stakes constraints. A recent review of AI for railway cybersecurity highlights anomaly modeling over time-series signaling and train-control telemetry as a promising path, particularly when GenAI models learn multi-modal dependencies beyond static, rule-based logic; for example, how speed profiles, signal timing, and network messages interact (Qi and Wang, 2025). Security also extends to the power interfaces that

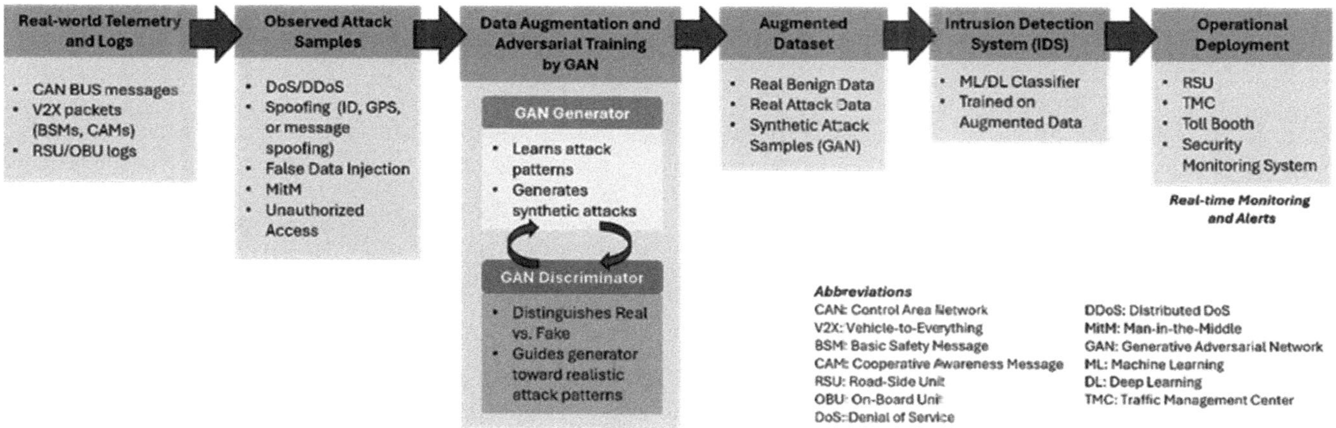

Fig. 9.1. GAN-based data augmentation for intrusion detection in transportation systems, where synthetic attack samples enhance IDS training prior to deployment.

keep electric rails and vehicles moving. Many traction substations, station auxiliaries, and depot fast chargers rely on International Electrotechnical Commission (IEC) 61850 substation-automation messaging. Therefore, applying GenAI-based anomaly detection to these message streams helps protect traction power, depot operations, and charging operations (Yigit *et al.*, 2025; Zaboli *et al.*, 2025). Complementary surveys on the Internet of Vehicles (IoV), a broader ecosystem of networked vehicles and services, advocate the use of GAN/variational autoencoder (VAE)/Transformer-based intrusion detection to address the challenges of scarce labels, evolving attack tooling, and adversarial evasion (Mahmoudi *et al.*, 2025). Scarce labels arise because real attack instances are rare, sensitive, and unevenly shared among operators and deployments. Evolving attack tooling refers to new tactics and malware variants that shift data distributions and invalidate fixed signatures. Adversarial evasion involves crafted packets, messages, or sensor inputs such as forged basic safety messages (BSMs), replay or Sybil identities, or subtle timing and feature perturbations that are designed to bypass detectors (Mahmoudi *et al.*, 2025).

9.2.2 *GenAI-Enabled Threat Simulation and Response*

In routine operations, large language models (LLMs) powered playbooks and retrieval-augmented generation (RAG) systems gather logs, tickets, and advisories from multiple transportation agencies and operators, summarize the most relevant information, and answer targeted questions. In effect, they turn scattered information into clear and actionable context for incident responders (Thomas *et al.*, 2025; Uddin *et al.*, 2025). One recent study outlines how a transportation-focused incident RAG can index past event reports and advisories to answer “what-if” and “where else” questions during a live cyber incident, thereby moving beyond static runbooks toward adaptive, context-aware decision support (Thomas *et al.*, 2025).

Proactive attack simulation also helps. Generative agents can play the attacker, emulating evolving tactics, techniques, and procedures so that defenders can test and tune controls before a real incident, as shown in recent frameworks that use LLMs, reinforcement learning, and GANs (Adeyinka and Adeyinka, 2025; Anis and Hammoudeh,

2025). At the cyber-physical edge, GAN-based defenses can clean perception inputs. For example, a single-stage GAN defense removes adversarial patches from traffic sign images before classification, adding a small computational cost in exchange for stronger robustness in the perception pipeline (Enan and Chowdhury, 2025).

Despite the benefits of GenAI-enabled threat simulation, automated incident response, and perception-level defenses, deploying GenAI-based threat simulation and response capabilities at the edge introduces important practical barriers. On-board units (OBUs) and RSUs host multiple safety- and security-critical functions that share limited computational, memory, and communication resources. Resource contention between GenAI workloads and existing real-time functions can lead to unpredictable execution delays and degraded system performance (Burns and Davis, 2022; Burns and Jones, 2025). From a functional safety perspective, this directly conflicts with the Freedom from Interference (FFI) requirement defined in ISO 26262, which mandates that software components of different criticality levels must not adversely affect one another's timing or behavior (ISO, 2018). Ensuring FFI in the presence of computationally intensive GenAI models remains a key challenge for edge deployment, requiring strict partitioning, scheduling controls, and architectural separation when integrating GenAI into safety-critical transportation systems (Burns and Jones, 2025).

9.2.3 *Edge and Physical Layer Security*

GenAI models offer data-driven wireless channel modeling and semantic-aware communication that can improve reliability and security for connected fleets and roadside infrastructure. Physical-layer security surveys report that GenAI can synthesize waveforms (wireless signal shapes) that are resilient to jamming and spoofing, improve channel estimation by inferring the time-varying wireless channel needed to decode messages, and support anomaly modeling directly at the radio interface (Van Huynh *et al.*, 2024; Zhao *et al.*, 2025). However, these works also caution that deploying models near the radio increases computational demands and privacy risk, because on-device inference operates under very short real-time control and scheduling deadlines and can process sensitive telemetry (Van Huynh *et al.*, 2024; Zhao *et al.*, 2025).

In 5G NR V2X and emerging 6G networks that support ITS, diffusion-guided policies are trained offline on historical logs and simulations and then propose how to allocate spectrum, transmission power, and processing resources for each cell or RSU. GAN-generated test cases create rare and stressful situations (for example, sudden traffic surges, fast-moving convoys, deep fading, or interference bursts) so engineers can verify that handovers stay stable, load balancing works, and integrated sensing and communication (ISAC) is tuned to detect objects and conditions while carrying data (Khoramnejad and Hossain, 2025; Vu *et al.*, 2025). A recent communications survey describes safe offline exploration of network resource allocation policies. The approach learns candidate configurations from logged telemetry (e.g., scheduler settings, power control, beamforming, and handover thresholds) and then validates them with generative scenarios before deployment to real-world systems (Khoramnejad and Hossain, 2025).

9.2.4 *System-Level Resilience in ITS*

In ITS ecosystems, AI-related cyber risks, such as data poisoning of routing and dispatch models, coordinated fraud across payment and identity systems, and privacy leakage from travel histories, can be mitigated by GenAI techniques for privacy preservation (for example, learning from masked or synthetic data) and anomaly modeling integrated into identity, billing, and operations-management services (Chu *et al.*, 2024). In smart routing and logistics, belief-desire-intention (BDI) architectures have been extended with generative decision modules that explore possible operational scenarios such as changing traffic and demand patterns, link failures, and attack conditions, and they adapt routing under uncertain network conditions and adversarial disruptions (Almadhor *et al.*, 2025). Adversarial disruptions include GPS spoofing or jamming that degrades localization, falsified or replayed V2X safety messages and Sybil identities that misdirect traffic flows, denial-of-service (DoS) against RSUs or back-end links that cause outages, false-data injection into sensor feeds or demand forecasts that skews routing decisions, and malware targeting dispatch, fleet-management, or charging-control systems (Almadhor *et al.*, 2025). Resource-constrained modalities, such as unmanned aerial vehicle-assisted ITS, can apply GenAI to produce workload-aware schedules and adaptive trajectories that comply with energy and latency budgets. This helps maintain service continuity during incidents (for example, communication link failures, sensor outages, or localized cyber-attacks) and during demand surges (for example, special events, extreme weather, or evacuation traffic) (Sharif *et al.*, 2025). In parallel, dual-framework approaches for ITS that address privacy and security together show how GenAI models can mask personally identifiable mobility patterns (for example, via aggregation or synthetic surrogates) while still detecting abnormal flows, suggesting architectures where privacy preservation and intrusion detection evolve together (Khan *et al.*, 2025).

9.3 Challenges of GenAI for Transportation Cybersecurity

As GenAI techniques mature within safety-critical transportation CPS, they introduce constraints that must be managed alongside their benefits. The following subsections present the principal challenges and their implications for deployment, operational resilience, and the assurance of security, safety, and regulatory compliance. Addressing these issues early is essential to realizing the benefits of GenAI in transportation CPS without compromising security and safety.

9.3.1 *Dual-Use Risk and Adversary Acceleration*

The same GenAI models that help defenders can also accelerate phishing and social engineering, generate exploit code, and fabricate believable telemetry that misleads detectors. A particularly critical dual-use risk is the data poisoning of generative models, in which adversaries manipulate training, fine-tuning, or instruction-tuning datasets to bias model behavior, degrade defensive performance, or cause the generation of unsafe outputs (Wan *et al.*, 2023). Reviews of generative AI for cybersecurity describe how instruction-tuned models, when guided by carefully crafted prompts or seeded with risky examples, produce unsafe guidance or insecure code. This underscores the need for systematic red team testing (structured exercises that examine models with attacker-style prompts and scenarios) and for refusal tuning (policies, training data, and guardrails that teach models to decline unsafe requests) (Alqahtani and Kumar, 2025; Metta *et al.*, 2024; Sai *et al.*, 2024; Uddin *et al.*, 2025; Wang, 2024;

Yigit *et al.*, 2025). Many surveys on LLMs for critical infrastructure present adversarial benchmarks showing that even small amounts of poisoning in the instruction-tuning data of LLMs can significantly increase attack success rates (Wan *et al.*, 2023; Yigit *et al.*, 2025). For example, Wan *et al.* show that inserting as few as 100 poisoned instruction-tuning examples can cause instruction-tuned LLMs to produce degenerate outputs or systematic errors across held-out downstream tasks whenever a trigger phrase appears (Wan *et al.*, 2023). Notably, the poisoned models behave normally on most benign inputs, demonstrating how small-scale data poisoning can substantially increase attack success rates while remaining difficult to detect in practice. Another survey also reports coding-assistant evaluations in which more capable GenAI models sometimes produce insecure code suggestions (Yigit *et al.*, 2025). These findings transfer directly to transportation operations, where LLM and RAG assistants and coding tools are being introduced into safety-critical development and incident workflows, and procurement and supplier audits should mandate LLM-specific evaluations for poisoning resistance and insecure-code generation (Arif *et al.*, 2025; Aung *et al.*, 2025; Chu *et al.*, 2024; Illiashenko *et al.*, 2023; Khoramnejad and Hossain, 2025; Uddin *et al.*, 2025).

9.3.2 *Model and Data-Pipeline Integrity*

Transportation threat models now need to cover GenAI supply chains, including prompt injection against LLM-enabled consoles, document-level prompt injection in the RAG knowledge base, data poisoning of fleet logs, and compromise of third-party model weights (Arif *et al.*, 2025; Mahmoudi *et al.*, 2025; Uddin *et al.*, 2025). Recent work on detecting adversarial and supply-chain attacks against GenAI models recommends end-to-end validation of dependencies and behavioral anomaly detection around model inputs and outputs (for example, monitoring for sudden drift in prompt patterns, embeddings, or output distributions) (Arif *et al.*, 2025; Uddin *et al.*, 2025). These controls fit naturally into ITS model registries and the security-informed safety case. In practice, the registry records each version of the GenAI models and datasets, verifies their development history, and stores the information needed to reproduce the models later. It also documents who trained each model, which data were used, and which hyperparameters were selected, while enforcing role-based access, approval gates, and preservation of test artifacts for audit. The safety case then ties this evidence to identified hazards, pre-deployment red-team exercises, in-service monitoring for drift and poisoning, and defined rollback procedures (Chu *et al.*, 2024; Illiashenko *et al.*, 2023; Radanliev *et al.*, 2025; Uddin *et al.*, 2025). At the system level, these controls are enforced through established change control and safety certification led by transportation agencies, state departments of transportation, and transit authorities (Chu *et al.*, 2024; Illiashenko *et al.*, 2023; Radanliev *et al.*, 2025; Uddin *et al.*, 2025). They are executed in partnership with operators, manufacturers, and service providers, such as rail infrastructure managers and train operating companies, metropolitan transit agencies, highway and toll operators, port and airport authorities, vehicle original equipment manufacturers (OEMs) and Tier-1 suppliers, on-board unit (OBU) and RSU manufacturers, mobile-network operators, cloud providers, and, where relevant, electric utilities and charging-network operators.

9.3.3 *Evaluation and Data Scarcity under Realistic CPS Conditions*

Surveys of railway and IoV intrusion detection repeatedly note the lack of open, high-fidelity datasets that capture closed-loop dynamics, for example, how controller commands (braking, speed profiles, interlocking states) change sensor readings and network traffic, and then feed back into subsequent control decisions across vehicles, infrastructure, and human operators (Mahmoudi *et al.*, 2025; Qi and Wang, 2025). Static benchmarks understate concept drift, meaning gradual or abrupt shifts in data distributions over time (for example, software updates, seasonal traffic changes, new sensors, or attacker adaptation), and they fail to represent coordinated, stealthy cyber-attack campaigns in which adversaries stage multi-step actions across vehicles, roadside units, and back-end services (Mahmoudi *et al.*, 2025; Qi and Wang, 2025). New benchmarks tailored to critical infrastructure are emerging, yet they also introduce issues in LLM-based operations tooling, including hallucination (confident but incorrect answers that appear authoritative), brittle retrieval in RAG pipelines (pulling irrelevant or non-authoritative evidence for answers), and fragile refusal boundaries (safety guardrails that can be inconsistently triggered or bypassed), all of which require careful management before field deployment (Yigit *et al.*, 2025).

9.3.4 *Edge Deployment under Resource and Timing Constraints*

Physical-layer and mobile-network surveys emphasize that large GenAI models strain computational, memory, and energy budgets in OBUs and RSUs, especially when inference is performed near the radio under millisecond-scale control and scheduling deadlines, and may increase privacy risks when sensitive vehicular or user data are processed close to the network edge (Khoramnejad and Hossain, 2025; Van Huynh *et al.*, 2024; Vu *et al.*, 2025; Zhao *et al.*, 2025). These millisecond-scale timing constraints are typical of safety-related control loops in vehicular systems (Karagiannis *et al.*, 2011), leaving little tolerance for variable or prolonged inference delays introduced by large generative models. To make deployment feasible, existing work recommends several edge-centric strategies: (i) distributed or federated training, which keeps raw telemetry local while enabling model updates across vehicles and RSUs; (ii) on-device distillation, often combined with quantization and pruning, to transfer a large "teacher" model into a compact "student" model suitable for embedded hardware; and (iii) architectural separation, in which lightweight generative components are deployed at the edge while computationally intensive training or simulation remains in the cloud. Looking forward, these approaches point toward hybrid architectures in which GenAI is used in advisory, monitoring, or anomaly-detection roles rather than within hard real-time control loops, allowing future systems to benefit from generative capabilities while preserving deterministic timing guarantees required for safe vehicular operation (Khoramnejad and Hossain, 2025; Van Huynh *et al.*, 2024; Vu *et al.*, 2025; Zhao *et al.*, 2025).

9.3.5 *Probabilistic GenAI vs. Deterministic Safety Requirements*

A fundamental challenge in deploying GenAI in transportation systems lies in the gap between the probabilistic nature of generative models and the deterministic requirements of automotive safety and security standards. GenAI models, including LLMs and other generative architectures, exhibit nondeterministic behavior, sensitivity to context, and performance variability under distribution shift, making their outputs difficult to bound and verify. In contrast, ISO 26262, the international standard for the functional safety of road vehicles, is built on deterministic assumptions, requiring traceable safety requirements, predictable execution, and evidence that safety goals are consistently met under defined operating conditions (ISO, 2018). ISO/PAS 8800, the international standard focused on safety assurance for AI-enabled road-vehicle functions, explicitly acknowledges this mismatch and provides guidance on incorporating AI-enabled functions into safety engineering processes by managing uncertainty, constraining operational domains, and establishing appropriate assurance arguments (ISO, 2024). This challenge is further compounded by cybersecurity considerations, as adversarial manipulation of data or inputs can amplify probabilistic behavior. ISO/SAE 21434, the international standard for cybersecurity engineering across the road-vehicle lifecycle, addresses this dimension by defining a lifecycle-based cybersecurity engineering framework that complements functional safety (ISO, 2021). Bridging this gap, therefore, requires a co-assurance approach, in which GenAI is architecturally constrained to advisory or monitoring roles, protected by runtime safety monitors and fallback mechanisms, and jointly assessed under functional safety and cybersecurity standards to preserve deterministic safety guarantees in the presence of probabilistic AI behavior.

9.4 GenAI-Based Cyber Resilient Traffic Sign Classification System for Autonomous Vehicles: A Case Study

In this section, we present a GenAI-based resilient solution for autonomous traffic sign classification systems as a case study. We begin with an explanation of the problem addressed using this resilient solution, followed by descriptions of the attack models, GenAI training framework, constituent neural networks, evaluation method, analyses, results, and study conclusions.

9.4.1 Problem Statement and Motivation

Autonomous vehicles (AVs) perform the autonomous driving task with the help of a suite of sensors and software. Sensors, including camera, light detection and ranging (LiDAR), and radio detection and ranging (Radar), help AVs perceive their surrounding environment (Marti *et al.*, 2019). The sensed data is fed to the AV perception module, where the relevant information for autonomous navigation is extracted, such as traffic signs, signals, lane markings, surrounding vehicles, pedestrians, and obstacles. Nowadays, even many human-driven vehicles have dashboard camera-based traffic sign classification systems. These systems are typically highly dependent on machine learning (ML) or deep learning (DL) models, especially deep neural networks (DNNs) (Wali *et al.*, 2019). However, since AVs rely on such systems to realize roadway regulations and maneuver, compromised information regarding roadway traffic signs can be hazardous for AVs. Thus, researchers have emphasized developing DNN-based accurate traffic sign classification systems over the past few years (Lim *et al.*, 2023).

However, DNN-based classification systems have some cybersecurity vulnerabilities. For example, an adversarial attack can introduce slight perturbations to the input images fed to a traffic sign classification system and cause the underlying DNN models to misclassify the signs on the roadway. These perturbations can be so minimal that they are imperceptible to regular human eyes. However, they can be effective in deceiving the DNN models used in AVs' traffic sign classification systems. To this end, we aim to develop an AV traffic sign classification system resilient to such adversarial attacks.

Adversarial attacks can be categorized based on the extent of knowledge the attack model or the attacker has about the DNN models responsible for classification (Khamaiseh *et al.*, 2022). If the attacker has no knowledge about the DNN-based classification models, e.g., their architecture and parameters, then it is called a black-box attack. If the attacker has partial knowledge about DNN-based classification models, e.g., the attacker may know the DNN architecture but may be unaware of its trained parameters, then it is known as a gray-box attack. If the attacker has full knowledge of the DNN-based classification models, then the attack is called a white-box attack. In a white-box attack, the victim, i.e., in this case, the DNN-based classification system, is at a maximum disadvantage. This is because the attacker can craft an adversarial attack in such a way that the DNN-based classification system and its defense methods remain utterly unaware of the attack. In this study, the goal is to develop an AV traffic sign classification system that is resilient to adversarial attacks, particularly to white-box attacks. In a connected and AV environment, it is reasonable to assume that an attacker could access sensitive system configurations or algorithms even without physically accessing the AV.

Different defense methods have been proposed by researchers over the past few years to protect image classification systems from adversarial attacks (Khamaiseh *et al.*, 2022), such as modification of the DNNs (Papernot *et al.*, 2016; Ross and Doshi-Velez, 2018), adversarial training (Bai *et al.*, 2021), input transformation (Liu *et al.*, 2019; Xu *et al.*, 2018), and input reconstruction (Jin *et al.*, 2019; Laykaviriyakul and Phaisangittisagul, 2023). The recent breakthroughs in GANs have opened opportunities to utilize GANs for defense against adversarial attacks. For example, Samangouei *et al.* (2018) introduced a Wasserstein GAN (WGAN)-based defense method, known as the Defense-GAN, which can protect image classification systems against known and unknown adversarial attacks by reconstructing the input images before feeding them to a classifier. The generator model in the Defense-GAN method was trained to generate samples similar to the unperturbed (legitimate) images given a random input latent vector. The random input latent vector is found by solving an optimization problem to minimize the reconstruction error of the generator. However, WGAN is known for suffering from issues associated with weight clipping, such as vanishing gradient and non-convergence of the discriminator, which makes the training of an appropriate WGAN model very difficult (Gulrajani *et al.*, 2017). In Jin *et al.* (2019) developed a defense method for image classification systems against adversarial attacks called the Adversarial Perturbation Elimination with GAN (APE-GAN). The generator of the APE-GAN was trained with adversarial examples to eliminate adversarial perturbations by making changes to the input images. However, the authors in Jin *et al.* (2019) also utilized the loss function of WGAN, which inherits the same issues as discussed earlier. Besides, adversarial training-based defense methods work well for known attacks only and are susceptible to unknown attacks (Zhang *et al.*, 2019). Laykaviriyakul and Phaisangittisagul (2023) presented an adversarial defense framework for image classification systems based on the

DiscoGAN architecture (Kim *et al.*, 2017), where the authors utilized an attacker model to create adversarial examples from the training data and a defender model to reconstruct unperturbed images from the adversarial images. These two models were trained in tandem to play a competitive game with each other. However, the authors in Laykaviriyakul and Phaisangittisagul (2023) also used an adversarial training-based defense approach, which does not ensure classification performance under unknown attacks on which the models were not trained. Besides, all these studies considered benchmark datasets like CIFAR-10, MNIST, and Fashion MNIST. Thus, their performance on real-world datasets has not been explored yet.

In this case study, our aim is to develop an attack-resilient GAN-based defense method for an AV traffic sign classification system, which we refer to as the AR-GAN, that can protect the perception module of an AV from unknown attacks. We used a WGAN-based loss function with gradient penalty (WGAN-GP) to train the GAN models, which was shown to overcome common issues with GAN/WGAN training. The GAN models and classifiers are trained on the unperturbed traffic sign images, so that all types of adversarial attacks are unknown to the models.

9.4.2 *Attack Models*

Numerous adversarial attack models have been developed by researchers in the past few years to target NN-based classifiers (Khamaiseh *et al.*, 2022). In this case study, we considered both black-box and white-box attacks. In both cases, the attacker aims to find a perturbation δ that will cause misclassification by the classifier when added to legitimate input $x \in \mathbb{R}^n$, i.e., $\tilde{x} = x + \delta$, where $\tilde{x}$ is the modified input or the adversarial example that may cause misclassification. In this section, we discuss the various adversarial attack models used in this study.

9.4.2.1 *Fast Gradient Sign Method (FGSM) Attack*

FGSM is a simple but effective adversarial attack proposed by Goodfellow *et al.* (2015). FGSM utilizes the gradient of the loss (cost) function to generate adversarial examples. Despite its simplicity, FGSM is one of the widely popular adversarial attacks due to its effectiveness in causing misclassification with high confidence (Goodfellow *et al.*, 2015). Given an input image x and a classifier with parameters θ, FGSM attack aims to generate an adversarial example $\tilde{x} = x + \delta$, where the added perturbation δ is determined by computing the gradient of the loss function with respect to the input x as follows (Goodfellow *et al.*, 2015),

$$\delta = \varepsilon \cdot \text{sign}(\nabla_x J(\theta, x, y)) \tag{9.1}$$

where, J denotes the loss function, y denotes the output class, ∇_x is a differential operator with respect to x, and ε denotes the magnitude of perturbation chosen by the attacker.

9.4.2.2 *DeepFool Attack*

DeepFool is another simple yet effective optimization-based iterative adversarial attack model proposed by Moosavi-Dezfooli *et al.* (2016), which was reported to be more effective than the FGSM attack on MNIST and CIFAR-10 datasets. Given a binary classifier model, this attack aims to find the minimum perturbation δ^* that would cause misclassification by shifting the input x to the other side of the decision boundary. The minimum perturbation δ^* is determined through an optimization problem as follows (Moosavi-Dezfooli *et al.*, 2016),

$$\begin{gathered} \delta^* = \arg\min_{\delta} \|\delta\|_2 \\ \text{subject to: } \text{sign}(f(x+\delta)) \neq \text{sign}(f(x)) \end{gathered} \tag{9.2}$$

where, f is an arbitrary binary classification model. At ith iteration, DeepFool updates δ by linearizing the classification boundary around the current point x_i. To ensure that the final perturbation $\hat{\delta}$ crosses the decision boundary and causes misclassification, δ^* is multiplied by a constant $(1+\eta)$, where $\eta \ll 1$. The authors in Moosavi-Dezfooli *et al.* (2016) extended this approach to multi-class classifiers as well.

9.4.2.3 *Carlini and Wagner (C&W) Attack*

Carlini and Wagner (2017) introduced an optimization-based iterative adversarial attack known as the C&W attack. The C&W attack is a powerful attack that has been reported to cause very low classification accuracy on benchmark datasets, such as MNIST and CIFAR datasets (Carlini and Wagner, 2017). In a C&W attack, given an input image $x \in \mathbb{R}^n$, the goal of the attack is to determine an optimal perturbation δ^* to yield misclassification of the adversarial example $\tilde{x} = x + \delta^*$ by a target classifier. In an ℓ_2-norm C&W attack, the optimal perturbation δ^* is determined through the following optimization problem (Carlini and Wagner, 2017),

$$\begin{gathered} \delta^* = \arg\min_{\delta} \|\delta\|_2 + c.f(x+\delta) \\ \text{subject to: } x + \delta \in [0,1]^n \end{gathered} \tag{9.3}$$

where, $\|\cdot\|_2$ denotes the ℓ_2-norm, $c > 0$ is an arbitrary constant, and f is an objective function that helps the misclassification, which is chosen based on the knowledge of the target classifier model. Apart from the ℓ_2-norm attack explained above, C&W attacks can also be performed using ℓ_0 and ℓ_∞ norms.

9.4.2.4 *Projected Gradient Descent (PGD) Attack*

PGD is a powerful adversarial attack model that also utilizes an optimization-based iterative approach. The authors in Madry *et al.* (2019) showed the effectiveness of PGD attacks on MNIST and CIFAR-10 datasets, in which PGD was able to yield lower classification accuracy on the datasets compared to FGSM and C&W attacks. Given an input image x, a classifier with parameters θ, and a perturbation δ to be optimized for a PGD attack, the optimization problem can be written as (Madry *et al.*, 2019),

$$\begin{gathered} \delta^* = \arg\min_{\delta} \|\delta\|_2 \\ \text{subject to: } x + \delta \in [x_{\min}, x_{\max}] \end{gathered} \tag{9.4}$$

where, $\|\cdot\|_2$ denotes the l_2-norm, and $x_{\min}$ and $x_{\max}$ are the minimum and the maximum values of each pixel. The perturbation δ is updated in each iteration as follows,

$$\delta_{t+1} = \delta_t + \varepsilon \cdot \operatorname{sign}(\nabla_x J(\theta, x+\delta, y)) \tag{9.5}$$

where, J denotes the loss function, y denotes the output class, ∇_x is a differential operator with respect to x, and ε denotes the magnitude of perturbation.

9.4.3 *AR-GAN Training Framework*

The training framework to obtain the models in the traffic sign classification system of AR-GAN is depicted in Fig. 9.2. We start by training a classifier model to classify unperturbed (i.e., legitimate or without attack) images in a traffic sign image dataset. In this study, we used a 9-layer deep residual learning architecture known as ResNet9 (He *et al.*, 2016). Once the classifier is trained to classify unperturbed traffic sign images with acceptable accuracy, we call it Classifier #1, which is used later to select the best generator model. Next, we train a set of GANs based on the WGAN architecture with a gradient penalty using the same unperturbed traffic sign images from the dataset. Once the GANs are trained, we utilize Classifier

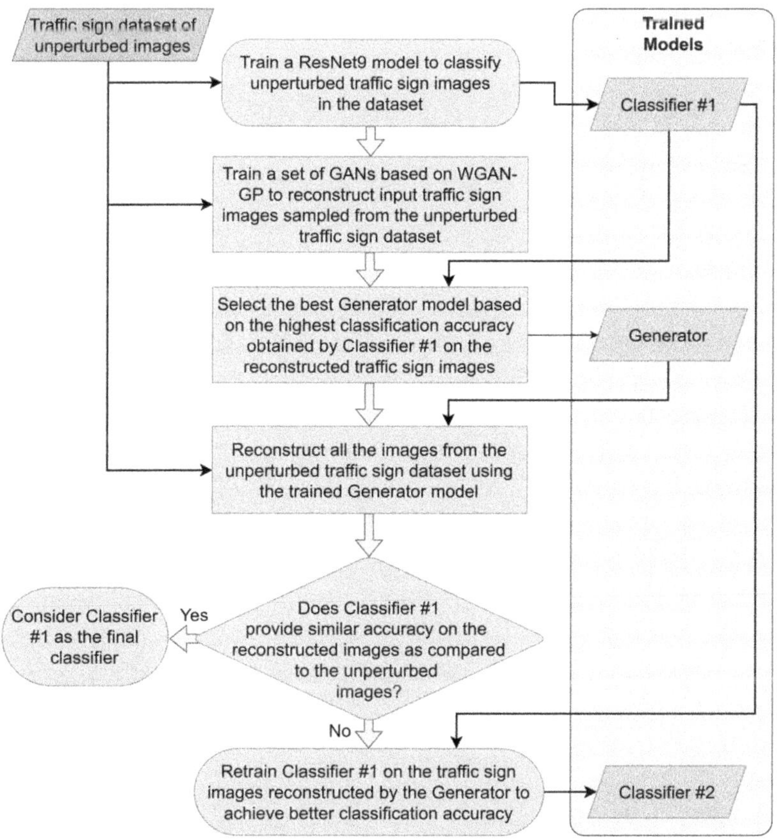

Fig. 9.2. AR-GAN training framework.

#1 to select the best generator model from the set of trained GANs that would yield the highest classification accuracy on the reconstructed traffic sign images of the test dataset. Then, we use the selected generator model to reconstruct all the unperturbed traffic sign images in the dataset. In an ideal case, if the GANs are trained to a point where the reconstructed traffic sign images look identical to the unperturbed traffic sign images, Classifier # 1 should be able to classify the reconstructed traffic sign images with similar accuracy to that of the unperturbed traffic sign images. However, if this level of accuracy is not achieved, we retrain Classifier #1 on the reconstructed traffic sign images to achieve better accuracy. We call this retrained classifier model Classifier #2. Finally, the AR-GAN traffic sign classification system is built with the best generator model, which reconstructs and denoises any input traffic sign images, and Classifier #2, which classifies these reconstructed images.

9.4.4 *AR-GAN Classifier Models*

The classifiers in the AR-GAN method are based on the ResNet9 architecture. ResNet9 is a 9-layer deep NN, including eight convolutional layers and one linear layer. We chose the ResNet9 architecture for developing the classifier models in this study because of the advantages of deep residual NNs over other CNNs and ResNet9's lightweight nature. Figure 9.3 presents the model architecture of the classifier used in the AR-GAN method, assuming the input traffic sign images have three color channels, i.e., red, green, and blue, each with 32×32 pixels. In Fig. 9.3, the output dimension of each layer is presented as $\mathrm{C} \times \mathrm{H} \times \mathrm{W}$, where C denotes the number of feature maps (for the input layer, it represents the number of color channels), and H and W denote the height and width of an output feature map, respectively.

9.4.5 *AR-GAN Generator and Discriminator Models*

GANs, first introduced by Goodfellow *et al.* (2014), consist of two NNs, known as the generator (G) and the discriminator (D). $G : \mathbb{R}^k \to \mathbb{R}^n$ takes a low-dimensional input latent vector $z \in \mathbb{R}^k$ and maps it to a higher dimensional sample space of $x \in \mathbb{R}^n$. The discriminator, D is a binary classifier that distinguishes real samples

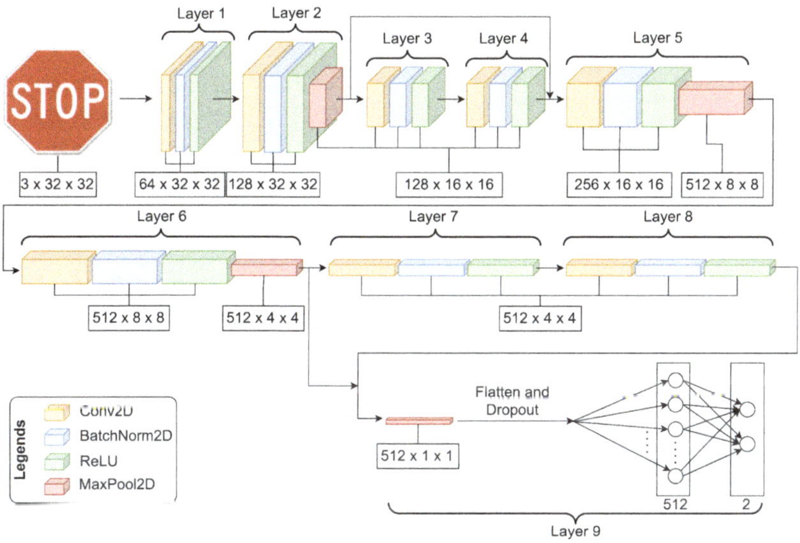

Fig. 9.3. AR-GAN classifier based on the ResNet9 architecture.

from fake (i.e., generated) samples. G and D are trained in tandem to optimize the following min-max loss function defined in Goodfellow *et al.* (2014),

$$\min_G \max_D V(D, G) = \mathrm{E}_{x \sim P_r(x)}[\log D(x)] + \mathrm{E}_{z \sim P_g(z)}[\log(1 - D(G(z)))] \tag{9.6}$$

where, $P_r(x)$ and $P_g(z)$ denote the real sample distribution and the generated sample distribution, respectively, and E denotes the expected value. The optimal GAN is obtained when these two distributions become the same. However, training GANs to optimality is difficult due to issues such as mode collapses and vanishing gradients. To resolve this, Arjovsky *et al.* (2017) proposed a variant of the GAN, known as the Wasserstein GAN (WGAN), that utilizes the concepts of Wasserstein distance and Kantorovich-Rubinstein duality (Rachev, 1990), with an alternative loss function given by,

$$\min_G \max_D V_W(D, G) = \mathrm{E}_{x \sim P_r(x)}[\log D(x)] - \mathrm{E}_{z \sim P_g(z)}[\log(D(G(z)))] \tag{9.7}$$

WGAN also removed the sigmoid function from the discriminator of the original GAN proposed in Goodfellow *et al.* (2014) to interpret the output of D in terms of probability to indicate how "real" the generated images are. However, WGAN still suffered from some convergence issues due to the hard constraints set by the weight clipping method to enforce the Lipschitz condition. To address these issues, Gulrajani *et al.* (2017) proposed an improved version known as WGAN with gradient penalty (WGAN-GP). WGAN-GP utilizes a soft version of the constraints by penalizing the model if the norm of the gradient deviates from its target norm value of 1 to meet the Lipschitz condition, as follows (Gulrajani *et al.*, 2017),

$$\min_G \max_D V_W(D, G) = \mathrm{E}_{x \sim P_r(x)}[\log D(x)] - \mathrm{E}_{z \sim P_g(z)}[\log(D(G(z)))] + \lambda \mathrm{E}_{\hat{x} \sim P_{\hat{x}}}[\|\nabla_{\hat{x}} D(\hat{x})\|_2 - 1]^2 \tag{9.8}$$

Here, λ is set to 10, $\nabla_{\hat{x}}$ is a differential operator with respect to $\hat{x}$, and $\hat{x}$ is sampled from x and $G(x)$ using the following linear equation,

$$\hat{x} = tG(x) + (1 - t)x \tag{9.9}$$

where, t is uniformly sampled between 0 and 1, i.e., $0 \leq t \leq 1$. The WGAN-GP, proposed in Gulrajani *et al.* (2017), removed the batch normalization steps from the discriminator as it affected the effectiveness of the gradient penalty.

However, to be able to reconstruct an input image with minimum reconstruction error, further extension is needed. The authors in Samangouei *et al.* (2018) proposed a simple extension to achieve this by optimizing the input latent vector z that will be fed to the generator to reconstruct an input image x, which we adopt in the AR-GAN method. The optimal latent vector z^* for reconstructing an input image x is obtained by solving the optimization problem given by,

$$z^* = \arg\min_z \|G(z) - x\|_2^2 \tag{9.10}$$

Equation (9.10) is solved in a gradient descent-based iterative approach. Because of its non-convex nature, the authors in Samangouei *et al.* (2018) utilized a fixed number of gradient descent steps

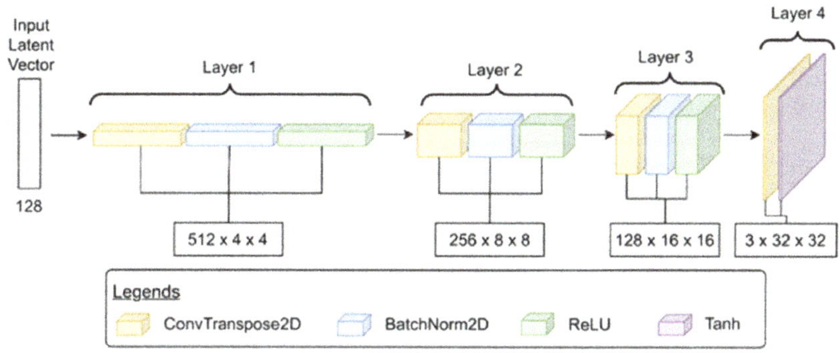

Fig. 9.4. AR-GAN generator based on DCGAN architecture.

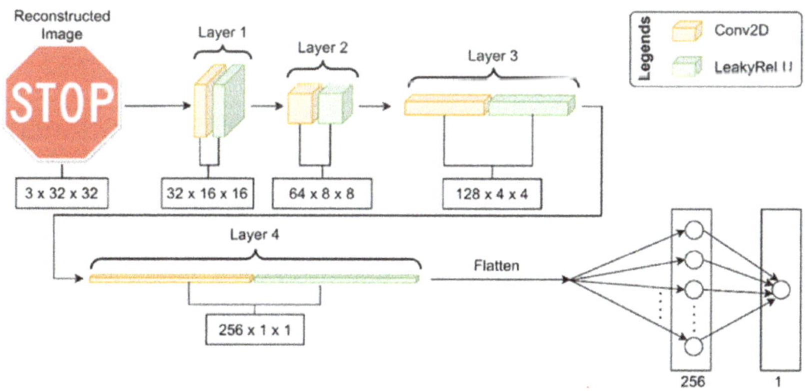

Fig. 9.5. AR-GAN discriminator based on WGAN architecture.

along with a given number of random initializations of the latent vector z.

Figures 9.4 and 9.5 present the architecture of the generator and the discriminator used in the AR-GAN method, respectively. The generator architecture is based on the DCGAN architecture (Radford *et al.*, 2016), whereas the discriminator architecture is kept the same as the WGAN architecture. In Figs. 9.4 and 9.5, the output dimension of each layer is presented as $\mathrm{C} \times \mathrm{H} \times \mathrm{W}$, where C denotes the number of feature maps, and H and W denote the height and width of each feature map, respectively.

Finally, the AR-GAN traffic sign classification system consists of the trained generator, the classifier, and the optimizer that optimizes the input latent vector applied to the generator to reconstruct an

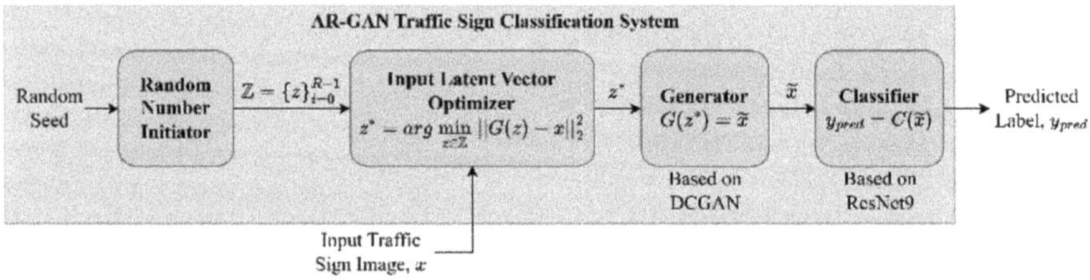

Fig. 9.6. AR-GAN discriminator based on WGAN architecture.

input traffic sign image. Figure 9.6 presents the AR-GAN traffic sign classification system with a flow diagram.

9.4.6 *Evaluation Method*

This section discusses the evaluation approach utilized in this study, outlining the traffic sign dataset and the traditional preprocessing-based defense methods we employed for comparison.

9.4.6.1 *Real-World Traffic Sign Dataset*

We reviewed the existing literature to select a comprehensive US traffic sign dataset and found the LISA traffic sign dataset (Mogelmose *et al.*, 2012) the most appropriate since it contains data collected from the real world. The LISA dataset covers 49 types of US traffic signs with 7,855 annotations on 6,610 frames. The traffic sign images in the LISA dataset were extracted from video frames captured by multiple vehicles' dashboard cameras while the vehicles roamed around San Diego, California. The LISA video frames exhibit varying resolutions, ranging from 640×480 to 1024×522 pixels. The traffic sign annotations have dimensions spanning from 6×6 to 167×168 pixels and include both color and grayscale images.

However, the LISA dataset does not contain enough images for each type of traffic sign to train GAN models. Therefore, we created a subset of the LISA dataset containing two traffic sign classes with the highest number of images, i.e., STOP signs and SPEED LIMIT signs. The original LISA dataset contains different types of SPEED LIMIT signs, such as 15, 25, 30, 35, 40, 45, 50, and 65 miles per hour (mph) signs. We combined all these SPEED LIMIT signs into one class to create a balanced dataset. The subset of

Fig. 9.7. Image samples from the LISA traffic sign dataset.

the LISA dataset used in this study for evaluation of the AR-GAN method contains a total of 1,562 traffic sign images, including 805 images under the STOP sign class and 757 images under the SPEED LIMIT sign class. We applied cropping and resizing to ensure that all the images in the dataset have the same dimension, i.e., each image has three channels for red, green, and blue colors, and each channel contains 32×32 pixels. Figure 9.7 presents some sample images from the dataset used in this study. As observed from the figure, the images are not very clean and contain some noise, which makes it even harder to classify them under adversarial perturbations.

9.4.6.2 *Traditional Preprocessing-Based Defense Methods as Baselines*

We selected several traditional preprocessing-based defense methods that can be used as benchmarks to compare with the classification performance of the AR-GAN traffic sign classification system. Table 9.2 highlights the traditional preprocessing-based adversarial attack defense methods considered in this study. Figure 9.8 shows the effect of these defense methods, including the AR-GAN method, on a set of sample traffic sign images used in this study.

9.4.7 *Analysis and Results*

This section presents the analysis and results based on the AR-GAN method and compares them with the traditional preprocessing-based defense methods. We divide the evaluation scenarios into two categories, i.e., (i) evaluation on unperturbed traffic sign images and

Table 9.2. Traditional preprocessing-based defense methods used in this study.

Defense Method	Method Highlights	Method Settings Used in This Study
Gaussian Augmentation (Grandvalet and Canu, 1997)	• Applies random noise to every pixel of an input image to grow robustness against adversarial attacks • Applies independent and identically distributed noise sampled from a zero-mean Gaussian distribution, $N(0, \sigma^2)$, to each pixel of an input image	Standard deviation σ is set to 1, as recommended in ... Grandvalet and Canu (1997)
JPEG Compression (Liu *et al.*, 2019)	• Eliminates high-frequency components from an input image that may cause the DNN-based classifiers to misclassify the image under an adversarial attack	Compressed image quality is set to 50%, as recommended in Liu *et al.* (2019)
Feature Squeezing (Xu *et al.*, 2018)	• Reduces the depth of each pixel's color bits of an input image that is necessary to represent the color value of that pixel • Transforms diverse feature vectors from the original space into more similar samples	Bit depth value is set to 4, as recommended in Xu *et al.* (2018)
Median Smoothing (Xu *et al.*, 2018)	• Reduces differences among the pixel values of an input image by moving a sliding window across the image while the window's center pixel value is replaced by its neighboring pixels' median value	Size of the sliding window is set to 3×3

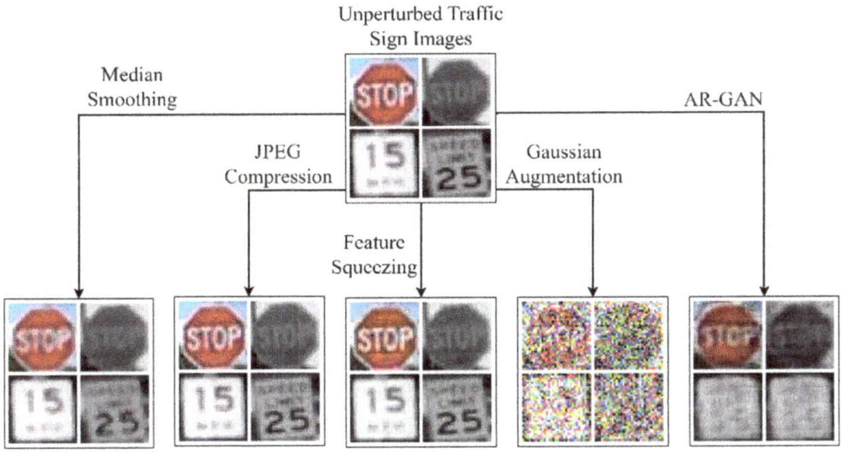

Fig. 9.8. Image samples from the LISA traffic sign dataset.

adversarial images, and (ii) evaluation under different perturbation magnitudes. The dataset was split into three subsets, i.e., train set (containing 60% of the images), validation set (containing 20% of the images), and test set (containing the remaining 20% of the images), after applying random shuffling on all the images in the dataset. The same train, validation, and test sets were used for all the evaluation scenarios to present a fair comparison among the different defense methods used in this study.

The AR-GAN traffic sign classification system utilizes a gradient descent-based optimization to determine the input latent vector that would minimize the reconstruction error for an input image. We conducted a sensitivity analysis of the AR-GAN traffic sign classification system's end-to-end delay (i.e., the time required to perform all the steps as presented in Fig. 9.6) and classification accuracy with respect to the number of gradient descent steps and the number of random initializations (see Fig. 9.9). For this analysis, we used all the images in the test set and considered the average values of delay and accuracy. As observed from Fig. 9.9(a), the end-to-end delay per image increases with the increase in the number of gradient descent steps as well as the number of random initializations. Liu and Deng (2021) reported the average delay for human drivers in recognizing traffic signs to range from 0.5 to 2.0 seconds, which sets a limit to the number of gradient descent steps as well as the number of random initializations we can allow. In addition, the lower the end-to-end

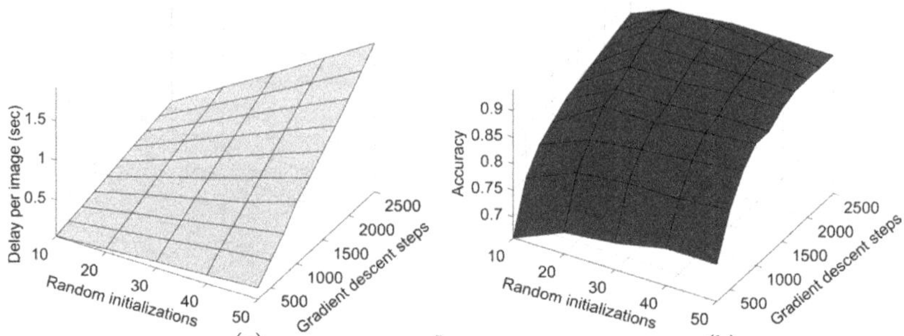

Fig. 9.9. Sensitivity analysis results of average (a) delay per image and (b) classification accuracy.

delay is, the better it is for real-world implementation in AVs. On the other hand, as observed from Fig. 9.9(b), the classification accuracy initially improves as the number of random initializations increases, but then it starts to decline. The highest traffic sign classification accuracy was achieved after 2,250 gradient descent steps and 20 random initializations, resulting in an end-to-end delay of 0.6 seconds. We considered this delay acceptable for real-world implementation as it is much lower than the human drivers' average traffic sign recognition delay reported in Liu and Deng (2021).

We used Pytorch packages[1] to implement the GAN and the classifier models, and the Adversarial Robustness Toolbox[2] to implement the attack models and traditional preprocessing-based defense methods. The source codes are provided in GitHub (Salek, 2023). All the NN models were trained using Nvidia Tesla A100 GPUs available in the Palmetto Cluster nodes at Clemson University.[3] These GPUs have a capacity of 312 trillion floating-point operations per second (TFLOPS).[4] One of the recent in-vehicle computational units developed by NVIDIA is the NVIDIA Drive Thor, offering a GPU-based

[1]PyTorch, https://www.pytorch.org.

[2]Adversarial Robustness Toolbox (ART) v1.15, https://github.com/Trusted-AI/adversarial-robustness-toolbox/tree/main.

[3]RCD Documentation. About the Palmetto Cluster, https://docs.rcd.clemson.edu/palmetto/about.

[4]NVIDIA A100 TENSOR CORE GPU, *NVIDIA Newsroom*, http://nvidianews.nvidia.com/news/nvidia-unveils-drive-thor-centralized-car-computer-unifying-cluster-infotainment-automated-driving-and-parking-in-a-single-cost-saving-system.

computational capacity of 2,000 TFLOPS,[5] which is well above the capacity of the A100 GPUs utilized in this study for training the NN models. Additionally, the in-vehicle computational units should only be responsible for running pre-trained models for traffic sign image classification, whereas the training task can be performed separately beforehand. Thus, the models developed under the AR-GAN method are considered feasible to be implemented in real-world AVs in terms of in-vehicle computational capacity.

9.4.7.1 *Evaluation with Unperturbed and Adversarial Traffic Sign Images*

Table 9.3 presents the evaluation results obtained using unperturbed traffic sign images. Tables 9.4 and 9.5 present the evaluation results obtained using black-box and white-box adversarial images, respectively. The adversarial images were generated under FGSM, DeepFool, C&W, and PGD attacks. We used precision, recall, F1-score, and accuracy for traffic sign classification performance comparison among the different defense methods. Among the performance metrics, accuracy was calculated globally for all the images in the test set, whereas the other metrics were calculated for each class, and then a weighted average was taken.

Table 9.3. Comparison of defense methods on unperturbed images.

Defense Method	**Precision** (%)	**Recall** (%)	**F1-Score** (%)	**Accuracy** (%)
Gaussian Augmentation	95.1	94.9	94.9	94.9
JPEG Compression	99.7	99.7	99.7	99.7
Feature Squeezing	99.7	99.7	99.7	99.7
Median smoothing	98.8	98.7	98.7	98.7
AR-GAN	**95.0**	**94.9**	**94.9**	**94.9**

[5]NVIDIA Unveils DRIVE Thor — Centralized car computer unifying cluster, infotainment, automated driving, and parking in a single, cost-saving system, *NVIDIA Newsroom*, http://nvidianews.nvidia.com/news/nvidia-unveils-drive-thor-centralized-car-computer-unifying-cluster-infotainment-automated-driving-and-parking-in-a-single-cost-saving-system.

Table 9.4. Comparison of defense methods on black-box adversarial images.

Attack Type	Defense Method	Precision (%)	Recall (%)	F1-Score (%)	Accuracy (%)
FGSM	Gaussian Augmentation	94.3	94.3	94.3	94.3
	JPEG Compression	99.7	99.7	99.7	99.7
	Feature Squeezing	99.4	99.4	99.4	99.4
	Median smoothing	97.6	97.4	97.4	97.4
	AR-GAN	**94.6**	**94.6**	**94.6**	**94.6**
DeepFool	Gaussian Augmentation	94.3	94.3	94.3	94.3
	JPEG Compression	99.7	99.7	99.7	99.7
	Feature Squeezing	99.7	99.7	99.7	99.7
	Median smoothing	98.8	98.7	98.7	98.7
	AR-GAN	**94.8**	**94.6**	**94.6**	**94.6**
C&W	Gaussian Augmentation	95.2	94.9	94.9	94.9
	JPEG Compression	99.7	99.7	99.7	99.7
	Feature Squeezing	99.7	99.7	99.7	99.7
	Median smoothing	98.8	98.7	98.7	98.7
	AR-GAN	**95.6**	**95.5**	**95.5**	**95.5**
PGD	Gaussian Augmentation	94.4	94.3	94.2	94.3
	JPEG Compression	99.7	99.7	99.7	99.7
	Feature Squeezing	99.7	99.7	99.7	99.7
	Median smoothing	98.8	98.7	98.7	98.7
	AR-GAN	**93.4**	**93.3**	**93.3**	**93.3**

As shown in Table 9.3, all the defense methods considered in this study achieved high precision, recall, F1-score, and accuracy on the unperturbed images. This proves that the classifier models used in all the defense methods in this study are well-trained to accurately classify the traffic sign images of the dataset. Although the AR-GAN method achieved approximately 95% classification accuracy on unperturbed images, it was lower than that of most other methods. This difference in performance is because, unlike the other preprocessing-based defense methods that transform or modify an input image, the AR-GAN completely reconstructs any input image. Gaussian augmentation achieved the second-lowest classification performance compared to the other defense methods, which is also expected because this defense method itself adds some Gaussian noise to the images as part of its adversarial defense strategy.

Table 9.5. Comparison of defense methods on white-box adversarial images.

Attack Type	Defense Method	Precision (%)	Recall (%)	F1-Score (%)	Accuracy (%)
FGSM	Gaussian Augmentation	81.1	80.5	80.4	80.5
	JPEG Compression	75.4	75.4	75.4	75.4
	Feature Squeezing	68.7	68.7	68.7	68.7
	Median smoothing	68.5	68.4	68.3	68.4
	AR-GAN	**92.8**	**92.7**	**92.6**	**92.7**
DeepFool	Gaussian Augmentation	74.0	73.8	73.8	73.8
	JPEG Compression	62.0	61.3	60.8	61.3
	Feature Squeezing	32.3	32.3	32.3	32.3
	Median smoothing	42.1	42.8	41.6	42.8
	AR-GAN	**92.7**	**92.7**	**92.7**	**92.7**
C&W	Gaussian Augmentation	79.0	78.6	78.5	78.6
	JPEG Compression	62.9	61.7	60.8	61.7
	Feature Squeezing	26.2	26.2	26.2	26.2
	Median smoothing	38.1	38.7	37.9	38.7
	AR-GAN	**93.4**	**93.3**	**93.3**	**93.3**
PGD	Gaussian Augmentation	75.6	75.4	75.4	75.4
	JPEG Compression	58.8	58.5	58.1	58.5
	Feature Squeezing	44.1	44.1	44.0	44.1
	Median smoothing	48.9	48.9	48.9	48.9
	AR-GAN	**91.7**	**91.7**	**91.7**	**91.7**

On the contrary, attackers can fine-tune their attack models using knowledge of the classifier's architecture and weights in a white-box attack. The novelty of the AR-GAN method lies in its consistent resilience to both black-box and white-box attacks. However, the performance gap between the traditional defense methods and the AR-GAN method becomes apparent under white-box attacks, as depicted in Table 9.5.

The FGSM attack was implemented with an $\varepsilon = 0.1$ perturbation magnitude, as recommended by Ye and Zhu (2018). The FGSM attack is not as powerful as the other attacks used in this study. As observed from Table 9.5, the performance metrics of the traditional preprocessing-based defense methods ranged from approximately 68% to 81%. However, the AR-GAN method improved the overall performance by about 10–12% compared to the second-best defense method, i.e., the Gaussian augmentation.

As observed from Table 9.5, the ℓ_2-norm DeepFool attack, with a perturbation magnitude of $\varepsilon = 0.1$, was more effective than the FGSM attack in degrading the classification performance. Feature squeezing and median smoothing performed the worst among all the defense methods. Gaussian augmentation was able to achieve about 74% classification accuracy, which was outperformed by the AR-GAN method with a classification accuracy of about 93%.

The ℓ_2-norm C&W attack was performed using a learning rate of 0.01 with a maximum of 10 iterations. As seen from Table 9.4, feature squeezing and median smoothing provided the worst traffic sign classification accuracies among all the defense methods. Gaussian augmentation achieved about 79% classification accuracy. However, the AR-GAN method outperformed all the other defense methods with a 93% classification accuracy under the C&W attack.

As observed from Table 9.5, the l_2-norm PGD attack, with a maximum iteration number of 100 and a perturbation magnitude of $\varepsilon = 0.1$, caused the traffic sign classification accuracies to drop under 60% for all the traditional defense methods, except for the Gaussian augmentation preprocessing, which achieved about 75% accuracy. Again, the AR-GAN method outperformed all the traditional defense methods with a classification accuracy of approximately 92%.

9.4.7.2 *Evaluation with Different Perturbation Magnitudes*

To evaluate how well the AR-GAN method performs under different perturbation magnitudes, we varied ε from 0.05 to 0.2 with a 0.05 step size for the FGSM, DeepFool, and PGD attacks, following the previous studies (Gao and Oates, 2019; Khan *et al.*, 2022; Pan *et al.*, 2019).

Figure 9.10 presents the results of this evaluation under different perturbation magnitudes. From Figs. 9.10(a)–9.10(c), it is observed that all the defense methods, including the AR-GAN, were able to achieve above 93% classification accuracy under the black-box FGSM, DeepFool, and PGD attacks with varied ε. However, the accuracy of the traditional preprocessing-based defense methods dropped abruptly as we increased the ε of the white-box FGSM and PGD attacks, as observed from Figs. 9.10(d) and 9.10(f), respectively. In Fig. 9.10(e), it is observed that these drops in traffic sign classification performance happen gradually for the traditional

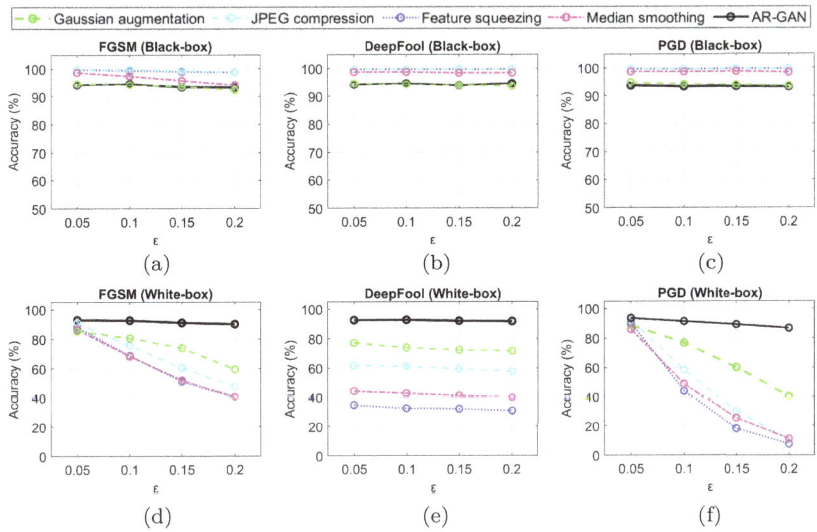

Fig. 9.10. Comparison of defense methods across varying perturbation magnitudes, ε, for: (a) FGSM (black-box), (b) DeepFool (black-box), (c) PGD (black-box), (d) FGSM (white-box), (e) DeepFool (white-box), and (f) PGD (white-box) attacks.

preprocessing-based defense methods under the white-box DeepFool attack. However, the AR-GAN method achieved above 90% classification accuracy in all these cases, except under the PGD attacks with $\varepsilon = 0.15$ and 0.2, where its accuracy dropped to about 87%. This consistency in traffic sign classification performance is achievable with the AR-GAN method because the generator in the AR-GAN method was trained to generate samples close to the distribution of unperturbed traffic sign images. Thus, the AR-GAN method developed in this study can effectively denoise the traffic sign images by reconstructing them with a generator trained on the unperturbed traffic sign images without any prior knowledge of adversarial attack types and samples.

9.4.8 *Case Study Conclusions*

In this case study, we developed a GAN-based adversarial defense method for an AV traffic sign classification system. We provided a framework for training a generator model and a classifier model, which comprise the attack-resilient traffic sign classification system of the AR-GAN method. The generator model was trained using a WGAN-GP-based loss function with a DCGAN architecture. The discriminator, used to support the training of the generator, was based on the WGAN architecture. The classifier was trained using the ResNet-9 architecture. All these models were trained only with unperturbed (i.e., legitimate) traffic sign images to ensure that any adversarial attacks are unknown to the trained models.

The traffic sign classification system of the AR-GAN defense method utilizes a generator-based image reconstruction to eliminate adversarial perturbations from an input traffic sign image. Once reconstructed, the image is fed to the classifier to identify the type of traffic sign. We evaluated the AR-GAN traffic sign classification system against widely used black-box and white-box adversarial attacks, including FGSM, DeepFool, C&W, and PGD attacks, and compared its performance with that of benchmark traditional adversarial defense methods, such as Gaussian augmentation, JPEG compression, feature squeezing, and median smoothing.

Under the black-box attacks considered in this study, the AR-GAN and the other traditional defense methods exhibited classification performance comparable to that on the unperturbed images. However, when encountered with white-box attacks, which assume that the attackers possess complete knowledge of the classification models, the AR-GAN method demonstrated superior resilience, outperforming all the traditional defense methods evaluated in this study. In addition, for the white-box adversarial images, the AR-GAN consistently achieved high traffic sign classification performance under various adversarial perturbation magnitudes, whereas the performance of the other traditional defense methods dropped abruptly at increased perturbation levels. This demonstrates the potential of the AR-GAN method to be deployed as a robust AV traffic sign classification system, achieving resiliency against various types of adversarial attacks. While white-box attacks were employed, the adversary is assumed to be non-adaptive with respect to the AR-GAN defense mechanism in this study. Although the attacker is assumed to have full access to the classification model, the attack generation process does not explicitly account for the presence of the generative reconstruction stage. As such, the adversarial perturbations are optimized to deceive the classifier alone, rather than being jointly optimized to survive or exploit AR-GAN's generator–classifier pipeline. Evaluating fully adaptive adversaries, incorporating knowledge of the AR-GAN generator and optimizing perturbations that remain effective after the reconstruction stage, represents a more stringent and challenging threat model, which was not considered in this study. Future work should extend the evaluation to adaptive white-box scenarios, enabling a more comprehensive assessment of AR-GAN's robustness under fully adaptive adversarial threats.

9.5 Summary

GenAI is rapidly reshaping the cybersecurity landscape and holds substantial potential for transportation systems and infrastructure, where the convergence of cyber and physical systems enhances both opportunities and risks. By modeling complex probability distributions and synthesizing high-fidelity data, GenAI can support anomaly detection, accelerate incident response, and provide resilience to operational technology (OT) networks. Compared with earlier ML approaches, GenAI provides not only predictive capability but also control over generation, enabling proactive defenses, such as synthetic data creation, red-teaming, and cyber-physical stress testing (i.e., a practice of deliberately subjecting a cyber-physical system to extreme or adversarial conditions). At the same time, challenges related to trustworthiness, governance, and adversarial misuse remain significant barriers to safe adoption of GenAI.

9.6 Promising Directions

For transportation, which is a safety-critical and highly regulated sector, the path forward requires a balance between innovation and assurance. Below, we present several directions for further research on GenAI-enabled cybersecurity and resiliency for transportation systems and infrastructure.

9.6.1 *GenAI-Augmented Digital Twins*

Digital twins, i.e., virtual replicas of physical infrastructure and processes, are increasingly employed to model traffic flows, railway control, or airport ground operations. By combining them with GenAI-driven scenario generation, researchers can create realistic cyberattack and failure conditions that would be unsafe or infeasible to test in a real-world setting. These digital twin-based experiments can support proactive stress-testing of transportation CPS and infrastructure, enabling pre-deployment validation of detection and mitigation strategies. Such approaches may also extend to simulating large-scale attacks; for example, a distributed DoS attack on intelligent transportation backbones and cellular vehicle-to-everything (C-V2X) communications.

9.6.2 *Synthetic Data Generation*

Another promising direction of GenAI-assisted cybersecurity is the generation of synthetic data for rare and safety-critical events. GenAI models can generate synthetic network traffic, control logs, or multimodal sensor data that preserve statistical distributions and structural integrity. This capability is particularly valuable for developing robust detection and resilient systems that can handle concept drift and emerging threats (Kholgh and Kostakos, 2023). For example, an AI model trained on traffic flow data may become obsolete as patterns evolve over time; however, incorporating GenAI-based synthetic traffic flows into training can help mitigate this issue if the models are designed to account for potential drift in data distribution for the foreseeable future. When combined with multimodal analyses, such as fusing cyber logs with multimodal data, like video and C-V2X messages, these synthetic datasets could support anomaly detection in transportation CPS.

9.6.3 *Large Language Models and Copilots for Enhanced Cybersecurity*

Security analysts in roadway, aviation, rail, or port network operation centers face overwhelming volumes of alerts and logs. LLM-based copilots could help them by summarizing incidents, correlating indicators of compromise, and even suggesting stepwise containment actions. Early prototypes in enterprise security have demonstrated the value of GenAI in incident response and threat intelligence workflows (Bono *et al.*, 2024; Google Cloud, 2023; Salek *et al.*, 2025). For transportation, such tools often require collecting and synthesizing heterogeneous information from different sources where GenAI or particularly LLM-based approaches are showing promise; for example, RAG-based pipelines to minimize hallucinations in analyzing legal cybersecurity-related documents applicable for transportation (Hockstad *et al.*, 2025).

9.6.4 *Multimodal Threat Detection and Software Supply Chain Assurance*

Security analysts in roadway, aviation, rail, or port network operation centers face overwhelming volumes of alerts and logs. LLM-based copilots could help them by summarizing incidents, correlating indicators of compromise, and even suggesting stepwise containment actions. Early prototypes in enterprise security have demonstrated the value of GenAI in incident response and threat intelligence workflows (Bono *et al.*, 2024; Google Cloud, 2023; Salek *et al.*, 2025). For transportation, such tools often require collecting and synthesizing heterogeneous information from different sources where GenAI or particularly LLM-based approaches are showing promise; for example, RAG-based pipelines to minimize hallucinations in analyzing legal cybersecurity-related documents applicable for transportation (Hockstad *et al.*, 2025).

9.6.5 *GenAI-Enhanced Transportation Supply Chain Management*

GenAI has shown promise to improve the resilience, efficiency, and adaptability of transportation supply chain management, which is increasingly exposed to disruptions, demand volatility, and cyber threats. Recent studies have highlighted the potential of GenAI, particularly LLMs, to support demand forecasting, inventory optimization, procurement, logistics coordination, and real-time decision-making by synthesizing heterogeneous operational data and unstructured information (Shekhar *et al.*, 2023). In transportation, GenAI is currently being explored by researchers to enable scenario-based disruption analysis, automated supplier risk assessment, and dynamic reconfiguration of multimodal logistics networks (Boone *et al.*, 2025). In addition to operational benefits, GenAI is expected to enhance sustainability by reducing waste, improving asset utilization, and supporting transparent supply chain governance. However, realizing these benefits requires addressing challenges related to data quality, integration with legacy systems, model governance, and trust in safety-critical decisions, which could be the focus of future research.

9.6.6 *Vision–Language Models for Semantic Consistency*

In transportation CPS, especially in vision-based autonomous driving, perception errors that alter semantic meaning can directly affect planning and control decisions. Vision–Language Models (VLMs) offer a practical approach to enforcing semantic consistency in GAN-based perception pipelines by jointly embedding generated images and their corresponding class descriptions in a shared multimodal space. Contrastive vision–language pretraining enables the detection of semantic drift, allowing the system to identify when a GAN hallucination changes the underlying image meaning despite visual plausibility (Jia *et al.*, 2021; Radford *et al.*, 2016). Prior work demonstrates that vision–language alignment enhances robustness to distribution shifts and adversarial perturbations, rendering VLM-based consistency checks potentially well-suited for safety-critical autonomous driving applications (Alayrac *et al.*, 2022; Zhou *et al.*, 2024), which require further investigations for vision-based transportation CPS applications.

In summary, because transportation is a safety-critical sector, GenAI deployments must be auditable, explainable, and subject to continuous oversight. This includes documenting hazards, such as model drift, misuse, and adversarial manipulation; designing human-in-the-loop refinement of automated decisions; and aligning with evolving regulatory frameworks, such as, the EU AI Act[6] as well as guidelines for best practices, like the Joint Guidance on Deploying AI Systems Securely (CISA, 2024). Only with such safeguards can GenAI realize its potential to strengthen the cybersecurity and resiliency of future transportation systems and infrastructure.

[6]EU Artificial Intelligence Act. Up-to-date developments and analyses of the EU AI act, *The EU AI Act Newsletter*, https://artificialintelligenceact.eu/.

Bibliography

Adeyinka, T. I. and Adeyinka, K. I. (2025). Leveraging generative AI for automated cyber threat simulation and response frameworks, https://doi.org/10.2139/ssrn.5334064.

Alayrac, J.-B., Donahue, J., Luc, P., Miech, A., Barr, I., Hasson, Y., Lenc, K., Mensch, A., Millican, K., Reynolds, M., Ring, R., Rutherford, E., Cabi, S., Han, T., Gong, Z., Samangooei, S., Monteiro, M., Menick, J. L., Borgeaud, S., Brock, A., Nematzadeh, A., Sharifzadeh, S., Bińkowski, M., Barreira, R., Vinyals, O., Zisserman, A., and Simonyan, K. (2022). Flamingo: A visual language model for few-shot learning, in *Advances in Neural Information Processing Systems*, Vol. 35, pp. 23716–23736.

Almadhor, A., Sampedro, G. A., Zaidi, M. M., Juanatas, R. A., and Hejaili, A. A. (2025). Generative AI-driven context-aware BDI-based smart routing protocol for intelligent transportation systems, *IEEE Transactions on Intelligent Transportation Systems*, pp. 1–10, https://doi.org/10.1109/TITS.2025.3570237.

Alqahtani, H. and Kumar, G. (2025). A comprehensive review of generative AI techniques and their impact on cybersecurity, *Soft Computing*, https://doi.org/10.1007/s00500-025-10702-z.

Anis, F. M. and Hammoudeh, M. (2025). AI-powered offensive security: Automating cyber attacks with large language models, reinforcement learning, and generative adversarial networks, https://doi.org/10.2139/ssrn.5366081.

Arif, H., Hussain, A. A., Nabi, H. A., and Ali, A. K. S. (2025). AI powered detection of adversarial and supply chain attacks on generative models, https://doi.org/10.5281/ZENODO.16080703.

Arjovsky, M., Chintala, S., and Bottou, L. (2017). Wasserstein generative adversarial networks, in *Proceedings of the 34th International Conference on Machine Learning*, pp. 214–223.

Aung, Y. L., Christian, I., Dong, Y., Ye, X., Chattopadhyay, S., and Zhou, J. (2025). Generative AI for Internet of Things security: Challenges and opportunities, https://doi.org/10.48550/ARXIV.2502.08886.

Bai, T., Luo, J., Zhao, J., Wen, B., and Wang, Q. (2021). Recent advances in adversarial training for adversarial robustness, https://doi.org/10.48550/arXiv.2102.01356.

Bilal, H., Rehman, A., Aslam, M. S., Ullah, I., Chang, W.-J., Kumar, N., and Almuhaideb, A. M. (2025). Hybrid TrafficAI: A generative AI framework for real-time traffic simulation and adaptive behavior modeling, *IEEE Transactions on Intelligent Transportation Systems*, pp. 1–17, https://doi.org/10.1109/TITS.2025.3571041.

Bono, J., Grana, J., and Xu, A. (2024). Generative AI and security operations center productivity: Evidence from live operations, https://doi.org/10.48550/arXiv.2411.03116.

Boone, T., Fahimnia, B., Ganeshan, R., Herold, D. M., and Sanders, N. R. (2025). Generative AI: Opportunities, challenges, and research directions for supply chain resilience, *Transportation Research Part E: Logistics and Transportation Review* **199**, p. 104135, https://doi.org/10.1016/j.tre.2025.104135.

Burns, A. and Davis, R. I. (2022). *Mixed Criticality Systems — A Review*. 13th edn.

Burns, A. and Jones, C. (2025). A specification framework for mixed-criticality scheduling protocols, *ACM Transactions on Embedded Computing Systems* **24**, 6, p. 166:1–166:24, https://doi.org/10.1145/3765522.

Carlini, N. and Wagner, D. (2017). Towards evaluating the robustness of neural networks, in *2017 IEEE Symposium on Security and Privacy (SP)*, pp. 39–57, https://doi.org/10.1109/SP.2017.49.

Chu, K.-F., Yuan, H., Yuan, J., Guo, W., Balta-Ozkan, N., and Li, S. (2024). A survey of artificial intelligence-related cybersecurity risks and countermeasures in mobility-as-a-service, *IEEE Intelligent Transportation Systems Magazine* **16**, 6, pp. 37–55, https://doi.org/10.1109/ MITS.2024.3427655.

CISA (2024). Joint guidance on deploying AI systems securely, https://www.cisa.gov/news-events/alerts/2024/04/15/joint-guidance-deploying-ai-systems-securely.

Enan, A. and Chowdhury, M. (2025). GAN-based single-stage defense for traffic sign classification under adversarial patch attack, https://doi.org/10.48550/ARXIV.2503.12567.

Gao, H. and Oates, T. (2019). Universal adversarial perturbation for text classification, https://doi.org/10.48550/arXiv.1910.04618.

Goodfellow, I., Pouget-Abadie, J., Mirza, M., Xu, B., Warde-Farley, D., Ozair, S., Courville, A., and Bengio, Y. (2014). Generative adversarial nets, in *Advances in Neural Information Processing Systems*, Vol. 27.

Goodfellow, I. J., Shlens, J., and Szegedy, C. (2015). Explaining and harnessing adversarial examples, https://doi.org/10.48550/arXiv.1412.6572.

Google Cloud (2023). How Google Cloud plans to supercharge security with generative AI, https://cloud.google.com/blog/products/identity-security/rsa-google-cloud-security-ai-workbench-generative-ai.

Grandvalet, Y. and Canu, S. (1997). Noise injection for inputs relevance determination, in *Advances in Intelligent Systems* (IOS Press, Netherlands), pp. 378–382.

Gulrajani, I., Ahmed, F., Arjovsky, M., Dumoulin, V., and Courville, A. C. (2017). Improved training of Wasserstein GANs, in *Advances in Neural Information Processing Systems* (Curran Associates, Inc.).

He, K., Zhang, X., Ren, S., and Sun, J. (2016). Deep residual learning for image recognition, in *Proceedings of the IEEE Conference on Computer Vision and Pattern Recognition*, pp. 770–778.

Hockstad, T. A., Rahman, M. M., Jones, S. L., Khan, L., Chowdhury, M. R., and Salek, M. S. (2025). Policy analysis and guidance to support secure transportation cyber-physical systems, https://rosap.ntl.bts.gov.

Illiashenko, O., Kharchenko, V., Babeshko, I., Fesenko, H., and Di Giandomenico, F. (2023). Security-informed safety analysis of autonomous transport systems considering AI-powered cyberattacks and protection, *Entropy* **25**, 8, p. 1123, https://doi.org/10.3390/e25081123.

ISO (2018). ISO 26262-6:2018, https://www.iso.org/standard/68388.html.

ISO (2021). ISO/SAE 21434:2021, https://www.iso.org/standard/70918.html.

ISO (2024). ISO/PAS 8800:2024, https://www.iso.org/standard/83303.html.

Jia, C., Yang, Y., Xia, Y., Chen, Y.-T., Parekh, Z., Pham, H., Le, Q., Sung, Y.-H., Li, Z., and Duerig, T. (2021). Scaling up visual and vision-language representation learning with noisy text supervision, in *Proceedings of the 38th International Conference on Machine Learning* (PMLR), pp. 4904–4916.

Jin, G., Shen, S., Zhang, D., Dai, F., and Zhang, Y. (2019). APE-GAN: Adversarial perturbation elimination with GAN, in *ICASSP 2019 - 2019 IEEE International Conference on Acoustics, Speech and Signal Processing (ICASSP)*, pp. 3842–3846, https://doi.org/10.1109/ICASSP.2019.8683044.

Karagiannis, G., Altintas, O., Ekici, E., Heijenk, G., Jarupan, B., Lin, K., and Weil, T. (2011). Vehicular networking: A survey and tutorial on requirements, architectures, challenges, standards and solutions, *IEEE Communications Surveys & Tutorials* **13**, 4, pp. 584–616, https://doi.org/10.1109/SURV.2011.061411.00019.

Khamaiseh, S. Y., Bagagem, D., Al-Alaj, A., Mancino, M., and Alomari, H. W. (2022). Adversarial deep learning: A survey on adversarial attacks and defense mechanisms on image classification, *IEEE Access* **10**, pp. 102266–102291, https://doi.org/10.1109/ACCESS.2022.3208131.

Khan, Z., Chowdhury, M., and Khan, S. M. (2022). A hybrid defense method against adversarial attacks on traffic sign classifiers in autonomous vehicles, *arXiv preprint* arXiv:2205.01225.

Khan, M. A., Alasiry, A., Marzougui, M., Bayhan, I., Kuna, S. S., Rao, G. S. N., Algamdi, S. A., and Aldossary, H. (2025). Securing intelligent transportation systems: A dual-framework approach for privacy protection and cybersecurity using generative AI, *IEEE Transactions on Intelligent Transportation Systems*, pp. 1–12, https://doi.org/10.1109/TITS.2025.3591007.

Kholgh, D. K. and Kostakos P. (2023). PAC-GPT: A novel approach to generating synthetic network traffic with GPT-3, *IEEE Access* **11**, pp. 114936–114951, https://doi.org/10.1109/ACCESS.2023.3325727.

Khoramnejad, F. and Hossain, E. (2025). Generative AI for the optimization of next-generation wireless networks: Basics, state-of-the-art, and open challenges, *IEEE Communications Surveys & Tutorials*, p. 1, https://doi.org/10.1109/COMST.2025.3535554.

Kim, T., Cha, M., Kim, H., Lee, J. K., and Kim, J. (2017). Learning to discover cross-domain relations with generative adversarial networks, in *Proceedings of the 34th International Conference on Machine Learning* (PMLR), pp. 1857–1865.

Laykaviriyakul, P. and Phaisangittisagul, E. (2023). Collaborative defense-GAN for protecting adversarial attacks on classification system, *Expert Systems with Applications* **214**, p. 118957, https://doi.org/10.1016/j.eswa.2022.118957.

Lim, X. R., Lee, C. P., Lim, K. M., Ong, T. S., Alqahtani, A., and Ali, M. (2023). Recent advances in traffic sign recognition: Approaches and datasets, *Sensors* **23**, 10, p. 4674, https://doi.org/10.3390/s23104674.

Liu, Z., Liu, Q., Liu, T., Xu, N., Lin, X., Wang, Y., and Wen, W. (2019). Feature distillation: DNN-oriented JPEG compression against adversarial examples, in *2019 IEEE/CVF Conference on Computer Vision and Pattern Recognition (CVPR)*, pp. 860–868, https://doi.org/10.1109/CVPR.2019.00095.

Liu, K., and Deng, H. (2021). The analysis of driver's recognition time of different traffic sign combinations on urban roads via driving simulation, *Journal of Advanced Transportation* **2021**, p. e8157293, https://doi.org/10.1155/2021/8157293.

Madry, A., Makelov, A., Schmidt, L., Tsipras, D., and Vladu, A. (2019). Towards deep learning models resistant to adversarial attacks, https://doi.org/10.48550/arXiv.1706.06083.

Mahmoudi, I., Boubiche, D. E., Athmani, S., Toral-Cruz, H., and Chan-Puc, F. I. (2025). Toward generative AI-based intrusion detection systems for the Internet of Vehicles (IoV), *Future Internet* **17**, 7, p. 310, https://doi.org/10.3390/fi17070310.

Marti, E., de Miguel, M. A., Garcia, F., and Perez, J. (2019). A review of sensor technologies for perception in automated driving, *IEEE Intelligent Transportation Systems Magazine* **11**, 4, pp. 94–108, https://doi.org/10.1109/MITS.2019.2907630.

Metta, S., Chang, I., Parker, J., Roman, M. P., and Ehuan, A. F. (2024). Generative AI in cybersecurity, https://doi.org/10.48550/ARXIV.2405.01674 .

Mogelmose, A., Trivedi, M. M., and Moeslund, T. B. (2012). Vision-based traffic sign detection and analysis for intelligent driver assistance systems: Perspectives and survey, *IEEE Transactions on Intelligent Transportation Systems* **13**, 4, pp. 1484–1497.

Moosavi-Dezfooli, S.-M., Fawzi, A., and Frossard, P. (2016). Deepfool: A simple and accurate method to fool deep neural networks, in *Proceedings of the IEEE Conference on Computer Vision and Pattern Recognition*, pp. 2574–2582.

Nie, J., Jiang, J., Li, Y., and Ercisli, S. (2025). Generative AI-enhanced autonomous driving: Innovating decision-making and risk assessment in multi-interactive environments, *IEEE Transactions on Intelligent Transportation Systems*, pp. 1–16, https://doi.org/10.1109/TITS.2025.3555774.

Pan, R., Islam, M. J., Ahmed, S., and Rajan, H. (2019). Identifying classes susceptible to adversarial attacks, https://doi.org/10.48550/arXiv.1905.13284.

Papernot, N., McDaniel, P., Wu, X., Jha, S., and Swami, A. (2016). Distillation as a defense to adversarial perturbations against deep neural networks, in *2016 IEEE Symposium on Security and Privacy (SP)*, pp. 582–597, https://doi.org/10.1109/SP.2016.41.

Qi, J. and Wang, J. (2025). Bridging artificial intelligence and railway cybersecurity: A comprehensive anomaly detection review, *Transportation Research Record: Journal of the Transportation Research Board* **2679**, 5, pp. 232–255, https://doi.org/10.1177/03611981241302335.

Rachev, S. T. (1990). *Duality Theorems for Kantorovich-Rubinstein and Wasserstein Functionals* (Instytut Matematyczny Polskiej Akademi Nauk, Warszawa).

Radanliev, P., Santos, O., and Ani, U. D. (2025). Generative AI cybersecurity and resilience, *Frontiers in Artificial Intelligence* **8**, p. 1568360, https://doi.org/10.3389/frai.2025.1568360.

Radford, A., Metz, L., and Chintala, S. (2016). Unsupervised representation learning with deep convolutional generative adversarial networks, https://doi.org/10.48550/arXiv.1511.06434.

Ross, A. and Doshi-Velez, F. (2018). Improving the adversarial robustness and interpretability of deep neural networks by regularizing their input gradients, *Proceedings of the AAAI Conference on Artificial Intelligence* **32**, 1, https://doi.org/10.1609/aaai.v32i1.11504.

Sai, S., Yashvardhan, U., Chamola, V., and Sikdar, B. (2024). Generative AI for cyber security: Analyzing the potential of ChatGPT, DALL-E, and other models for enhancing the security space, *IEEE Access* **12**, pp. 53497–53516, https://doi.org/10.1109/ACCESS.2024.3385107.

Salek, M. S. (2023). Msabbirsalek/AR-GAN, https://github.com/msabbirsalek/AR-GAN.

Salek, M. S., Chowdhury, M., Munir, M. B., Cai, Y., Hasan, M. I., Tine, J.-M., Khan, L., and Rahman, M. (2025). A large language model-supported threat modeling framework for transportation cyber-physical systems, *IEEE Access* **13**, pp. 163046–163070, https://doi.org/10.1109/ACCESS.2025.3603580.

Samangouei, P., Kabkab, M., and Chellappa, R. (2018). Defense-GAN: Protecting classifiers against adversarial attacks using generative models, https://doi.org/10.48550/arXiv.1805.06605.

Sharif, S., Zeadally, S., and Ejaz, W. (2025). Resource optimization in UAV-assisted IoT networks: The role of generative AI, *IEEE Internet of Things Magazine* **8**, 1, pp. 34–41, https://doi.org/10.1109/IOTM.001.2400045.

Shekhar, A., Prabhat, P., Yandrapalli, V., Umar, S., Abdul, F., and Wakjira, W. D. (2023). Generative AI in supply chain management, *International Journal on Recent and Innovation Trends in Computing and Communication* **11**, 9, pp. 4179–4185.

Somvanshi, S., Liu, J., and Das, S. (2025). A survey on generative AI in transportation systems management and operation, in *2025 IEEE Conference on Artificial Intelligence (CAI)*, pp. 829–832, https://doi.org/10.1109/CAI64502.2025.00148.

Sun, Y. and Ortiz, J. (2024). GenAI-driven cyberattack detection in V2X networks for enhanced road safety and autonomous vehicle defense, *International Journal of Advance in Applied Science Research* **3**, pp. 67–75.

Thomas, O., Munir, M. B., Tine, J.-M., Rahman, M., Cai, Y., Akbar, K. A., Uddin, M. N., Khan, L., Hockstad, T., and Chowdhury, M. (2025). Transportation cyber incident awareness through generative AI-based incident analysis and retrieval-augmented question-answering systems, https://doi.org/10.48550/ARXIV.2508.02523.

Tremblay, J., Prakash, A., Acuna, D., Brophy, M., Jampani, V., Anil, C., To, T., Cameracci, E., Boochoon, S., and Birchfield, S. (2018). Training deep networks with synthetic data: Bridging the reality gap by domain randomization, in *Proceedings of the IEEE Conference on Computer Vision and Pattern Recognition Workshops*, pp. 969–977.

Uddin, M., Irshad, M. S., Kandhro, I. A., Alanazi, F., Ahmed, F., Maaz, M., Hussain, S., and Ullah, S. S. (2025). Generative AI revolution in cybersecurity: A comprehensive review of threat intelligence and operations, *Artificial Intelligence Review* **58**, 8, p. 236, https://doi.org/10.1007/s10462-025-11219-5.

Van Huynh, N., Wang, J., Du, H., Hoang, D. T., Niyato, D., Nguyen, D. N., Kim, D. I., and Letaief, K. B. (2024). Generative AI for physical layer communications: A survey, *IEEE Transactions on Cognitive Communications and Networking* **10**, 3, pp. 706–728, https://doi.org/10.1109/TCCN.2024.3384500.

Vu, T.-H., Kumar Jagatheesaperumal, S., Nguyen, M.-D., Van Huynh, N., Kim, S., and Pham, Q.-V. (2025). Applications of generative AI (GAI) for mobile and wireless networking: A survey, *IEEE Internet of Things Journal* **12**, 2, pp. 1266–1290, https://doi.org/10.1109/JIOT.2024.3487627.

Wali, S. B., Abdullah, M. A., Hannan, M. A., Hussain, A., Samad, S. A., Ker, P. J., and Mansor, M. B. (2019). Vision-based traffic sign detection and recognition systems: Current trends and challenges, *Sensors* **19**, 9, p. 2093, https://doi.org/10.3390/s19092093.

Wan, A., Wallace, E., Shen, S., and Klein, D. (2023). Poisoning language models during instruction tuning, https://doi.org/10.48550/arXiv.2305.00944.

Wang, M. (2024). Generative AI: A new challenge for cybersecurity, *Journal of Computer Science and Technology Studies* **6**, 2, pp. 13–18, https://doi.org/10.32996/jcsts.2024.6.2.3.

Xu, W., Evans, D., and Qi, Y. (2018). Feature squeezing: Detecting adversarial examples in deep neural networks, in *Proceedings 2018 Network and Distributed System Security Symposium*, https://doi.org/10.14722/ndss.2018.23198.

Ye, N., and Zhu, Z. (2018). Bayesian adversarial learning, in *Advances in Neural Information Processing Systems* (Curran Associates, Inc.).

Yigit, Y., Ferrag, M. A., Ghanem, M. C., Sarker, I. H., Maglaras, L. A., Chrysoulas, C., Moradpoor, N., Tihanyi, N., and Janicke, H. (2025) Generative AI and LLMs for critical infrastructure protection: Evaluation benchmarks, agentic AI, challenges, and opportunities, *Sensors* **25**, 6, p. 1666, https://doi.org/10.3390/s25061666.

Zaboli, A., Kim, Y.-H., and Hong, J. (2025). An advanced generative AI-based anomaly detection in IEC61850-based communication messages in smart grids, *IEEE Access* **13**, pp. 89997–90016, https://doi.org/10.1109/ACCESS.2025.3571881.

Zhang, H., Chen, H., Song, Z., Boning, D., Dhillon, I. S., and Hsieh, C.-J. (2019). The limitations of adversarial training and the blind-spot attack, https://doi.org/10.48550/arXiv.1901.04684.

Zhao, C., Du, H., Niyato, D., Kang, J., Xiong, Z., Kim, D. I., Shen, X., and Letaief, K. B. (2025). Generative AI for secure physical layer communications: A survey, *IEEE Transactions on Cognitive Communications and Networking* **11**, 1, pp. 3–26, https://doi.org/10.1109/TCCN.2024.3438379.

Zhou, W., Bai, S., Mandic, D. P., Zhao, Q., and Chen, B. (2024). Revisiting the adversarial robustness of vision language models: A multimodal perspective, https://doi.org/10.48550/arXiv.2404.19287.

Zou, Y., Xu, Z., Wang, T., Xiong, G., Lin, Z., and Li, D. (2025). Generative AI-driven dynamic information prioritization for enhanced autonomous driving, *IEEE Transactions on Intelligent Transportation Systems*, pp. 1–14, https://doi.org/10.1109/TITS.2025.3552795.

https://doi.org/10.1142/9789819831548_0010

Chapter 10

Architecting Resilience: GenAI-Enhanced Threat Modeling in ITS

Muhaimin Bin Munir*, **Yuchen Cai**†, **Bhavani Thuraisingham**‡, **and Latifur Khan**§

The University of Texas at Dallas, Richardson, Texas, USA
**muhaimin.binmunir@utdallas.edu*
†yuchen.cai@utdallas.edu
‡bxt043000@utdallas.edu
§lkhan@utdallas.edu

Abstract
Modern Intelligent Transportation Systems (ITS) face significant cybersecurity challenges due to their complex cyber-physical architectures and extensive connectivity, making traditional manual threat modeling approaches impractical for systems with over 1,000 information flows. This chapter introduces TraCR-TMF (Transportation Cyber Resilience-Threat Modeling Framework), which leverages Large Language Models to automate the mapping of ITS information flows to MITRE ATT&CK adversarial techniques and NIST Cybersecurity Framework controls. We systematically evaluate multiple AI approaches including zero-shot learning, retrieval-augmented generation, multimodal processing, in-context learning, and supervised fine-tuning using ModernBERT on 433 real-world transportation flows from 26 ARC-IT service packages. Supervised fine-tuning achieves the highest performance with 72% F1-score, representing a 300% improvement over zero-shot approaches (18%), while expert validation confirms 73% precision for predicted attack techniques. The framework demonstrates 98% reduction in threat

assessment time compared to manual approaches and successfully identifies 23 previously unrecognized threats in pilot deployments with transportation agencies. This research provides the first comprehensive automated threat modeling framework for ITS, along with empirical benchmarks that establish a foundation for AI-enhanced transportation cybersecurity.

Keywords: LLM, Intelligent transportation system security, Cyber physical system, RAG, Multimodal RAG, NIST control, MITRE ATT&CK, In context learning, BERT, Supervised learning, Threat mapping.

10.1 Introduction

Modern transportation systems have evolved into complex cyber-physical ecosystems that integrate vehicles, infrastructure, sensors, and control systems to enhance safety, efficiency, and mobility. The emergence of Intelligent Transportation Systems (ITS) represents a paradigm shift toward highly interconnected and automated transportation networks, encompassing connected and autonomous vehicles (CAVs), smart traffic management systems, and comprehensive sensor networks. While these technological advances offer unprecedented opportunities for transportation optimization, they simultaneously introduce significant cybersecurity challenges that threaten the integrity, availability, and confidentiality of critical transportation services.

The cybersecurity landscape for ITS is particularly complex due to the heterogeneous nature of the underlying cyber-physical systems (CPS). Transportation networks comprise diverse components including roadside units, traffic control centers, vehicle onboard units, communication networks, and sensor arrays, each presenting unique attack surfaces and vulnerabilities (Checkoway *et al.*, 2011; Koscher *et al.*, 2010). The increasing connectivity and automation in transportation systems exponentially expand these attack surfaces, creating new pathways for adversaries to exploit system vulnerabilities and potentially cause catastrophic consequences ranging from traffic disruptions to safety-critical incidents.

Traditional cybersecurity methods are not suitable for the unique challenges of Intelligent Transportation Systems (ITS) due to their real-time, safety-critical nature and distributed architecture. The complexity of modern ITS makes manual threat analysis too slow and prone to errors, necessitating specialized threat modeling methodologies to identify and mitigate cybersecurity risks (Boyes *et al.*, 2018; Humayed *et al.*, 2017). Recent research has explored LLM-supported frameworks for transportation cyber-physical systems threat modeling (Salek *et al.*, 2025), while multimodal approaches have shown promise for enhancing cyber attack detection in transit systems (Munir *et al.*, 2024).

To address these challenges, this chapter presents a comprehensive framework that leverages Large Language Models (LLMs) and other generative AI tools to automate and enhance threat modeling processes for intelligent transportation systems. Our approach systematically maps information flows within ITS service packages to relevant adversarial techniques catalogued in the most recent MITRE ATT&CK knowledge base. It is also capable of mapping the information flows to NIST control IDs with proper justification. By automating this critical mapping process, our framework enables security analysts to efficiently identify potential attack scenarios and design appropriate countermeasures and mitigation strategies for complex transportation systems.

10.1.1 *Research Motivation and Objectives*

The primary motivation for this research stems from the critical gap between the rapidly evolving threat landscape in transportation cybersecurity and the limited scalability of traditional threat modeling approaches. Current threat modeling practices for ITS rely heavily on manual analysis by cybersecurity experts, a process that becomes increasingly impractical and economically challenging as system complexity grows. The Architecture Reference for Cooperative and Intelligent Transportation (ARC-IT) framework, which serves as the de facto standard for ITS architecture design in the United States, encompasses over 150 service packages and approximately 1,000 information flows, each requiring individual security assessment (US Department of Transportation, 2022).

The manual process of associating each information flow with relevant attack techniques from frameworks such as MITRE ATT&CK is not only labor-intensive but also prone to inconsistencies and omissions. Security experts must possess deep knowledge of both transportation systems and cybersecurity to accurately identify potential threats, making the process dependent on scarce specialized expertise. They may have differences of opinion that are conflicting. This bottleneck significantly limits the ability of transportation agencies and system integrators to conduct comprehensive security assessments, particularly for large-scale deployments.

Our research addresses these limitations by pursuing the following key objectives:

Objective 1: Automated Multi-Framework Threat-Flow Mapping. Develop an intelligent system that automatically maps ITS information flows to relevant MITRE ATT&CK techniques (see Figure 10.1) and other cybersecurity frameworks, such as the NIST Cybersecurity Framework (CSF). This system will reduce the manual effort of threat modeling and enable comprehensive security assessments that address both tactical threats and strategic security functions.

Objective 2: Analysis Using AI Tools. Investigate and compare various LLM-based approaches for cybersecurity knowledge extraction and application, including zero-shot learning, few shot learning, retrieval-augmented generation, multimodal analysis, and supervised fine-tuning methodologies.

10.1.2 *Key Contributions*

This chapter makes several significant contributions to the intersection of transportation cybersecurity and artificial intelligence:

Novel Framework Architecture: We introduce TraCR-TMF, the first comprehensive framework specifically designed to automate threat modeling for intelligent transportation systems using advanced LLM techniques. The framework integrates multiple AI approaches and provides a systematic methodology for mapping ITS information flows to cybersecurity threats.

Comprehensive Empirical Evaluation: We present an extensive comparative analysis of multiple LLM-based approaches for automated threat modeling, including zero-shot learning, retrieval-augmented generation (RAG), multimodal RAG, in-context learning

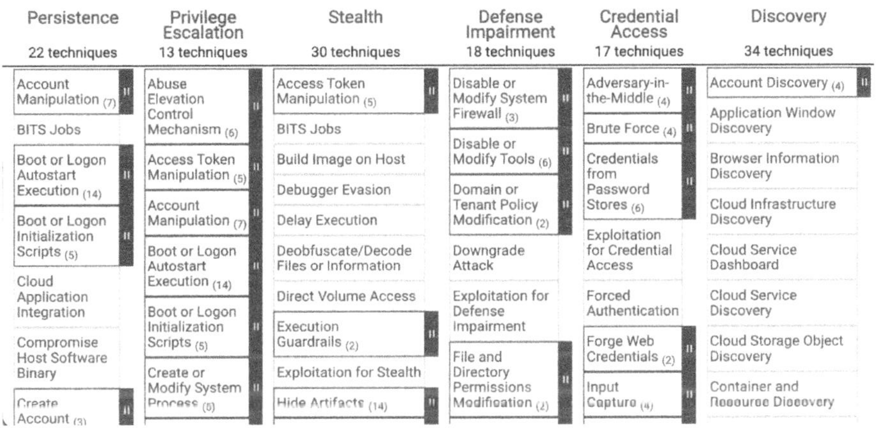

Fig. 10.1. Overview of the MITRE ATT&CK Enterprise Systems matrix, highlighting some tactics and techniques relevant to transportation cyber-physical systems. The framework provides a structured approach for categorizing adversarial behaviors in industrial environments.

(ICL), and supervised fine-tuning. Our evaluation provides practical insights into the strengths and limitations of each approach for cybersecurity applications. We also demonstrate how active learning techniques can significantly reduce the annotation burden for training effective threat modeling systems.

Multi-Framework Mapping: Our work extends beyond traditional MITRE ATT&CK mapping to include automated alignment with NIST CSF functions, providing a more comprehensive security assessment capability that addresses both threat identification and security management requirements.

Validated Dataset and Benchmarks: We contribute a labeled dataset of ITS information flows mapped to cybersecurity frameworks, along with evaluation benchmarks that can support future research in this domain.

10.1.3 *Chapter Organization*

This chapter is organized as follows: Section 10.2 provides essential background on the MITRE ATT&CK framework and ITS information flow architectures. Section 10.3 presents a comprehensive overview of the machine learning and NLP techniques employed, including zero-shot learning and retrieval-augmented generation. The framework architecture and implementation details for automated threat-flow mapping and multi-framework integration are described in Section 10.4. Section 10.5 compiles and analyzes experimental results, demonstrating the effectiveness of different approaches. Finally, Section 10.6 summarizes the key findings, discusses limitations, and outlines future research directions for LLM-enhanced cybersecurity in transportation systems.

As ITS continue to evolve toward greater connectivity and autonomy, the ability to automatically and accurately assess cybersecurity risks becomes essential for maintaining safety, security, and reliability. This chapter provides both theoretical foundations and practical methodologies for achieving this goal, contributing to the broader effort of building resilient and secure intelligent transportation systems.

10.2 Background: MITRE ATT&CK Framework and ITS Information Flows

This section establishes the foundational concepts underlying our LLM-enhanced threat modeling framework. We begin with a comprehensive overview of the MITRE ATT&CK knowledge base, examining its structure, applicability to industrial control systems, and relevance to transportation cybersecurity. Subsequently, we explore the Architecture Reference for Cooperative and Intelligent Transportation (ARC-IT) framework, focusing on service packages, functional objects, and information flows that form the basis for our automated threat modeling approach (see Figure 10.2).

10.2.1 *MITRE ATT&CK Framework*

The MITRE ATT&CK (Adversarial Tactics, Techniques, and Common Knowledge) framework represents one of the most comprehensive and widely adopted knowledge bases for understanding cybersecurity threats (MITRE Corporation, 2021; Strom *et al.*, 2018). Originally developed by MITRE Corporation, ATT&CK provides a structured taxonomy of adversarial behaviors observed

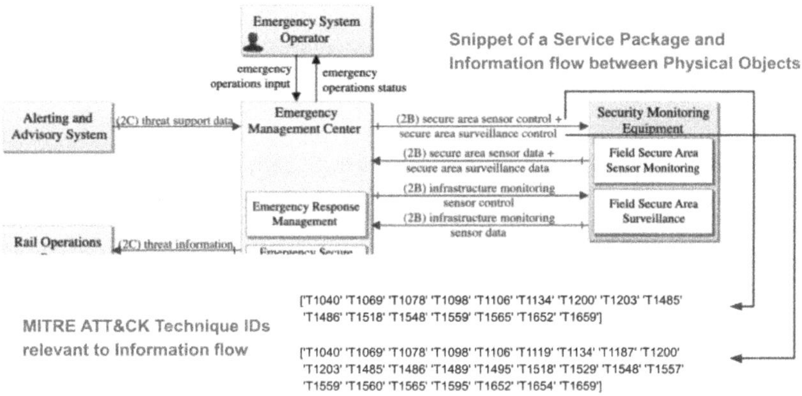

(a)

ID	Name	ID	Name
T1040	Network Sniffing	**T1489**	Service Stop
T1069	Permission Groups Discovery	**T1495**	Firmware Corruption
T1078	Valid Accounts	**T1518**	Software Discovery
T1098	Account Manipulation	**T1529**	System Shutdown/Reboot
T1659	Content Injection	**T1548**	Abuse Elevation Control Mechanism
T1119	Automated Collection	**T1557**	Adversary-in-the-Middle
T1134	Access Token Manipulation	**T1559**	Inter-Process Communication
T1187	Forced Authentication	**T1560**	Archive Collected Data
T1200	Hardware Additions	**T1565**	Data Manipulation
T1203	Exploitation for Client Execution	**T1595**	Active Scanning
T1485	Data Destruction	**T1652**	Device Driver Discovery
T1486	Data Encrypted for Impact	**T1654**	Log Enumeration

(b)

Fig. 10.2. Overview of the ITS system architecture (a) High-level architecture of a service package, Transit Security of Intelligent Transportation Systems, showing the interconnection of various physical objects, functional objects, and information flows. The complexity of these interconnections necessitates automated approaches for comprehensive threat modeling. (b) Mapped technique IDs and their names for better readability.

in real-world cyber attacks, organized according to the tactics and techniques employed by threat actors throughout the attack lifecycle.

10.2.1.1 *Framework Structure and Organization*

The ATT&CK framework organizes adversarial behaviors into a matrix structure comprising tactics (the "why" of an attack) and techniques (the "how" of an attack). Tactics represent the tactical objectives that adversaries seek to achieve during an attack, such as initial access, persistence, privilege escalation, and exfiltration. Techniques describe the specific methods used to accomplish these tactical objectives, with many techniques further subdivided into sub-techniques that provide more granular detail about implementation approaches (Applebaum *et al.*, 2016; MITRE Corporation, 2021).

The framework's hierarchical structure enables security analysts to understand both the strategic intent behind attacks and the specific technical methods employed.

10.2.1.2 *Enterprise Matrix*

Of particular relevance to transportation cybersecurity is the MITRE ATT&CK Enterprise matrix, which specifically addresses threats to operational technology (OT) environments (Assante and Lee, 2015). The Enterprise matrix recognizes that operating systems, network devices, containers, including transportation infrastructure face unique threats that differ significantly from traditional enterprise IT environments. These systems prioritize availability and safety over confidentiality, operate with real-time constraints, and often employ legacy technologies with limited security capabilities.

As of 2025, the Enterprise ATT&CK matrix comprises:

- 14 tactics, spanning the full adversary lifecycle from initial access to impact (MITRE Corporation, 2021).
- 211 techniques, each mapped to one or more tactics and describing high-level offensive behaviors.
- 468 subtechniques, which elaborate on specific ways techniques may be implemented in various environments (MITRE Corporation, 2021).

The Enterprise matrix lists the techniques most extensively and covers any system. It is to be noted that cybersecurity experts and professionals constantly update this knowledge base with newer techniques, as cyber-criminals also add weapons to their arsenal. It is important to address the newer updates with any system that references the MITRE knowledge base to include the latest cyber threats and their mitigation strategies.

10.2.1.3 *Application to Transportation Systems*

Transportation systems share many characteristics with Industrial Control Systems (ICS), making the ICS ATT&CK matrix a highly relevant tool for cybersecurity analysis (Checkoway *et al.*, 2011). Modern transportation infrastructure, including traffic management and railway signaling networks, relies on industrial components and protocols like MODBUS, DNP3, and IEC 61850, which are specifically addressed in the ICS ATT&CK matrix (Hurd and Smith, 2008; Morris *et al.*, 2012). The safety-critical nature of these systems makes them prime targets for adversaries, as seen in attacks on other critical infrastructure, reinforcing the real-world applicability of these techniques to the transportation sector (Liang *et al.*, 2017).

The integration of connected and autonomous vehicles complicates this threat landscape by introducing new attack vectors through vehicle-to-infrastructure (V2I) and vehicle-to-vehicle (V2V) communications (Mundhenk *et al.*, 2017; Parkinson *et al.*, 2017). These hybrid threats, which bridge traditional automotive and infrastructure-based concerns, can be systematically analyzed using the ICS ATT&CK matrix. This makes the framework essential for understanding and mitigating the evolving cybersecurity risks in transportation.

10.2.2 *ITS Information Flows and ARC-IT Architecture*

The Architecture Reference for Cooperative and Intelligent Transportation (ARC-IT) serves as the authoritative framework for ITS architecture design and implementation in the United States (US Department of Transportation, 2020, 2022). Developed and maintained by the US Department of Transportation, ARC-IT provides a comprehensive reference architecture that standardizes the description of ITS functionality, interfaces, and information exchanges.

10.2.2.1 *Service Packages and Functional Decomposition*

ARC-IT structures Intelligent Transportation Systems (ITS) into 156 discrete service packages across 12 functional areas, including Data Management and Public Safety. Each package is a building block

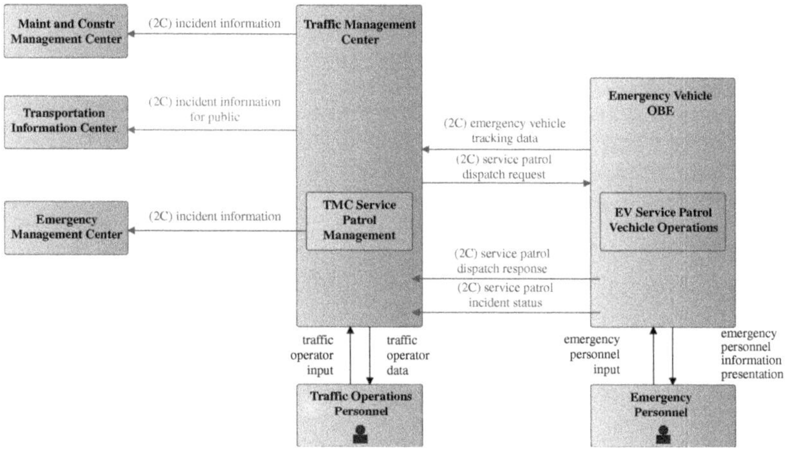

Fig. 10.3. Example of an ITS service package Roadway Service Patrol showing the relationship between physical objects, functional objects, and information flows. The diagram illustrates how multiple systems collaborate through standardized interfaces to deliver transportation services.

that combines physical objects, functional objects, and information flows to deliver specific transportation services (see Figure 10.3).

Each service package is documented with multiple architectural views, including physical view (showing systems and their interfaces), functional view (depicting functional objects and data flows), and enterprise view (illustrating organizational relationships and responsibilities) (US Department of Transportation, 2012).

10.2.2.2 *Information Flows and Data Exchanges*

Information flows represent the structured data exchanges between functional objects within and across service packages. These flows define the specific data elements, formats, and exchange patterns required to support ITS operations. ARC-IT catalogs approximately 1,000 distinct information flows, each characterized by its source and destination, data content, communication characteristics, and security attributes.

The security attributes associated with each information flow provide crucial context for threat modeling activities. ARC-IT assigns CIA ratings to information flows using a standardized scale that indicates the potential impact of compromising each security property. These ratings serve as input to risk assessment processes and help prioritize security investments.

10.2.3 *Integration Challenges and Opportunities*

Mapping ARC-IT information flows to MITRE ATT&CK techniques presents significant challenges, primarily due to:

- Domain expertise requirements, as it demands a rare combination of transportation systems and cybersecurity knowledge.
- The immense scale and complexity of over 150 service packages and 1,000 information flows, which makes manual mapping efforts overwhelming.
- The need to adapt to the dynamic evolution of both the ARC-IT and MITRE ATT&CK frameworks.
- The financial constraints associated with the high cost of manual expert labor required for consistent annotation.

This challenge has been recognized in recent work developing LLM-supported threat modeling frameworks for transportation CPS (Salek *et al.*, 2025).

Despite these challenges, integrating ARC-IT and MITRE ATT&CK offers substantial benefits by enabling more comprehensive risk assessments, supporting the development of targeted security controls, and improving communication between transportation and cybersecurity communities (Humayed *et al.*, 2017; Yaacoub *et al.*, 2020). The application of advanced artificial intelligence, particularly large language models, offers a promising path to overcome these challenges and automate the threat modeling process for intelligent transportation systems.

10.3 Gen-AI Techniques for Automated Threat Modeling

This section provides a comprehensive technical overview of the machine learning and natural language processing methodologies employed in our LLM-enhanced threat modeling framework. We examine the theoretical foundations, algorithmic implementations, and domain-specific adaptations of each technique, with particular emphasis on their application to cybersecurity knowledge extraction and automated threat-flow mapping in intelligent transportation systems.

10.3.1 *Zero-Shot Learning for Threat Classification*

Zero-shot learning (ZSL) represents a paradigm where models perform classification tasks on previously unseen classes without explicit training examples (Wang *et al.*, 2019; Xian *et al.*, 2018). In the context of threat modeling, zero-shot approaches leverage pre-trained language models' inherent knowledge to classify information flows according to relevant MITRE ATT&CK techniques without requiring domain-specific training data.

10.3.1.1 *Semantic Embedding Approaches*

Contemporary zero-shot learning approaches rely heavily on semantic embeddings to bridge the gap between visual/textual features and class semantics. For threat modeling applications, we employ transformer-based language models that encode both information flow descriptions and ATT&CK technique descriptions into high-dimensional semantic spaces (Brown *et al.*, 2020; Radford *et al.*, 2019).

The semantic similarity between an information flow x and an ATT&CK technique c is computed using cosine similarity in the embedded space:

$$\text{sim}(x, c) = \frac{\phi(x) \cdot \psi(c)}{||\phi(x)|| \cdot ||\psi(c)||} \tag{10.1}$$

where $\phi(x)$ represents the embedding of information flow description x, and $\psi(c)$ represents the embedding of ATT&CK technique description c. The classification decision is made by selecting the technique with highest semantic similarity:

$$\hat{y} = \arg\max_{c \in \mathcal{C}_u} \text{sim}(x, c) \tag{10.2}$$

10.3.1.2 *Prompt Engineering for Cybersecurity Domains*

Zero-shot learning's effectiveness in cybersecurity threat modeling relies on carefully designed prompts. We developed a prompt structure that includes a clear task definition, domain context on ITS and ATT&CK framework knowledge, and a specific output format to facilitate automated parsing (Dong *et al.*, 2022; Liu *et al.*, 2023).

10.3.2 *Few-Shot and In-Context Learning*

Few-shot learning and in-context learning (ICL) extend zero-shot approaches by providing a small number of labeled examples to guide model predictions (Brown *et al.*, 2020; Dong *et al.*, 2022). These techniques are particularly valuable for specialized domains like cybersecurity, where domain-specific patterns and relationships may not be fully captured in pre-training data.

10.3.2.1 *Example Selection Strategies*

The effectiveness of few-shot learning critically depends on example selection strategies. For threat modeling applications, we investigate several approaches:

Random Sampling: Baseline approach that randomly selects examples from the labeled dataset, providing an unbiased but potentially suboptimal context.

Similarity-Based Selection: Examples are selected based on semantic similarity to the query instance, computed using pre-trained embeddings.

10.3.2.2 *Chain-of-Thought Reasoning*

Chain-of-thought (CoT) reasoning is a prompt engineering technique that enhances a model's ability to solve complex tasks by guiding it to generate a series of intermediate reasoning steps (Kojima *et al.*, 2022; Wei *et al.*, 2022). For threat modeling, this process involves four key steps: analyzing the information flow, assessing the attack surface, mapping vulnerabilities to specific ATT&CK techniques, and providing justifications for each selection. This structured approach mimics human-like reasoning, leading to more accurate and reliable threat analysis.

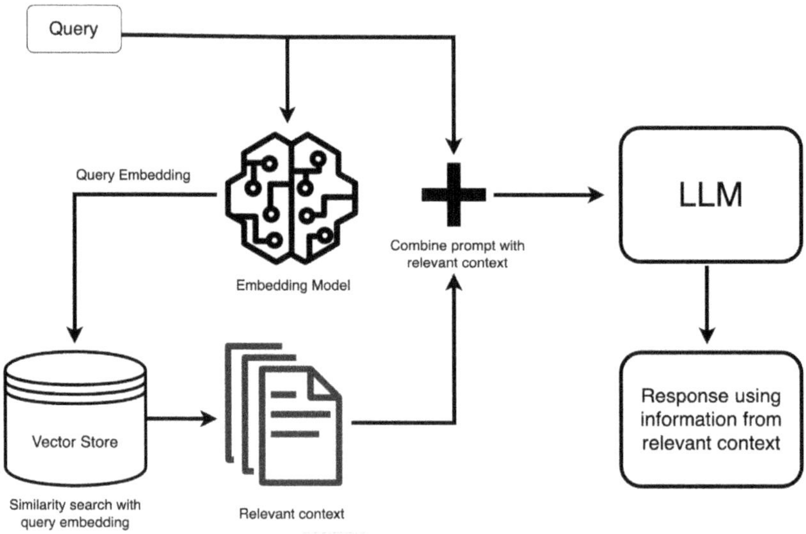

Fig. 10.4. The generic RAG architecture.

10.3.3 *Retrieval-Augmented Generation*

Retrieval-Augmented Generation (RAG) addresses the knowledge limitations of parametric language models by dynamically incorporating relevant external information during generation (Guu *et al.*, 2020; Lewis *et al.*, 2020) (see Figure 10.4). For cybersecurity applications, RAG enables models to access up-to-date threat intelligence, detailed technique descriptions, and domain-specific knowledge that may not be present in pre-training data.

10.3.3.1 *Architecture and Components*

The RAG framework consists of three primary components: a retriever, a knowledge base, and a generator. The mathematical formulation of RAG can be expressed as:

$$P(y|x) = \sum_{z \in \text{top-k}(x)} P(z|x) \cdot P(y|x, z) \tag{10.3}$$

where x represents the input query (information flow description), y represents the output (ATT&CK technique predictions), and z represents retrieved knowledge base entries. The retriever computes $P(z|x)$ to identify relevant knowledge, while the generator computes

$P(y|x, z)$ to produce the final output conditioned on both input and retrieved knowledge (Lewis *et al.*, 2020).

10.3.4 *Multimodal Retrieval-Augmented Generation*

Traditional RAG approaches focus exclusively on textual information, potentially overlooking valuable visual information present in technical diagrams, network schematics, and security matrices. Multimodal RAG extends the framework to incorporate visual information alongside textual content. Adding these visual diagrams in the context of the LLM helps the model to reason better, as it gives a holistic picture of the information flow with other components in the service package, which the text-only prompt cannot provide.

10.3.5 *Visual Information Processing*

Intelligent Transportation Systems (ITS) documentation often contains network diagrams, data flow diagrams, and security matrices that convey critical security information. We employ vision-language models (VLMs) like GPT-4V and CLIP to extract semantic information from these visual elements, assisting in a more comprehensive understanding of the system's security posture (Achiam *et al.*, 2023; Li *et al.*, 2022; Radford *et al.*, 2021).

10.3.5.1 *Multimodal Fusion Strategies*

The integration of visual and textual information requires sophisticated fusion approaches:

Early Fusion: Visual and textual features are combined at the input level, creating joint representations that capture cross-modal relationships:

$$h_{\text{joint}} = f_{\text{fusion}}([h_{\text{text}}; h_{\text{visual}}]) \tag{10.4}$$

Late Fusion: Separate processing pipelines handle textual and visual information, with results combined at the decision level:

$$P(y|x_{\text{text}}, x_{\text{visual}}) = \alpha P(y|x_{\text{text}}) + (1-\alpha)P(y|x_{\text{visual}}) \tag{10.5}$$

Attention-Based Fusion: Cross-modal attention mechanisms enable dynamic weighting of textual and visual information based on query relevance (Lu *et al.*, 2019; Tan and Bansal, 2019).

10.3.6 *Supervised Fine-Tuning for Multi-Label Classification*

Supervised fine-tuning represents the most sophisticated approach within the TraCR-TMF framework for mapping information flows to MITRE ATT&CK techniques. This approach leverages the power of pre-trained transformer-based models and treats the threat identification task as a multi-label classification problem, where each information flow can be associated with multiple relevant attack techniques.

10.3.6.1 *Model Architecture Selection*

We selected ModernBERT over traditional architectures like BERT due to its ability to handle significantly longer input sequences, a crucial feature for complex transportation cybersecurity texts. The base version of ModernBERT, with its 22 layers and 149 million parameters, was chosen for its optimal balance of performance and computational efficiency. Furthermore, its open-source nature allows for local deployment, making it suitable for organizations with data privacy concerns or limited cloud computing resources.

10.3.6.2 *Transfer Learning Approach*

To perform supervised fine-tuning of the ModernBERT model, we follow a transfer learning approach, starting with a pre-trained model and fine-tuning it for our specific threat identification task. The primary advantage of transfer learning over training a model from scratch is leveraging the knowledge already embedded in the pre-trained model from its initial training on large online datasets. This approach enables the model to avoid learning basic language features from the ground up, yielding improved performance especially when the target dataset is comparatively smaller.

10.3.6.3 *Multi-Label Classification Formulation*

In the supervised learning context, our approach can be used when the training dataset contains ground-truth labels provided by cybersecurity experts. In our multi-label classification problem, each information flow instance can be associated with more than one MITRE ATT&CK technique.

Consider the dataset formulation:

$$\mathcal{D} = \{(x_i, y_i)\}_{i=1}^{N} \tag{10.6}$$

where x_i represents the basic input features for information flow i, and $y_i = \{y_{i,1}, y_{i,2}, y_{i,3}, \ldots, y_{i,L}\}$ is a binary vector with $y_{i,j} = 1$ indicating that technique j is associated with flow i based on the ground-truth annotations, and $y_{i,j} = 0$ otherwise.

Our goal is to train a multi-label classifier based on ModernBERT that can perform the mapping from x_i to y_i. In the context of mapping transportation CPS information flows to relevant MITRE ATT&CK techniques, x_i represents the basic input containing flow descriptions, security attributes, and STRIDE threats, while y_i represents the relevance of the MITRE ATT&CK techniques present in the expert-annotated ground-truth list.

10.3.6.4 *Training Configuration*

To ensure the model's robustness and generalizability, we employed a five-fold cross-validation strategy, utilizing the MultilabelStratifiedKFold method to create balanced train (80%) and test (20%) splits. The model was fine-tuned using the Hugging Face Transformers' Trainer API, with all layers updated by the AdamW optimizer. To prevent overfitting, early stopping was implemented, monitoring the validation loss with a tolerance of five epochs and a minimum improvement threshold of 0.001. The training was conducted with a maximum of 50 epochs, a batch size of 8, a learning rate of 0.00002, and a weight decay of 0.01, with the batch size specifically chosen to fit the memory capacity of an A100 GPU.

10.3.6.5 *Threshold Optimization*

For each individual ATT&CK technique, we optimized the binary classification threshold to maximize the F1-score on the validation sets. This process involved analyzing receiver operating characteristic (ROC) curves to derive a range of true positive rates and false positive rates. The optimal operating point was selected by computing the F1-score for each potential threshold, with ties resolved by prioritizing the highest true positive rate. This supervised fine-tuning method achieved a 72% F1-score on the evaluation dataset, demonstrating its suitability for production environments where accuracy is the primary concern and sufficient expert-annotated data is available.

10.3.7 *Active Learning for Efficient Data Annotation*

The development of high-quality labeled datasets for cybersecurity applications faces significant challenges due to the specialized expertise required for accurate annotation and the time-intensive nature of threat analysis. Active learning provides a systematic approach to minimize annotation costs while maximizing model performance by strategically selecting the most informative examples for labeling (Ren *et al.*, 2021; Settles, 2009).

10.3.7.1 *Query Strategy Design*

For multi-label threat modeling, we implement several query strategies adapted from single-label approaches, including:

- **Entropy-Based Selection:** This method measures prediction uncertainty across all labels and selects instances with the highest average entropy.
- **Monte Carlo Dropout:** This strategy estimates model uncertainty by conducting multiple forward passes with dropout enabled during inference.
- **Temporal Prediction Discrepancy (TPD):** Specifically designed for multi-label scenarios, TPD measures the disagreement between predictions at different stages of training.

10.3.7.2 *Batch Selection Algorithms*

Selecting batches of examples for annotation requires balancing informativeness with diversity to avoid redundant selections:

Uncertainty Sampling: Selects the top-k most uncertain examples based on individual uncertainty scores.

Diversity-Based Sampling: Uses clustering or maximum marginal relevance to ensure selected examples cover different regions of the feature space (Sener and Savarese, 2017).

Expected Model Change: Selects examples that are expected to cause the largest changes in model parameters, approximated through gradient-based methods (Settles and Craven, 2008).

10.3.8 *NIST Cybersecurity Framework Integration*

The National Institute of Standards and Technology (NIST) Cybersecurity Framework provides a comprehensive approach to managing cybersecurity risks through five core functions: Identify, Protect, Detect, Respond, and Recover (National Institute of Standards and Technology, 2018). Integrating NIST CSF mapping with ATT&CK technique identification enables a more holistic approach to cybersecurity planning that addresses both tactical threats and strategic security management.

10.3.8.1 *Framework Alignment Methodology*

We used the RAG framework for the MITRE techniques mapping to map the NIST IDs with the information flow. In addition to that, the model also provides justification, as we don't have enough ground truth annotated to evaluate the results. The equations for this model are similar to those we used for the MITRE technique mapping. Just instead of the MITRE techniques, the model will recommend NIST control families and, more specifically, the NIST Control IDs related to the information flow.

The techniques presented in this section form the foundation for our comprehensive threat modeling framework, each contributing unique capabilities for automated cybersecurity analysis. The following section details how these techniques are integrated into a cohesive architecture for practical threat modeling applications.

10.4 TraCR-TMF: Architecture and Implementation Methods

This section presents the detailed architecture and implementation methodology of the Transportation Cyber Resilience-Threat

Modeling Framework (TraCR-TMF). We provide comprehensive descriptions of the system components, data processing pipelines, and algorithmic implementations that enable automated mapping between ITS information flows and cybersecurity frameworks. The architecture is designed to be modular, scalable, and extensible to accommodate evolving threat landscapes and expanding transportation system complexity.

10.4.1 *System Architecture Overview*

The TraCR-TMF framework employs a layered architecture that separates concerns between data ingestion, knowledge representation, machine learning inference, and result synthesis. Figure 10.5 illustrates the high-level system design, consisting of five primary layers:

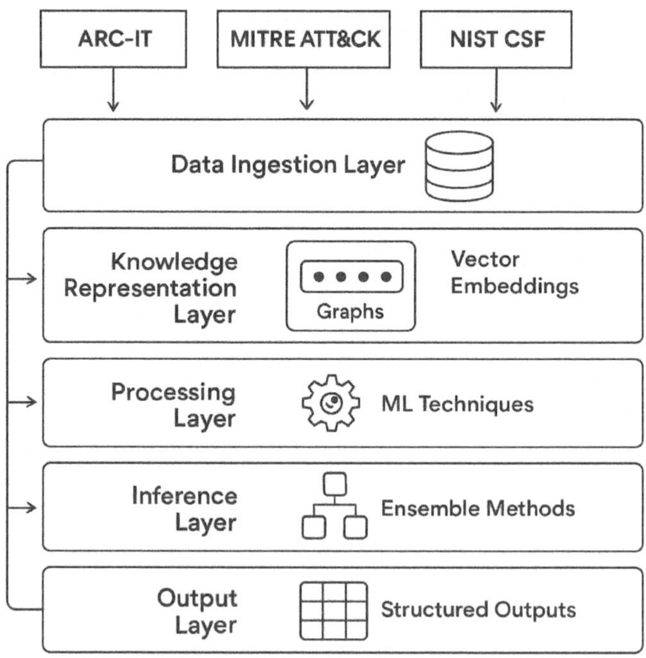

Fig. 10.5. Detailed architecture of the TraCR-TMF framework showing the five-layer design with data flow between components. The modular architecture enables flexible integration of multiple machine learning techniques while maintaining consistent interfaces and evaluation protocols.

Data Ingestion Layer: Responsible for extracting, parsing, and normalizing information from ARC-IT service packages, MITRE ATT&CK knowledge bases, and NIST CSF documentation.

Knowledge Representation Layer: Constructs structured representations of domain knowledge, including vector embeddings, semantic indexes, and relationship graphs.

Processing Layer: Implements the various machine learning techniques described in Section 10.3, providing unified interfaces for different analytical approaches.

Inference Layer: Orchestrates the application of multiple techniques, manages ensemble methods, and coordinates active learning workflows.

Output Layer: Synthesizes results from multiple techniques, performs post-processing validation, and generates structured outputs for downstream applications.

This layered approach enables flexible experimentation with different techniques while maintaining consistent data formats and evaluation protocols across the framework.

10.4.2 *Data Processing*

The effectiveness of automated threat modeling depends critically on the quality and completeness of input data representation. Our data processing pipeline transforms raw ARC-IT information flows into structured feature vectors that capture both syntactic and semantic characteristics relevant to cybersecurity analysis.

10.4.2.1 *Information Flow Parsing and Normalization*

ARC-IT information flows are defined within service package XML documents that specify functional objects, physical objects, and data exchange relationships. The parsing module extracts the following components for each information flow:

Flow Metadata: Including flow identifier, name, description, source functional object, destination functional object, and associated physical objects.

Security Attributes: CIA (Confidentiality, Integrity, Availability) ratings specified on a scale from "Not Applicable" to "High," with intermediate values of "Low" and "Moderate."

Contextual Information: Service package classification, functional area designation, and related flows within the same service package.

Technical Specifications: Protocol requirements, message formats, timing constraints, and performance requirements where available.

The normalization process standardizes terminology, resolves acronyms and abbreviations, and creates consistent representations across different service packages. This preprocessing step is critical for enabling effective semantic matching between information flows and threat descriptions.

10.4.3 *Knowledge Base Construction and Management*

The TraCR-TMF framework maintains multiple specialized knowledge bases optimized for different retrieval and analysis tasks. Each knowledge base is designed to support specific components of the threat modeling pipeline while enabling efficient cross-referencing and relationship discovery.

10.4.3.1 *MITRE ATT&CK Knowledge Base*

The ATT&CK knowledge base contains comprehensive information about adversarial techniques, tactics, and procedures, structured to support both keyword-based and semantic retrieval:

Technique Descriptions: Complete textual descriptions of each ATT&CK technique, including behavioral characteristics, implementation methods, and detection strategies.

10.4.3.2 *NIST Control Framework*

The framework incorporates relevant standards and regulatory guidance to ensure alignment with established best practices:

Cybersecurity Frameworks: Complete documentation of NIST CSF, ISO 27001, and other relevant cybersecurity frameworks with structured mappings between framework elements.

Regulatory Guidance: Transportation cybersecurity policies, guidelines, and requirements from federal, state, and local agencies.

10.4.4 *Zero-Shot and ICL Implementation Architecture*

The zero-shot learning component of TraCR-TMF leverages large language models to perform threat classification without requiring task-specific training data. Using the same prompt, we implement the few-shot learning by adding examples at the bottom of the prompt. The implementation addresses several key challenges, including prompt optimization, response parsing, and confidence estimation.

10.4.4.1 *Prompt Engineering Framework*

Effective zero-shot classification requires carefully crafted prompts that provide sufficient context while avoiding information overload. Our prompt engineering framework employs a template-based approach with dynamic content insertion. Figure 10.6 illustrates the exact prompt structure provided to the LLM agent.

The framework supports multiple prompt variations optimized for different model architectures and capabilities. Dynamic prompt construction adapts content based on information flow characteristics, emphasizing relevant domain knowledge and providing appropriate examples.

10.4.4.2 *Response Processing and Validation*

Zero-shot model responses require sophisticated parsing and validation to ensure reliability and consistency:

Structured Response Parsing: Regular expressions and natural language processing techniques extract technique identifiers, confidence scores, and justification text from model responses.

Technique Validation: Cross-referencing against the ATT&CK knowledge base ensures that predicted techniques are valid and properly categorized.

General Input (System Role):
"I am working on a multilabel classification of data flow descriptions from ARC-IT to MITRE ATT&CK techniques. The information includes the data flow name, its initiator and acceptor, a functional object description associated with it, and the data flow's description itself. An attacker may attempt to compromise the confidentiality, integrity, or availability of the data flows in various ways. Identify all relevant MITRE ATT&CK techniques that an attacker might use to target the data flow."

Instructions:
Follow the instructions below carefully.

- We have a predefined list of MITRE ATT&CK techniques. You must select only the relevant MITRE ATT&CK techniques from this list that pertain to the given data flow.
- Understand the entire context and generate a sublist of MITRE ATT&CK techniques from the provided list.
- Do not include any additional descriptions in your response.
- Only return the technique IDs in Python list format.

Context Data:
Given MITRE ATT&CK technique list = ['T1495', 'T1485', 'T1595', ..., 'T1556', 'T1589']

¡Basic Input¿
What relevant MITRE ATT&CK techniques from the given list might an attacker use to compromise the data flow? Return the technique IDs in Python list format.

¡Examples included for one-shot or few-shot learning¿

Fig. 10.6. Structure of the prompt injected into the LLM. The "General Input" section designates the system's role and follows it with the specific question.

Consistency Checking: Multiple prompt variations and response aggregation improve reliability and identify potential inconsistencies in model predictions.

Confidence Calibration: Statistical models map model response characteristics to calibrated confidence scores that reflect true prediction accuracy.

10.4.5 *Retrieval-Augmented Generation Implementation*

The RAG implementation in TraCR-TMF combines dense vector retrieval with large language model generation to produce accurate and well-grounded threat assessments. The system architecture supports multiple retrieval strategies and enables dynamic knowledge base updates. Figure 10.7 shows the architecture of our model.

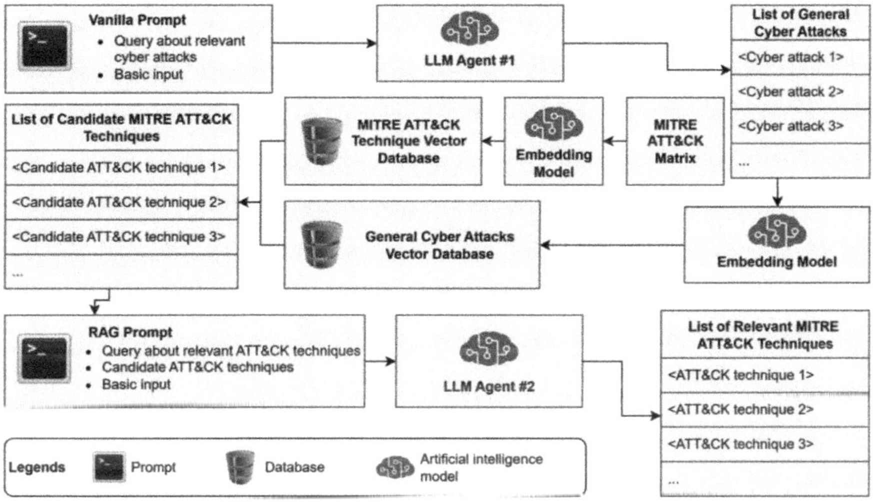

Fig. 10.7. RAG-based LLM architecture for ATT&CK technique mapping.

10.4.5.1 *Vector Index Construction*

Dense vector retrieval requires high-quality embeddings that capture semantic relationships between information flows and threat descriptions. We employ a multi-stage indexing process:

Embedding Generation: Pre-trained language models (e.g., BERT, RoBERTa, SecBERT) generate dense vector representations for all knowledge base entries and information flows.

Hybrid Indexing: Combination of dense and sparse (BM25) indexes supports both semantic and keyword-based retrieval strategies.

10.4.5.2 *Dynamic Retrieval Pipeline*

The retrieval pipeline adapts to query characteristics and available context to optimize relevance and coverage.

The pipeline incorporates multiple retrieval strategies to maximize coverage while maintaining efficiency. Query expansion techniques automatically add related terms, synonyms, and domain-specific vocabulary to improve retrieval effectiveness.

10.4.6 *Multimodal RAG Architecture*

The multimodal extension of RAG processes both textual and visual information from ITS documentation to provide comprehensive threat analysis. The architecture integrates vision-language models with traditional text processing pipelines.

10.4.6.1 *Visual Information Extraction*

ITS documentation contains various types of visual information that provide valuable context for threat modeling:

Diagram Processing: Computer vision techniques extract entities, relationships, and annotations from network diagrams, system architectures, and data flow diagrams.

Table Analysis: Optical character recognition (OCR) and structured extraction techniques process security matrices, configuration tables, and parameter specifications.

Image-Text Alignment: Vision-language models establish correspondences between visual elements and textual descriptions to create unified representations.

The visual processing pipeline employs multiple specialized models optimized for different document types and visual layouts:

$$\mathbf{v}_{\text{extracted}} = \text{VisionModel}(\text{Image}_{\text{input}}) \oplus \text{OCR}(\text{Image}_{\text{input}}) \quad (10.7)$$

where $\oplus$ represents feature fusion combining visual understanding with extracted text content.

10.4.6.2 *Cross-Modal Fusion and Reasoning*

The integration of visual and textual information requires sophisticated fusion mechanisms that preserve the unique contributions of each modality:

Attention-Based Fusion: Cross-modal attention mechanisms enable dynamic weighting of information sources based on relevance to the specific threat modeling query.

The fusion architecture employs transformer-based models with specialized attention mechanisms for cross-modal reasoning:

$$\mathbf{A}_{\text{text}\rightarrow\text{vision}} = \text{Attention}\left(\mathbf{Q}_{\text{text}}, \mathbf{K}_{\text{vision}}, \mathbf{V}_{\text{vision}}\right) \tag{10.8}$$

$$\mathbf{A}_{\text{vision}\rightarrow\text{text}} = \text{Attention}\left(\mathbf{Q}_{\text{vision}}, \mathbf{K}_{\text{text}}, \mathbf{V}_{\text{text}}\right) \tag{10.9}$$

$$\mathbf{h}_{\text{fused}} = \text{Combine}\left(\mathbf{A}_{\text{text}\rightarrow\text{vision}}, \mathbf{A}_{\text{vision}\rightarrow\text{text}}\right) \tag{10.10}$$

10.4.7 *Supervised Fine-Tuning Implementation*

The supervised fine-tuning component of the TraCR-TMF framework represents the most accurate approach for automated threat-flow mapping, leveraging transfer learning principles to adapt pre-trained transformer models for cybersecurity-specific multi-label classification tasks.

10.4.7.1 *ModernBERT Architecture and Selection Rationale*

Our implementation utilizes ModernBERT as the foundation model for supervised fine-tuning due to several compelling advantages over traditional BERT architectures. ModernBERT is a state-of-the-art encoder-only transformer model specifically designed to handle extended input sequences while maintaining computational efficiency.

The architectural advantages of ModernBERT for our cybersecurity application include:

Extended Context Window: ModernBERT can process significantly longer input sequences compared to BERT's 512-token limit. This capability is essential for our application, as the longest input sequence in our transportation cybersecurity dataset requires nearly seven times BERT's maximum token length. This extended capacity allows the model to process complete information flow descriptions, security attributes, STRIDE threat categories, and contextual information without truncation.

Optimized Architecture: The base version of ModernBERT features 22 transformer layers and 149 million parameters, providing an optimal balance between model expressiveness and computational requirements. This configuration delivers superior performance compared to similarly-sized models while maintaining reasonable inference times for practical deployment.

Open-Source Accessibility: Unlike proprietary models such as GPT-4, ModernBERT is open-source and can be deployed locally, addressing data privacy concerns and reducing operational costs for organizations with sensitive cybersecurity information.

10.4.7.2 *Transfer Learning Methodology*

The supervised fine-tuning process follows established transfer learning principles, beginning with the pre-trained ModernBERT model and adapting it for our specific threat identification task. This approach leverages the extensive linguistic knowledge embedded in the pre-trained model from its training on large-scale text corpora, eliminating the need to learn fundamental language representations from scratch.

The transfer learning advantages in our cybersecurity context include:

Knowledge Transfer: The pre-trained model already understands complex linguistic patterns, technical terminology, and contextual relationships that are valuable for interpreting cybersecurity documentation and threat descriptions.

Sample Efficiency: Transfer learning enables effective training with relatively modest amounts of domain-specific labeled data, as the model only needs to learn the mapping between cybersecurity concepts and MITRE ATT&CK techniques rather than basic language understanding.

Faster Convergence: Fine-tuning converges more rapidly than training from scratch, reducing computational costs and enabling iterative model improvements.

10.4.7.3 *Multi-Label Classification Framework*

The threat identification task is formulated as a multi-label classification problem where each information flow can be simultaneously associated with multiple MITRE ATT&CK techniques. This formulation accurately reflects the reality that transportation cybersecurity threats often involve multiple attack vectors and techniques.

The mathematical formulation defines the dataset as:

$$\mathcal{D} = \{(x_i, y_i)\}_{i=1}^{N} \tag{10.11}$$

where x_i represents the feature vector for information flow i containing:

- Information flow name and description
- Source and destination functional objects
- Associated processes and security attributes
- STRIDE-based threat categories
- Additional contextual information

The label vector $y_i = \{y_{i,1}, y_{i,2}, \ldots, y_{i,L}\}$ is binary, where $y_{i,j} = 1$ indicates that MITRE ATT&CK technique j is relevant to flow i according to expert cybersecurity annotations, and $y_{i,j} = 0$ otherwise.

10.4.7.4 *Loss Function and Optimization Strategy*

The model optimization employs binary cross-entropy loss, which is well-suited for multi-label classification scenarios:

$$\mathcal{L} = -\frac{1}{N}\sum_{i=1}^{N}\sum_{j=1}^{L}\left[y_{i,j}\log(\hat{y}_{i,j}) + (1 - y_{i,j})\log(1 - \hat{y}_{i,j})\right] \tag{10.12}$$

where $\hat{y}_{i,j}$ represents the predicted probability that technique j applies to information flow i. This loss function treats each technique independently while allowing the model to learn correlations between techniques through the shared representation layers.

The optimization process utilizes the AdamW optimizer with carefully tuned hyperparameters:

- **Learning Rate:** 2e-5 (empirically optimized for cybersecurity domain)
- **Weight Decay:** 0.01 (regularization to prevent overfitting)
- **Batch Size:** 8 (constrained by GPU memory limitations)
- **Maximum Epochs:** 50 with early stopping.

10.4.7.5 *Cross-Validation and Model Robustness*

To ensure model generalization and robustness, we implement five-fold cross-validation using MultilabelStratifiedKFold, which maintains balanced label distributions across folds despite the multi-label nature of the problem. This approach ensures that each fold contains representative samples of both common and rare MITRE ATT&CK techniques.

The training process incorporates several robustness mechanisms:

Early Stopping: Training terminates when validation loss fails to improve by at least 0.001 over 5 consecutive epochs, preventing overfitting while maximizing learning from available data.

Validation Monitoring: Separate validation sets (randomly selected from training data) enable continuous performance monitoring and hyperparameter adjustment during training.

Threshold Optimization: Post-training threshold optimization maximizes F1-scores for each technique individually, accounting for class imbalance and varying technique importance.

The supervised fine-tuning approach provides the most accurate and reliable technique identification within the TraCR-TMF framework, making it particularly suitable for production deployments where accuracy is paramount and sufficient expert-labeled training data is available.

10.4.8 *Active Learning Implementation*

The active learning component minimizes annotation costs by strategically selecting the most informative examples for expert labeling. The implementation supports multiple query strategies and integrates seamlessly with the supervised learning pipeline.

10.4.8.1 *Query Strategy Implementation*

Multiple query strategies are implemented to address different aspects of uncertainty and informativeness:

Entropy-Based Selection: Computes prediction entropy across all labels and selects instances with highest average uncertainty:

$$\text{Score}_{\text{entropy}}(\mathbf{x}) = -\frac{1}{L}\sum_{j=1}^{L}\sum_{c\in\{0,1\}} p(y_j = c|\mathbf{x}) \log p(y_j = c|\mathbf{x}) \quad (10.13)$$

Monte Carlo Dropout: Estimates predictive uncertainty through multiple forward passes with dropout enabled:

$$\text{Score}_{\text{MC}}(\mathbf{x}) = \frac{1}{L} \sum_{j=1}^{L} \text{Var}_{t=1}^{T}[\hat{y}_j^{(t)}] \tag{10.14}$$

Temporal Prediction Discrepancy: Measures disagreement between predictions at different training stages:

$$\text{Score}_{\text{TPD}}(\mathbf{x}) = \frac{1}{L} \sum_{j=1}^{L} |\hat{y}_j^{(\text{current})} - \hat{y}_j^{(\text{previous})}| \tag{10.15}$$

10.4.8.2 *Batch Selection Optimization*

Selecting batches of examples requires balancing informativeness with diversity to maximize learning efficiency:

Greedy Selection: Iteratively selects examples that maximize a utility function combining uncertainty and diversity measures.

Submodular Optimization: Formulates batch selection as a submodular maximization problem, enabling principled approximation algorithms with theoretical guarantees.

Clustering-Based Diversity: Uses k-means clustering in feature space to ensure selected examples cover different regions of the input distribution.

The batch selection algorithm optimizes the following objective:

$$\mathcal{B}^* = \underset{\mathcal{B} \subset \mathcal{U}, |\mathcal{B}|=k}{\arg\max} \sum_{\mathbf{x} \in \mathcal{B}} \text{Uncertainty}(\mathbf{x}) - \lambda \sum_{\mathbf{x}_i, \mathbf{x}_j \in \mathcal{B}} \text{Similarity}(\mathbf{x}_i, \mathbf{x}_j) \tag{10.16}$$

where λ controls the trade-off between uncertainty and diversity.

10.4.9 *NIST Framework Integration Architecture*

The NIST Cybersecurity Framework integration extends the threat modeling capabilities to include strategic security management functions alongside tactical threat identification. The implementation supports automated mapping to NIST control families, IDs, and control enhancements.

10.4.9.1 *Framework Alignment Pipeline*

The NIST integration follows a similar architectural pattern to ATT&CK mapping, with specialized components for framework-specific requirements:

Function Classification: High-level mapping of information flows to NIST control families based on operational role and security characteristics.

Category Mapping: More granular classification to specific Control IDs such as Access Enforcement, Access Control for Mobile Devices, Event Logging, and Impact Analyses.

Subcategory Alignment: Detailed mapping to specific CSF subcategories that provide actionable implementation guidance.

The mapping pipeline employs similar machine learning techniques adapted for the NIST framework structure.

10.4.9.2 *Cross-Framework Correlation Analysis*

The simultaneous mapping to both ATT&CK and NIST frameworks enables sophisticated correlation analysis that provides strategic insights:

Statistical Correlation: Pearson and Spearman correlation coefficients identify relationships between attack techniques and security functions.

Co-Occurrence Analysis: Frequent pattern mining identifies common combinations of attack techniques and security controls.

Coverage Assessment: Gap analysis identifies information flows that lack adequate security coverage or are vulnerable to specific attack patterns.

Priority Ranking: Risk-based prioritization combines threat likelihood with security impact to guide resource allocation decisions.

This comprehensive architecture enables TraCR-TMF to leverage multiple complementary approaches while providing robust, accurate, and interpretable threat modeling results. The following section presents detailed experimental evaluation of these architectural components and their performance characteristics.

10.5 Experimental Results and Performance Analysis

This section presents a comprehensive evaluation of the TraCR-TMF framework across multiple dimensions, including individual technique performance, comparative analysis, and practical deployment considerations. We analyze results from extensive experiments conducted on real-world ITS information flows, demonstrating the effectiveness of different machine learning approaches for automated threat modeling and providing insights into optimal configuration strategies for practical implementations.

10.5.1 *Experimental Setup and Dataset Construction*

The evaluation of automated threat modeling techniques requires high-quality ground truth data that accurately reflects expert knowl edge and real-world threat scenarios. This section describes the dataset construction process, evaluation metrics, and experimental protocols employed throughout our analysis.

10.5.1.1 *Dataset Characteristics*

Our evaluation dataset was constructed from 847 distinct information flow extracted from 25 representative ARC-IT service packages covering various ITS functional areas. A team of five cybersecurity experts with extensive experience independently annotated each flow, mapping it to relevant MITRE ATT&CK techniques from the ICS matrix and providing NIST CSF alignments. This process resulted in a multi-label dataset with an average of 3.2 techniques per flow, ensuring a comprehensive and robust foundation for our evaluation.

10.5.1.2 *Evaluation Metrics*

Multi-label classification evaluation requires specialized metrics that account for partial correctness and label interdependencies. We employed label-based evaluation approaches.

Label-Based Metrics: Evaluate performance for individual ATT&CK techniques:

$$\text{Precision}_{\text{label}} = \frac{1}{L}\sum_{j=1}^{L}\frac{\text{TP}_j}{\text{TP}_j + \text{FP}_j} \tag{10.17}$$

$$\text{Recall}_{\text{label}} = \frac{1}{L}\sum_{j=1}^{L}\frac{\text{TP}_j}{\text{TP}_j + \text{FN}_j} \tag{10.18}$$

$$\text{F1}_{\text{label}} = \frac{1}{L}\sum_{j=1}^{L}\frac{2 \cdot \text{TP}_j}{2 \cdot \text{TP}_j + \text{FP}_j + \text{FN}_j} \tag{10.19}$$

10.5.2 *Zero-Shot Learning Performance Analysis*

Zero-shot learning approaches provide an important baseline for threat modeling applications, as they require no task-specific training data and can be rapidly deployed in new domains. However, our experiments reveal significant limitations in accuracy and reliability for this specialized cybersecurity task.

10.5.2.1 *Model Comparison*

We evaluated the performance of several large language models in a zero-shot configuration. Among the models tested, GPT-4 demonstrated the highest performance with an F1-score of 0.301, outperforming GPT-3.5-turbo (0.236), Claude-2 (0.263), and Llama-2-70B (0.218). Furthermore, the application of different prompt engineering strategies on GPT-4 significantly improved its performance. The use of a basic prompt yielded an F1-score of 0.301, while adding domain context increased it to 0.325. The inclusion of ATT&CK examples and the use of a Chain-of-Thought strategy further boosted the F1-score to 0.348 and 0.361, respectively.

10.5.2.2 *Error Analysis and Limitations*

Analysis of zero-shot predictions revealed several systematic error patterns that underscore the limitations of using pre-trained models for this specialized task. Models frequently suffered from **technique hallucination**, generating non-existent ATT&CK technique identifiers, and **overgeneralization**, often predicting common techniques regardless of specific flow characteristics, leading to a high rate of false positives. Furthermore, the models struggled with **context misunderstanding**, failing to differentiate between general IT environments and the specialized context of industrial control systems, which resulted in **incomplete coverage** for critical, specialized ICS techniques. These findings demonstrate that the specialized nature of cybersecurity threat modeling requires more than a general knowledge base, highlighting the need for domain-specific knowledge integration.

10.5.3 *Retrieval-Augmented Generation Evaluation*

RAG approaches demonstrate substantial improvements over zero-shot baselines by providing explicit access to curated cybersecurity knowledge during the generation process. Our evaluation examines multiple RAG configurations and retrieval strategies to identify optimal deployment parameters.

10.5.3.1 *Retrieval Strategy Comparison*

Table 10.1 presents performance results for different retrieval strategies within the RAG framework.

10.5.4 *Multimodal RAG Performance Assessment*

Multimodal RAG extends traditional text-based approaches by incorporating visual information from ITS documentation. Our evaluation examines the added value of visual processing and optimal fusion strategies for transportation cybersecurity applications.

10.5.4.1 *Visual Information Contribution*

Table 10.2 quantifies the performance impact of different types of visual information.

We notice that when we add mode context in the RAG prompt with visual information, we see some improvements in the F1-score.

Table 10.1. RAG retrieval strategy performance analysis.

Retrieval Method	Precision	Recall	F1-Score	MAP@5
Dense (text-embedding-3-small)	0.524	0.367	0.432	0.398
Dense (text-embedding-3-large)	0.547	0.389	0.456	0.421
Sparse (BM25)	0.489	0.421	0.452	0.387
Hybrid (Dense + Sparse)	0.573	0.445	0.501	0.467
Query Expansion + Hybrid	0.591	0.467	0.522	0.489

Table 10.2. Multimodal information contribution analysis.

Information Type	Precision	Recall	F1-Score	Improvement
Text-only RAG	0.591	0.467	0.522	—
+ Network Diagrams	0.608	0.485	0.540	+3.4%
+ Security Matrices	0.621	0.501	0.555	+6.3%
+ Protocol Diagrams	0.598	0.473	0.529	+1.3%
+ All Visual Types	0.634	0.518	0.570	+9.2%

10.5.4.2 *Fusion Strategy Evaluation*

Different approaches to combining textual and visual information yield varying performance characteristics.

Early Fusion: Concatenating visual and textual embeddings at the input level achieves modest improvements (+4.1% F1) but requires careful dimensionality management.

Attention-Based Fusion: Cross-modal attention mechanisms achieve the best performance (+9.2% F1) by dynamically weighting information sources based on relevance, but require additional computational resources.

10.5.5 *In-Context Learning Analysis*

In-context learning demonstrates significant performance improvements over zero-shot approaches by providing carefully selected examples within the prompt context. Our evaluation examines example selection strategies and optimal context configurations.

10.5.5.1 *Example Selection Strategy Impact*

Random sampling the examples yields the lowest F1-score. However, if we carefully choose intelligent examples, like diverse information flows from different packages and so on, it improves the F1-score significantly. So for our final experiments, we went with the intelligent example selection strategy.

10.5.5.2 *Context Length Optimization*

The number of examples included in the context significantly affects performance. Table 10.3 compares different number of examples in the prompt.

Table 10.3. In-context learning performance comparison for MITRE ATT&CK technique identification.

ICL Approach	Precision	Recall	F1-Score
Zero-shot	0.18	0.18	0.18
One-shot	0.38	0.31	0.34
Few-shot (8 examples)	0.47	0.48	0.48
Comparison with Other Approaches			
RAG-based	0.33	0.20	0.25
Supervised Fine-tuning	0.70	0.75	0.72

Single Example (1-Shot): Provides substantial improvement over zero-shot baselines with minimal context overhead.

Few Examples (3–5 Shot): Achieves optimal performance with diminishing returns beyond 5 examples due to context window limitations.

Many Examples (10+ Shot): Performance degrades due to context dilution and potential confusion from conflicting examples.

10.5.6 *Supervised Fine-Tuning Performance Analysis*

The supervised fine-tuning approach represents the most effective technique within the TraCR-TMF framework, achieving superior performance compared to both zero-shot learning and retrieval-augmented generation approaches. This section presents the comprehensive evaluation results based on our empirical analysis of the IEEE Access paper implementation.

10.5.6.1 *Overall Performance Metrics*

The supervised fine-tuning approach, utilizing ModernBERT as the foundation model, achieved the following performance metrics across the 433 transportation CPS data flows from 26 different ARC-IT applications.

These results demonstrate that the supervised approach significantly outperforms alternative methods, achieving 72% F1-score compared to 25% for RAG-based approaches and 48% for the best few-shot in-context learning configuration.

10.5.6.2 *Expert Validation Results*

To assess the true effectiveness of the supervised model beyond initial ground-truth limitations, we conducted additional expert validation on 50 randomly selected data flows. Cybersecurity experts with decades of combined academic and industrial experience evaluated each technique predicted by the model, including those initially classified as a false positives.

Table 10.4 shows that the expert validation revealed significant insights. The precision improvement from 64% to 73% (a 14% relative improvement) indicates that approximately 73% of the attack techniques identified by the supervised model were deemed correct by cybersecurity experts. This demonstrates the model's ability to identify relevant techniques beyond the original ground-truth annotations, suggesting that the initial expert-generated ground truth was incomplete rather than the model producing false positives.

10.5.6.3 *Model Generalization Capabilities*

The supervised model demonstrated remarkable generalization capabilities, identifying relevant MITRE ATT&CK techniques beyond its training domain. Analysis of the expert validation revealed two key findings:

Technique Discovery: The model correctly identified techniques such as T1136 (Create Account) for 11 out of 50 evaluated flows, all of which were confirmed as relevant by experts despite not being present in the original ground-truth annotations. This capability stems from the model's extensive pre-trained knowledge and contextual reasoning abilities.

Systematic Gaps: The model consistently missed certain techniques, such as T1098 (Account Manipulation), which was absent from predictions for 13 out of 50 flows where experts deemed it

Table 10.4. Expert validation comparison.

Evaluation Basis	Precision	Recall	F1-Score
Initial ground truth	0.64	0.67	0.65
Expert-validated predictions	0.73	0.66	0.69

relevant. These patterns provide insights for threshold adjustment and model improvement strategies.

10.5.6.4 *Cross-Validation Analysis*

The five-fold cross-validation process revealed consistent performance across different data partitions, with training and validation loss curves demonstrating proper convergence without overfitting. The MultilabelStratifiedKFold approach ensured balanced label distributions across folds, addressing the multi-label nature of the classification problem.

Training characteristics observed include:

- Stable convergence within 20–30 epochs across all folds.
- Validation loss reduction consistent with training loss trends.
- Early stopping activation preventing overfitting.
- Minimal variance in performance across folds (standard deviation < 0.02 for F1-scores).

10.5.6.5 *Comparative Analysis with Alternative Approaches*

The supervised fine-tuning approach substantially outperformed other TraCR-TMF techniques.

The supervised approach achieves a 50% relative improvement in F1-score compared to the best alternative (few-shot learning), validating the investment in expert-annotated training data and domain-specific model adaptation (see Table 10.5).

The supervised fine-tuning approach is highly suitable for production due to its high accuracy (72% F1-score), which ensures reliable threat identification. This method allows for local deployment

Table 10.5. Performance comparison across TraCR-TMF approaches.

Approach	Precision	Recall	F1-Score
Zero-shot	0.18	0.18	0.18
One-shot	0.38	0.31	0.34
Few-shot (8 examples)	0.47	0.48	0.48
RAG-based	0.33	0.20	0.25
Supervised Fine-tuning	**0.70**	**0.75**	**0.72**

using open-source models like ModernBERT, addressing data privacy concerns. It is also **scalable**, capable of rapidly processing new data without real-time expert input, and offers threshold flexibility to meet varying risk tolerances. While it requires moderate expert intervention for initial training and periodic updates, this is a necessary trade-off for its superior accuracy and reliability.

The comprehensive evaluation confirms that supervised fine-tuning represents the most effective approach for organizations requiring high-accuracy automated threat modeling, particularly when sufficient expert-labeled training data is available and accuracy is prioritized over complete automation.

10.5.7 *Active Learning Effectiveness*

Active learning techniques significantly reduce annotation requirements while maintaining high performance levels. Our evaluation demonstrates the practical value of strategic example selection for cybersecurity applications.

10.5.7.1 *Query Strategy Performance*

We compared the performance of several active learning query strategies and found that the Temporal Prediction Discrepancy (TPD) strategy, designed for multi-label scenarios, achieved the highest F1-scores across all evaluated data budgets. At a 25% data budget, TPD reached an F1-score of 0.663, outperforming Random Sampling (0.612), Entropy-Based (0.651), and Monte Carlo Dropout (0.648) strategies. A hybrid strategy, which combines multiple approaches, further improved performance slightly, achieving a final F1-score of 0.775, the highest among all tested methods.

10.5.7.2 *Annotation Efficiency Analysis*

Active learning provides substantial efficiency gains in practical deployment scenarios, achieving 88.9% of full-dataset performance with only 25% of the labeled examples. This translates to a four-fold efficiency improvement and saves an estimated 600–800 hours of expert annotation time for a typical deployment. Furthermore, the active learning curves show rapid initial performance improvements, with diminishing returns beyond 60–70% of the dataset, which suggests a clear optimal stopping point for practical applications.

10.5.8 *NIST Framework Integration Results*

The integration of NIST Cybersecurity Framework mapping provides complementary strategic insights alongside tactical ATT&CK technique identification. Our evaluation examines cross-framework correlation patterns and practical utility for cybersecurity planning.

10.5.8.1 *Cross-Framework Correlation Analysis*

Statistical analysis reveals significant correlations between specific ATT&CK techniques and NIST CSF functions:

- **Identify Function:** Strong correlation ($r = 0.73$) with reconnaissance techniques (e.g., T0840, T0842), indicating systematic relationships between asset discovery and attack preparation phases.
- **Protect Function:** Significant negative correlation ($r = -0.68$) with impact techniques (e.g., T0809, T0815), suggesting that robust protection measures effectively mitigate destructive attack outcomes.
- **Detect Function:** Moderate correlation ($r = 0.54$) with persistence and privilege escalation techniques, highlighting detection system effectiveness for identifying advanced persistent threats.
- **Respond Function:** Strong correlation ($r = 0.79$) with techniques affecting availability (e.g., T0814, T0816), reflecting incident response priorities for maintaining operational continuity.

10.5.8.2 *Framework Mapping Accuracy*

Table 10.6 presents performance metrics for NIST CSF mapping accuracy.

Table 10.6. NIST CSF mapping performance analysis.

CSF Level	Precision	Recall	F1-Score	Agreement
Functions (5 classes)	0.847	0.823	0.835	0.89
Categories (23 classes)	0.721	0.698	0.709	0.76
Subcategories (108 classes)	0.634	0.587	0.609	0.68

10.5.9 *Computational Performance Analysis*

Practical deployment requires consideration of computational requirements and scalability characteristics. This section analyzes processing times, memory requirements, and optimization strategies for different approaches.

10.5.9.1 *Processing Time Analysis*

Table 10.7 presents detailed timing analysis for different techniques.

10.5.9.2 *Scalability Considerations*

Different approaches exhibit varying scalability characteristics for large-scale deployments:

Supervised Models: Excellent scalability with linear processing time growth and minimal per-query overhead. GPU acceleration provides additional 10-15x speedup for batch processing.

RAG Systems: Moderate scalability limited by retrieval index size and language model inference. Caching and approximate retrieval provide optimization opportunities.

Generative Methods: Poor scalability due to sequential generation requirements and API rate limits for commercial models.

This comprehensive evaluation demonstrates the effectiveness of the TraCR-TMF framework while identifying areas for future improvement and optimization. The results provide strong evidence

Table 10.7. Computational performance characteristics.

Method	Inference (ms)	Memory (GB)	Throughput (flows/sec)	Scalability
Zero-Shot (GPT-4)	2,340	0.5	0.43	Poor
RAG (Local)	487	2.1	2.05	Moderate
Multimodal RAG	1,023	3.7	0.98	Moderate
ICL (5-shot)	1,876	0.8	0.53	Poor
Supervised (ModernBERT)	23	1.4	43.5	Excellent

for the practical value of LLM-enhanced threat modeling in intelligent transportation systems, with significant performance improvements over traditional manual approaches across multiple evaluation dimensions.

10.6 Conclusion and Future Directions

This chapter has presented TraCR-TMF, a comprehensive framework for automating cybersecurity threat modeling in intelligent transportation systems through advanced large language model techniques. The framework addresses critical scalability and accuracy challenges in traditional threat assessment methodologies while providing practical solutions for real-world ITS deployments. Our research makes several key contributions, including a novel automated threat modeling system, a comprehensive empirical evaluation of its performance, and a high-quality expert-annotated dataset for future research. TraCR-TMF delivers substantial practical benefits by reducing threat analysis time by over 98%, identifying previously unrecognized threats, and offering a cost-effective solution for security practitioners.

10.6.1 *Limitations and Constraints*

While TraCR-TMF demonstrates substantial improvements over manual approaches, several limitations must be acknowledged for practical deployment. The supervised models exhibit performance degradation on information flows significantly different from training examples and are susceptible to biases from expert annotation variability. The framework also faces challenges from the continuous evolution of both the MITRE ATT&CK and ARC-IT frameworks, which necessitate ongoing model retraining. Technical constraints include limited context windows for large language models, lower-than-human accuracy in processing complex technical diagrams, and significant computational requirements for advanced features like multimodal processing.

10.6.2 *Future Research Directions*

We identify three high-priority areas for continued investigation:

- **Advanced AI and Adaptive Capabilities:** Future work should investigate domain-specific pre-training for transportation cybersecurity, graph neural networks for modeling ITS architectural relationships, and federated learning for privacy-preserving model training. Integration with real-time threat intelligence feeds and continuous learning systems could enable dynamic, adaptive threat assessments.
- **Expanded Framework and Domain Integration:** The framework should be extended to include integration with standards such as ISO 27001 and COBIT, and quantitative risk assessment methodologies. Adapting the framework to other critical infrastructure sectors (e.g., energy, water) and emerging domains like autonomous vehicle systems would address broader security challenges.
- **Enhanced Explainability and Collaboration:** Future research should focus on developing interactive visualization tools and explainable AI integration to improve analyst trust and facilitate effective human-AI collaboration. Advanced attack path visualization and other tools could support strategic security planning.

Acknowledgments

This work is based upon the work supported by the National Center for Transportation Cybersecurity and Resiliency (TraCR) (a US Department of Transportation National University Transportation Center) headquartered at Clemson University, Clemson, South Carolina, USA. Any opinions, findings, conclusions, and recommendations expressed in this material are those of the author(s) and do not necessarily reflect the views of TraCR, and the US Government assumes no liability for the contents or use thereof.

Bibliography

Achiam, J., Adler, S., Agarwal, S., Ahmad, L., Akkaya, I., Aleman, F. L., Almeida, D., Altenschmidt, J., Altman, S., Anadkat, S., *et al.* (2023). GPT-4 technical report, *arXiv preprint* arXiv:2303.08774.

Applebaum, A., Miller, D., Strom, B., Korban, C., and Wolf, R. (2016). Intelligent, automated red team emulation, in *Proceedings of the 32nd Annual Conference on Computer Security Applications*, pp. 363–373.

Assante, M. J., Conway, T., and Lee, R. M. (2015). Analysis of the cyber attack on the Ukrainian power grid, Electricity Information Sharing and Analysis Center (E-ISAC).

Boyes, H., Hallaq, B., Cunningham, J., and Watson, T. (2018). The industrial internet of things (IIoT): An analysis framework, *Computers in Industry* **101**, pp. 1–12.

Brown, T., Mann, B., Ryder, N., Subbiah, M., Kaplan, J. D., Dhariwal, P., Neelakantan, A., Shyam, P., Sastry, G., Askell, A., *et al.* (2020). Language models are few-shot learners, *Advances in Neural Information Processing Systems*, Vol. 33, pp. 1877–1901.

Checkoway, S., McCoy, D., Kantor, B., Anderson, D., Shacham, H., Savage, S., Koscher, K., Czeskis, A., Roesner, F., Kohno, T., *et al.* (2011). Comprehensive experimental analyses of automotive attack surfaces, in *USENIX Security Symposium*, Vol. 4 (San Antonio, TX), pp. 447–462.

Dong, Q., Li, L., Dai, D., Zheng, C., Wu, Z., Chang, B., Sun, X., Xu, J., and Sui, Z. (2022). A survey on in-context learning, *arXiv preprint* arXiv:2301.00234.

Guu, K., Lee, K., Tung, Z., Pasupat, P., and Chang, M.-W. (2020). Retrieval augmented language model pre-training, in *International Conference on Machine Learning* (PMLR), pp. 3929–3938.

Humayed, A., Lin, J., Li, F., and Luo, B. (2017). Cyber-physical systems security—A survey, *IEEE Communications Surveys & Tutorials* **19**, 2, pp. 1802–1831.

Hurd, S. and Smith, R. (2008). DNP3 security: Improving security of the distributed network protocol, *DigitalBond.*

Kojima, T., Gu, S. S., Reid, M., Matsuo, Y., and Iwasawa, Y. (2022). Large language models are zero-shot reasoners, in *Advances in Neural Information Processing Systems*, Vol. 35, pp. 22199–22213.

Koscher, K., Czeskis, A., Roesner, F., Patel, I., Kohno, T., Checkoway, S., McCoy, D., Kantor, B., Anderson, D., Shacham, H., *et al.* (2010). Experimental security analysis of a modern automobile, in *2010 IEEE Symposium on Security and Privacy* (IEEE), pp. 447–462.

Lewis, P., Perez, E., Piktus, A., Petroni, F., Karpukhin, V., Goyal, N., Küttler, H., Lewis, M., Yih, W.-t., Rocktäschel, T., *et al.* (2020). Retrieval-augmented generation for knowledge-intensive NLP tasks, in *Advances in Neural Information Processing Systems*, Vol. 33, pp. 9459–9474.

Li, J., Li, D., Xiong, C., and Hoi, S. (2022). BLIP: Bootstrapping language-image pre-training for unified vision-language understanding and generation, in *International Conference on Machine Learning* (PMLR), pp. 12888–12900.

Liang, G., Zhao, J., Luo, F., Weller, S. R., and Dong, Z. Y. (2017). A review of false data injection attacks against modern power systems, *IEEE Transactions on Smart Grid* **8**, 4, pp. 1630–1638.

Liu, P., Yuan, W., Fu, J., Jiang, Z., Hayashi, H., and Neubig, G. (2023). Pre-train, prompt, and predict: A systematic survey of prompting methods in natural language processing, *ACM Computing Surveys* **55**, 9, pp. 1–35.

Lu, J., Batra, D., Parikh, D., and Lee, S. (2019). ViLBERT: Pretraining task-agnostic visiolinguistic representations for vision-and-language tasks, in *Advances in Neural Information Processing Systems*, Vol. 32.

MITRE Corporation (2021). MITRE Attack, https://attack.mitre.org/.

Morris, T. H., Thornburgh, B., and Turnipseed, I. (2012). Industrial control system network intrusion detection by combining multiple machine learning algorithms and feature selection methods, in *Critical Infrastructure Protection V*, pp. 49–63.

Mundhenk, P., Paverd, A., Mrowca, A., Steinhorst, S., Lukasiewycz, M., Fahmy, S. A., and Chakraborty, S. (2017). Security in automotive networks: Lightweight authentication and authorization, *ACM Transactions on Design Automation of Electronic Systems (TODAES)* **22**, 2, pp. 1–27.

Munir, M. B., Cai, Y., Khan, L., and Thuraisingham, B. (2024). Leveraging multimodal retrieval-augmented generation for cyber attack detection in transit systems, in *2024 IEEE 6th International Conference on Trust, Privacy and Security in Intelligent Systems, and Applications (TPS-ISA)*, pp. 341–350, doi:10.1109/TPS-ISA62245.2024.00046.

National Institute of Standards and Technology (2018). Framework for improving critical infrastructure cybersecurity, NIST Cybersecurity Framework.

Parkinson, S., Ward, P., Wilson, K., and Miller, J. (2017). Cyber threats facing autonomous and connected vehicles: Future challenges, *IEEE Transactions on Intelligent Transportation Systems* **18**, 11, pp. 2898–2915.

Radford, A., Kim, J. W., Hallacy, C., Ramesh, A., Goh, G., Agarwal, S., Sastry, G., Askell, A., Mishkin, P., Clark, J., *et al.* (2021). Learning transferable visual models from natural language supervision, in *International Conference on Machine Learning* (PMLR), pp. 8748–8763.

Radford, A., Wu, J., Child, R., Luan, D., Amodei, D., Sutskever, I., *et al.* (2019). Language models are unsupervised multitask learners, *OpenAI blog* **1**, 8, p. 9.

Ren, P., Xiao, Y., Chang, X., Huang, P.-Y., Li, Z., Gupta, B. B., Chen, X., and Wang, X. (2021). A survey of deep active learning, *ACM Computing Surveys* **54**, 9, pp. 1–40.

Salek, M. S., Chowdhury, M., Munir, M. B., Cai, Y., Hasan, M. I., Tine, J.-M., Khan, L., and Rahman, M. (2025). A large language model-supported threat modeling framework for transportation cyber-physical systems, *arXiv preprint* arXiv:2506.00831.

Sener, O. and Savarese, S. (2017). Active learning for convolutional neural networks: A core-set approach, in *International Conference on Learning Representations.*

Settles, B. (2009). Active learning literature survey, University of Wisconsin-Madison Department of Computer Sciences.

Settles, B. and Craven, M. (2008). An analysis of active learning strategies for sequence labeling tasks, in *Proceedings of the 2008 Conference on Empirical Methods in Natural Language Processing*, pp. 1070–1079.

Strom, B. E., Applebaum, A., Miller, D. P., Nickels, K. C., Pennington, A. G., and Thomas, C. B. (2018). MITRE att&ck: Design and philosophy, Technical Report.

Tan, H. and Bansal, M. (2019). LXMERT: Learning cross-modality encoder representations from transformers, in *Proceedings of the 2019 Conference on Empirical Methods in Natural Language Processing and the 9th International Joint Conference on Natural Language Processing (EMNLP-IJCNLP)*, pp. 5100–5111.

US Department of Transportation (2012). Regional its architecture guidance document, Federal Highway Administration.

US Department of Transportation (2020). Arc-it version 8.3, https://www.arc-it.net/.

US Department of Transportation (2022). Architecture reference for cooperative and intelligent transportation, https://www.arc-it.net/.

Wang, W., Zheng, V. W., Yu, H., and Miao, C. (2019). A survey on zero-shot learning, *ACM Transactions on Intelligent Systems and Technology (TIST)* **10**, 2, pp. 1–35.

Warner, B., Chaffin, A., Clavié, B., Weller, O., Hallström, O., Taghadouini, S., Gallagher, A., Biswas, R., Ladhak, F., Aarsen, T., *et al.* (2025). Smarter, Better, Faster, Longer: A Modern Bidirectional Encoder for Fast, Memory Efficient, and Long Context Finetuning and Inference, in *Proceedings of the 63rd Annual Meeting of the Association for Computational Linguistics* (Volume 1: Long Papers) (Vienna, Austria), pp. 2526–2547.

Wei, J., Wang, X., Schuurmans, D., Bosma, M., Xia, F., Chi, E., Le, Q. V., Zhou, D., *et al.* (2022). Chain-of-thought prompting elicits reasoning in large language models, in *Advances in Neural Information Processing Systems*, Vol. 35, pp. 24824–24837.

Xian, Y., Lampert, C. H., Schiele, B., and Akata, Z. (2018). Zero-shot learning—A comprehensive evaluation of the good, the bad and the ugly, *IEEE Transactions on Pattern Analysis and Machine Intelligence* **41**, 9, pp. 2251–2265.

Yaacoub, J.-P., Salman, O., Noura, H. N., Kaaniche, N., Chehab, A., and Malli, M. (2020). Cybersecurity analysis of cyber-physical systems against stealthy deception attacks, *Sensors* **20**, 6, p. 1537.

https://doi.org/10.1142/9789819831548_0011

Chapter 11

Quantum Computing: Threats or Opportunities to Cybersecurity of Transportation Systems?

Abyad Enan* and M. Sabbir Salek†

**Glenn Department of Civil Engineering*
Clemson University, Clemson, SC
†*USDOT National Center for Transportation Cybersecurity and Resiliency (TraCR), Greenville, SC*

Abstract

Quantum computing is rapidly gaining momentum across various domains, leveraging its principles to enhance computational capabilities and efficiency beyond the limits of classical computing techniques. Quantum computers perform tasks with a level of speed and complexity unattainable by today's classical computing architectures. This emerging technology introduces transformative opportunities in numerous fields, particularly in cybersecurity, through enhanced cryptographic techniques, rapid threat detection, and resilient defense mechanisms. Building on these advances, quantum artificial intelligence (QAI) integrates quantum computing with machine-learning paradigms to enable faster optimization, improved pattern recognition, and more efficient learning in high-dimensional spaces, offering new capabilities for intelligent and adaptive cybersecurity solutions.

This chapter explores the evolving role of quantum computing, with a particular emphasis on QAI applications in cybersecurity, and highlights its advantages over classical techniques for predicting, detecting, and mitigating cyberattacks in transportation systems. The chapter begins by introducing the foundations of quantum computing, explaining how quantum bits (qubits) differ fundamentally from classical bits. Concepts, such as superposition and entanglement, are introduced alongside foundational quantum gates to illustrate the core distinctions between classical and quantum computing. The discussion further covers essential elements of quantum computation, including quantum data encoding, variational quantum circuits (VQCs), quantum machine learning strategies, and quantum-enhanced optimization techniques.

Subsequently, the chapter examines QAI and quantum optimization in depth, contrasting their capabilities with those of classical approaches and showcasing their applicability across diverse domains. It then transitions to cybersecurity challenges within transportation systems, outlining the unique complexities that make securing these systems particularly demanding. Recent studies are presented to demonstrate how quantum computing enhances security through improved cyber-attack prediction, anomaly detection, and real-time threat mitigation, achieving superior accuracy and efficiency compared to conventional methods. At the same time, the discussion acknowledges that quantum computing may also introduce new vulnerabilities, offering a balanced perspective on its dual-edged impact.

Finally, the chapter explores potential threats posed by quantum computing itself and outlines future directions for integrating quantum-based solutions into real-world security infrastructures for transportation systems. By discussing theoretical advancements with practical implementations, this chapter provides an overview of leveraging quantum computing to strengthen cybersecurity and enhance the resiliency of transportation systems.

11.1 Introduction to Quantum Computing

Over the past few decades, quantum computing has rapidly gained attention as a powerful emerging computational paradigm with the potential to outperform classical systems for specific problem classes. Its foundation lies in fundamental quantum-mechanical phenomena, such as superposition, entanglement, and measurement uncertainty, which enable information to be encoded and processed in fundamentally different ways. In contrast to classical computers that rely on binary bits restricted to either 0 or 1, quantum computers employ quantum bits (qubits), which can exist in a superposition of multiple basis states simultaneously. This capability allows quantum systems to evaluate large solution spaces in parallel, offering significant computational advantages over classical methods for certain complex tasks. This section introduces the foundational concepts behind quantum computation, focusing on its key principles, mathematical representations, and operational mechanisms.

11.1.1 *From Classical Bits to Qubits*

The fundamental unit for classical computing is classical bits, where each bit can be in either state 0 or 1. The manipulations of the classical bits are performed through classical logic gates, NAND, NOR, NOT, etc., that are deterministic as well. Classical computing relies on deterministic bits and logic gates that produce identical outputs for identical inputs. While this determinism ensures reliability, it inherently limits the classical computing paradigm's capacity to efficiently explore large and/or complex solution spaces.

Quantum computing extends this paradigm by introducing the qubit, the basic unit of quantum information. Unlike classical bits, a qubit is a two-level quantum state that can exist in a superposition of both states (0 and 1) simultaneously until it is measured. These states, often defined by electron spin, photon polarization, or atomic energy levels, serve as fundamental properties that can be precisely adjusted to encode and process information. Mathematically, a qubit's state is expressed as a superposition of its two basis states, typically denoted $|0\rangle$ and $|1\rangle$:

$$|\psi\rangle = \alpha|0\rangle + \beta|1\rangle \tag{11.1}$$

where, α and β represent the complex amplitudes associated with the computational basis states $|0\rangle$ and $|1\rangle$, respectively, that satisfies the following condition:

$$|\alpha|^2 + |\beta|^2 = 1 \tag{11.2}$$

where, $|\alpha|^2$ and $|\beta|^2$ are the probabilities of getting 0 and 1, respectively, when measured, and hence they must sum up to 1. The geometric representation of the state of a qubit is captured by the Bloch sphere, presented in Fig. 11.1. Any single-qubit state corresponds to a point on the surface of a unit sphere, parameterized by two angles θ and ϕ. In the Bloch sphere form, a qubit can be represented by replacing α and β by $\cos(\theta/2)$ and $e^{i\phi}\sin(\theta/2)$, respectively, and can be written as:

$$|\psi\rangle = \cos\left(\frac{\theta}{2}\right)|0\rangle + e^{i\phi}\sin\left(\frac{\theta}{2}\right)|1\rangle \tag{11.3}$$

where, $\theta \in [0, \pi]$ corresponds to the polar angle, while $\phi \in [0, 2\pi]$ corresponds to the azimuthal angle. In the Bloch sphere representation,

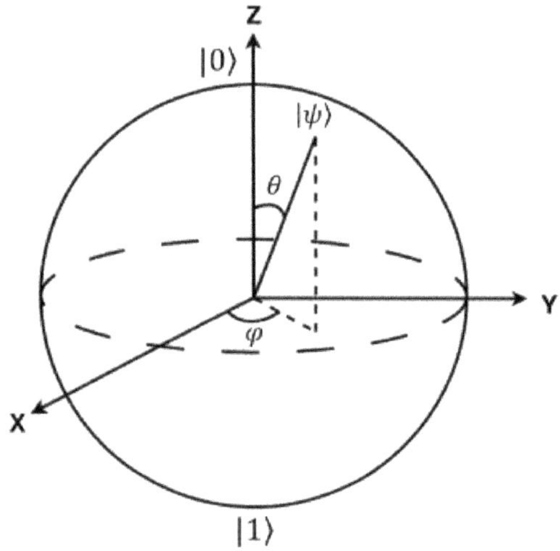

Fig. 11.1. Geometrical representation of a state of a qubit.

$|0\rangle$ is located at the north pole and $|1\rangle$ at the South Pole. Any qubit state corresponds to a point on the sphere's surface, defined by its coordinates ($x = \sin\theta\cos\phi$, $y = \sin\theta\sin\phi$, $z = \cos\theta$). The Bloch sphere provides an intuitive visualization of qubit states, illustrating how quantum operations can manipulate these states through rotations of the Bloch sphere.

11.1.2 *Quantum Measurement and Probabilistic Outcomes*

One of the most perplexing aspects of quantum mechanics is the role of measurement. Before measurement, a qubit exists in a blend of both states, as mentioned in Equation (11.1), a phenomenon with no classical analogue. Upon measurement on the $|0\rangle$ and $|1\rangle$ computational basis, the outcome is inherently probabilistic, where the probabilities of observing 0 and 1 are $|\alpha|^2$ and $|\beta|^2$, respectively. A simple example includes both α and β equal $1/\sqrt{2}$, Equation (11.1) becomes $|\psi\rangle = (|0\rangle + |1\rangle)/\sqrt{2}$, where the probabilities of 0 and 1 are $|\alpha|^2 = |\beta|^2 = (1/\sqrt{2})^2 = 1/2$.

After measurement, the qubit collapses irreversibly into the observed state. The probabilistic nature of quantum measurement contrasts sharply with classical determinism, introducing new challenges in algorithm design. To gain reliable information, quantum algorithms often rely on repeated preparation and measurement of identically initialized qubits. Because measurement of a qubit only provides random outcomes according to the probability amplitudes, quantum computing relies on running the same quantum circuit multiple times to extract reliable information, unlike classical computing, where one readout is always enough.

The measurements on the computational basis $|0\rangle$ and $|1\rangle$ only tell us about the qubit along the z-axis of the Bloch sphere. However, the measurement need not be limited to the computational basis; one can choose to measure on a different basis, defined by other orthogonal states. By appropriately redefining the basis, information can be extracted from a qubit in different directions on the Bloch sphere. As an illustration, measurements on the basis of $|+\rangle = (|0\rangle + |1\rangle)/\sqrt{2}$ and $|-\rangle = (|0\rangle - |1\rangle)/\sqrt{2}$ corresponds to checking the state along the x-axis of the Bloch sphere.

This flexibility is useful because, depending on the algorithm, the information of interest may not be apparent on the computational basis. For instance, a qubit in the state $|+\rangle$ appears completely random if measured in the $|0\rangle$ and $|1\rangle$ basis, yielding 0 and 1 with equal 50% probability. However, if the same qubit is measured in the $|+\rangle$ and $|-\rangle$ basis, the outcome will always be $|+\rangle$. Thus, by carefully choosing the measurement basis, hidden structure in the quantum state can be revealed that would remain inaccessible, enabling more powerful algorithmic strategies.

11.1.3 *Multiple Qubits and Entanglement*

While a single qubit exhibits remarkable behavior, the true power of quantum computing arises when multiple qubits are combined. An n-qubit system corresponds to a Hilbert space of dimension 2^n, meaning that the system's state can encode exponentially many amplitudes:

$$|\psi\rangle = \sum_{x=0}^{2^n-1} a_x |x\rangle \tag{11.4}$$

where, a_x are complex probability amplitudes associated with each computational basis state $|x\rangle$. In other words, an n-qubit system is capable of encoding 2^n amplitudes simultaneously, which is the source of the exponential representational power of quantum machines. As an example, a 2-qubit system can be presented as:

$$|\psi\rangle = a_0|00\rangle + a_1|01\rangle + a_2|10\rangle + a_3|11\rangle \tag{11.5}$$

where, a 2-qubit system can encode four amplitudes simultaneously, in contrast to a classical system that can only represent one of the four possible states at a time.

A central feature that arises uniquely in multi-qubit systems is quantum entanglement. Entanglement is the phenomenon where qubits exhibit correlations that defy explanation within the framework of classical probability. In an entangled state, the description of one qubit inherently depends on the others, no matter how far apart they are. One example is the Bell state:

$$|\Phi^+\rangle = \frac{|00\rangle + |11\rangle}{\sqrt{2}} \tag{11.6}$$

An entangled state cannot be factored as a product of two individual qubit states, known as a non-separability characteristic of an entangled state, where the state of one qubit is inherently connected to the other, highlighting its non-classical nature. The outcome of the second qubit is immediately known upon measuring the first. For an example of the Bell state in Equation (11.6), neither qubit has a definite value before measurement; the joint system is in a superposition of $|00\rangle$ and $|11\rangle$. Upon measuring the first qubit, the superposition collapses into one of the two possible outcomes: either $|00\rangle$ or $|11\rangle$. Due to entanglement, the measurement of the first qubit completely determines the state of the second qubit. If the first collapses to $|0\rangle$, the second is guaranteed to also be $|0\rangle$. If the first collapses to $|1\rangle$, the second is guaranteed to also be $|1\rangle$. This does not mean that information is transmitted instantaneously between the qubits. Rather, the two qubits share a non-classical correlation established when they are entangled. The outcomes are random (each pair of outcomes occurs with 50% probability), but once one outcome is realized, the other is fixed. Entanglement is a cornerstone resource in quantum algorithms, quantum communication, and emerging technologies like quantum teleportation.

11.1.4 *Quantum Gates and Circuits*

Just as classical computations are built from circuits of logic gates, quantum computations are constructed from quantum circuits composed of unitary gates. A unitary gate is mathematically described by a unitary matrix U, which satisfies the condition $U^\dagger U = I$, where $U^\dagger$ is the conjugate transpose of U, and I is the identity matrix. This condition guarantees that quantum evolution is reversible and that the norm of the quantum state vector is preserved throughout computation. Preservation of the state vector's norm is equivalent to the conservation of total probability, since the squared magnitudes of the probability amplitudes in a quantum state must always sum to 1. Thus, applying a unitary transformation corresponds to rotating the state vector in Hilbert space without altering its overall length.

11.1.4.1 *Single Qubit Gates*

Fundamental to quantum computing, single-qubit gates operate on one qubit at a time, modifying its state via a unitary transformation. Important examples include Pauli-X gate (X), Pauli-Y gate (Y), Pauli-Z gate (Z), Hadamard gate (H), etc. For example, the X gate performs a bit flip, exchanging the computational basis states, which is directly comparable to the classical NOT gate. The Z gate applies a phase flip, leaving $|0\rangle$ unchanged but mapping $|1\rangle$ to $-|1\rangle$. Here, while the computational basis probabilities are unaffected, the relative phases between states are modified. The Hadamard gate creates equal superpositions of basis states. By generating superposition states, the Hadamard gate plays a key role in facilitating quantum parallelism in algorithms. All operations on a single qubit can be visualized as rotations of its state vector on the Bloch sphere, and together they provide the building blocks for more complex quantum circuits.

11.1.4.2 *Multi-Qubit Gates*

Multi-qubit gates extend quantum operations beyond individual qubits, enabling entanglement and conditional logic. The Controlled-NOT (CNOT) gate serves as the fundamental example.

When applied to qubits in superposition, CNOT produces entangled states, making it essential for quantum computation and communication protocols. A critical insight in quantum computing is that the combination of single-qubit gates and the CNOT gate is universal, meaning that any quantum operation can be constructed from sequences of these gates, forming the essential primitives of quantum computation.

11.1.5 *Quantum Parallelism*

Quantum computation differs from classical computation through its use of superposition, enabling the powerful effect of quantum parallelism. In this paradigm, a quantum computer can, in effect, evaluate a function on multiple inputs simultaneously within a single computational step, as proven by the Deutsch algorithm. In quantum computing, a superposition of all possible inputs encodes the corresponding outputs in parallel across the amplitudes of the quantum state. This exponential parallelism is unattainable in classical computing, where each function evaluation must be performed sequentially. However, this parallelism does not directly allow the extraction of all results at once, since measurement collapses the state to a single outcome. The power of quantum algorithms lies in leveraging quantum interference: by carefully designing circuits, amplitudes corresponding to correct solutions are amplified, while those corresponding to incorrect solutions interfere destructively and diminish. This principle forms the foundation of several landmark quantum algorithms:

- Shor's factoring algorithm exploits quantum parallelism and interference in the quantum Fourier transform to efficiently determine periodicities, enabling polynomial-time integer factorization.
- Grover's search algorithm enhances the amplitude of the desired solution within an unstructured database, providing a quadratic speedup compared to classical search methods.

11.1.6 *Physical Realizations of Qubits*

Qubits are not just abstract mathematical constructs; they require physical realization, and several competing architectures are being actively pursued. Superconducting circuits, used by companies such as Google and IBM, enable fast operations. Trapped ions, such as those developed by IonQ, offer exceptionally high fidelity and long coherence. Photonic qubits are naturally suited for quantum communication, though scaling them for large-scale computation remains a significant challenge. Spin qubits, which exploit electron spins in semiconductor devices, hold promise for integration with existing electronic technologies. Each approach embodies a distinct balance of scalability, error rates, and operational stability, underscoring the formidable engineering challenges of building practical quantum computers.

11.2 Quantum Artificial Intelligence and Quantum Optimizations

Quantum Artificial Intelligence (Quantum AI) and quantum-based optimization techniques represent two transformative paradigms that leverage the principles of quantum mechanics to enhance computational efficiency and decision-making. While classical AI relies on conventional computing architectures, Quantum AI exploits phenomena such as superposition, entanglement, and quantum interference to process information in fundamentally different ways. Similarly, quantum optimization approaches use quantum algorithms and hardware to solve large-scale, combinatorial, and constraint-heavy optimization problems more efficiently than traditional methods. Together, these domains promise significant breakthroughs across fields where complex decision-making, high-dimensional data, and computational bottlenecks are prevalent.

11.2.1 *How Are They Different from Classical*

11.2.1.1 *Quantum AI*

The distinction between classical AI and Quantum AI arises from how information is represented and manipulated. In classical AI, data is encoded in binary bits (0s and 1s) and processed sequentially or in parallel through deterministic algorithms. By contrast, Quantum AI encodes information into quantum states, where qubits can represent multiple states simultaneously due to superposition. This allows

Quantum AI models to explore exponentially large feature spaces without explicitly computing all possibilities, offering advantages in tasks such as classification, clustering, and generation.

Furthermore, quantum entanglement introduces correlations between qubits that classical systems cannot easily replicate. This property enables quantum-enhanced learning models to capture intricate relationships within data, thereby improving accuracy when dealing with highly nonlinear or interdependent features. Quantum algorithms such as the Variational Quantum Eigensolver (VQE) and the Quantum Approximate Optimization Algorithm (QAOA) leverage these principles to achieve better convergence properties and solution diversity compared to purely classical optimizers (Schuld and Killoran, 2019).

Another benefit lies in quantum parallelism. Classical AI models must sequentially evaluate hypotheses or configurations, whereas Quantum AI inherently evaluates multiple possibilities in parallel through the evolution of quantum states. This makes QAI especially promising for search and optimization problems, where the solution space grows exponentially with problem size. While fully scalable quantum advantage is still under investigation, hybrid architectures combining quantum subroutines with classical AI pipelines already demonstrate improvements in solution quality, sample efficiency, and uncertainty estimation in specific contexts (Huang *et al.*, 2021).

The differences between Quantum AI (QAI) and classical AI can be perceived by studying how they process and encode data, train models, and perform inference.

Data/Problem Encoding: Quantum AI focuses on leveraging quantum computing principles to enhance machine learning models. In QAI, data/problem encoding plays a critical role, as classical data must be represented within quantum states (Chatterjee and Yu, 2017). This is typically achieved through various encoding strategies, such as angle encoding, amplitude encoding, or problem-inspired encodings. Since encoding impacts the efficiency and effectiveness of the model, the encoding cost can dominate computational resources. A well-designed QAI model aligns the encoding scheme with the task's structure, allowing quantum interference to contribute constructively to the learning process.

Training: In terms of training, QAI models, particularly variational quantum circuits (VQCs), are trained using gradient-based or gradient-free optimization techniques (Schuld *et al.*, 2019). A key innovation enabling efficient training is the parameter-shift rule, which allows exact computation of gradients on quantum hardware by executing circuits with slightly shifted parameters (Wierichs *et al.*, 2022). Recent advances extend this method to broader gate sets and even continuous-variable quantum systems. However, due to noise and finite measurement shots, variance in gradient estimates can be significant, necessitating techniques such as batching, careful optimizer selection, and hybrid quantum-classical strategies to stabilize convergence.

Inference: QAI predictions involve executing the trained quantum circuit and measuring its outputs. The latency of inference depends on the circuit's depth and the number of shots required to obtain statistically reliable results. Popular tooling ecosystems such as Qiskit, PennyLane, and Cirq provide frameworks to design, train, and deploy variational quantum models effectively across simulators and quantum hardware.

11.2.1.2 *Quantum Optimization*

Quantum optimization, on the other hand, focuses on solving combinatorial and constrained optimization problems using quantum-enhanced techniques. Here, problem encoding typically involves mapping the optimization task to a Quadratic Unconstrained Binary Optimization (QUBO) (Glover *et al.*, 2018) or Ising model formulation (Lucas, 2014). This process represents the objective function using linear and quadratic coefficients, often achieved through standard reductions or dedicated libraries. Constraints are incorporated either through penalty terms or by using constrained model variants.

Quantum optimization transforms a discrete problem into a physical phenomenon. Many NP-hard tasks, such as routing, scheduling, and portfolio selection, maps to minimize an Ising or QUBO objective; the cost function becomes a Hamiltonian, and the device searches for its low-energy states (Lucas, 2014). The advantage of this approach is that many different problems can be written in the same form and solved using a single framework. Research in this area primarily follows two paradigms:

Quantum Annealing/Adiabatic Optimization: In quantum annealing, the evolution begins from a readily preparable reference state and gradually transitions toward the Hamiltonian encoding the target problem, where quantum tunneling facilitates transitions across energy landscapes that are challenging for classical optimization methods. Recent research shows that quantum annealers can outperform classical methods for approximate optimization, and how the difficulty of a problem affects this advantage. Hybrid quantum-classical solvers make it possible to tackle even larger or more complex problems (Munoz-Bauza and Lidar, 2025).

Gate-Model (Circuit-Based Quantum Computing): This approach utilizes quantum gates, basic operations applied to qubits, to construct circuits, much like classical computers use logic gates. By arranging these gates in different ways, one can run algorithms that exploit superposition and entanglement. Prominent examples include Shor's algorithm, which efficiently factors large numbers, and Grover's algorithm, which enables faster searching. Today's noisy devices can already try out methods like the VQE and the Quantum Approximate Optimization Algorithm (QAOA), while future advances in error correction aim to make large-scale, fault-tolerant quantum computing possible (Farhi *et al.*, 2014).

Solving an optimization problem typically involves re-running the quantum solver whenever input conditions change. Quantum annealing or QAOA techniques often generate ensembles of near-optimal solutions, which are particularly useful for downstream decision-making when multiple feasible solutions exist. Quantum devices can rapidly sample diverse high-quality solutions (useful for multi-start metaheuristics and robust planning), navigate through rugged landscapes via tunneling, and serve as specialized subroutines within hybrid pipelines that decompose large instances. Yet classical solvers remain formidable; empirically demonstrated quantum advantages are emerging but still selective, particularly in approximate regimes and hybrid workflows (Munoz-Bauza and Lidar, 2025; McGeoch, 2020).

11.2.2 *Use Cases*

Quantum computing has emerged as a revolutionary paradigm that leverages the principles of quantum mechanics to address complex computational challenges across diverse domains. Its ability to perform operations at scales and speeds unattainable by classical systems enables groundbreaking applications in fields such as materials science, healthcare, finance, energy, transportation, artificial intelligence, cybersecurity, and others. Figure 11.2 illustrates the major application areas of quantum computing, highlighting its transformative potential across industries, and the applications are discussed in this subsection.

Quantum computing plays a transformative role in chemistry and material science, particularly in modeling molecular systems and predicting chemical properties that are extremely difficult to simulate using classical techniques. Classical computational approaches struggle with solving the Schrödinger equation for complex molecules due to the exponential scaling of resources with system size. Quantum algorithms, such as the VQE and Quantum Phase Estimation (QPE), allow direct simulation of quantum phenomena, providing substantial computational efficiency. This capability accelerates the discovery of new materials, including catalysts, alloys, sorbents, and high-density energy storage systems. In the energy sector specifically, quantum computing facilitates the design of next-generation hydrogen carriers, CO_2 capture technologies, and advanced battery chemistries. Moreover, quantum simulations enable improvements in

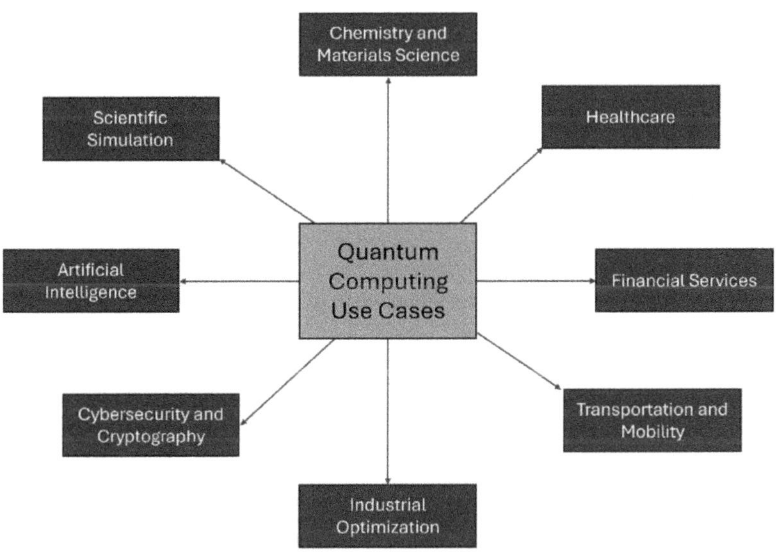

Fig. 11.2. Quantum computing applications in different domains.

seismic imaging, reservoir modeling, and pipeline network design, allowing engineers to optimize the placement of wind farms, solar arrays, and fossil or nuclear power plants for maximum efficiency and minimal environmental impact (Paudel *et al.*, 2022).

The healthcare and life sciences sectors are also rapidly integrating quantum computing to address challenges in drug discovery, clinical research, diagnostics, and personalized medicine. Recent proof-of-concept studies demonstrate that Quantum Machine Learning (QML) can outperform or match classical models when dealing with high-dimensional biomedical datasets (Flöther, 2023). In genomics, quantum computing accelerates genome sequencing pipelines and enhances the understanding of gene-disease relationships, enabling faster identification of potential therapeutic targets. Within diagnostics, quantum-enhanced algorithms improve the classification of complex medical images, supporting early detection of tumors and other pathologies. In treatment planning, quantum computing facilitates the optimization of radiation therapy schedules and tailors interventions to individual patient profiles, improving clinical outcomes. Furthermore, quantum-inspired techniques have been applied to drug discovery, where quantum computing accelerates simulations of molecular binding affinities and metabolic pathways, potentially reducing the time and cost associated with pharmaceutical development (Flöther, 2023; Marchetti *et al.*, 2022).

In the domain of finance, quantum computing is driving significant innovation in stochastic modeling, risk management, portfolio optimization, and fraud detection. Financial systems inherently involve highly complex and multidimensional datasets, making them an ideal testing ground for quantum algorithms. One of the most promising advancements is the application of Quantum Monte Carlo Integration (QMCI), which achieves a quadratic speedup compared to classical Monte Carlo methods, enabling faster pricing of derivatives and improved risk modeling under uncertain market conditions (Herman *et al.*, 2023). Similarly, variational quantum algorithms and quantum annealing techniques are applied to optimize large-scale financial portfolios, ensuring better trade-offs between risk and returns in scenarios involving hundreds or thousands of assets. Different institutions have already begun deploying hybrid quantum-classical frameworks to simulate market dynamics, evaluate counterparty risks, and optimize trading strategies more efficiently than classical approaches (Herman *et al.*, 2023).

Transportation systems and intelligent mobility planning are also being reshaped by quantum computing. Urban transportation networks face significant challenges in real-time optimization, traffic congestion management, and multi-modal logistics planning, which involve large-scale combinatorial problems that are computationally demanding for classical solvers. Quantum computing provides a promising alternative by enabling multi-objective route optimization and dynamic activity-based modeling to simulate travel behaviors, predict congestion patterns (Cooper, 2022), and traffic control optimization (Enan *et al.*, 2024). Through algorithms like the QAOA and quantum annealing, quantum computing has been applied to the Vehicle Routing Problem (VRP) and dynamic traffic signal control, allowing real-time updates to traffic management strategies and improving system efficiency (Somvanshi *et al.*, 2025). Quantum computing, with the development of QAI, contributes to detecting traffic incidents (Khan *et al.*, 2023) in transportation systems. Furthermore, quantum computing supports simulation-driven approaches in Connected and Automated Vehicle (CAV) frameworks, enabling faster responses to evolving traffic conditions and promoting efficient mobility solutions in smart cities.

QML integrates quantum algorithms with classical models to accelerate tasks such as classification, clustering, regression and generation. Unlike classical neural networks, Quantum Neural Networks (QNNs) offer superior storage capacity and computational efficiency, enabling scalable solutions to high-dimensional problems. In natural language processing, QC has enabled the development of meaning-aware algorithms that understand the semantics of words and sentences, improving real-time machine translation and conversational AI (Rawat *et al.*, 2022). Hybrid quantum-classical architectures are increasingly deployed to combine the strengths of deep learning with quantum-enhanced feature extraction, supporting breakthroughs in diverse fields (Valdez and Melin, 2023).

Finally, cybersecurity represents a unique domain where quantum computing simultaneously introduces threats and solutions. Shor's algorithm illustrates the power of quantum computing to factor large integers exponentially faster than classical algorithms, thereby undermining the security of cryptographic systems such as RSA and

ECC. On the other hand, quantum computing enables the design of quantum-resistant cryptographic techniques, including lattice-based encryption and Quantum Key Distribution (QKD), which provide provably secure communication channels. Governments and industries are investing heavily in post-quantum cryptography to prepare critical infrastructure, such as healthcare records, financial transactions, and transportation data, for a future that is quantum secure.

11.3 Quantum Computing for Cybersecurity of Transportation Systems

To set the stage for understanding the role of quantum computing in transportation, this section first discusses the cybersecurity and resiliency challenges currently confronting transportation systems and considers the future risks posed by quantum technologies. It then examines how quantum computing innovations can be applied to improve their cybersecurity.

11.3.1 *Cybersecurity and Resiliency Challenges of Transportation Systems*

The rise of Intelligent Transportation Systems (ITS) and CAVs marks a paradigm shift in modern mobility, enabling improved safety, efficiency, and reduced emissions. These systems integrate cyber-physical infrastructures in real-time, leveraging advanced sensing, communications, edge computing, and AI-based decision-making to manage traffic safely and efficiently, optimize routing, and enhance user experiences. However, this interconnectedness particularly expands the attack surface and introduces numerous vulnerabilities, creating complex cybersecurity and resiliency challenges that must be addressed to ensure system trustworthiness. The key cybersecurity and resiliency challenges of transportation systems are illustrated in Fig. 11.3. In this subsection, the key cybersecurity and resiliency challenges of modern transportation systems are discussed.

11.3.1.1 *Shift toward Increasingly Connected Systems*

The growing integration of ITS with advanced sensing, computing, and communication technologies has brought transformative benefits

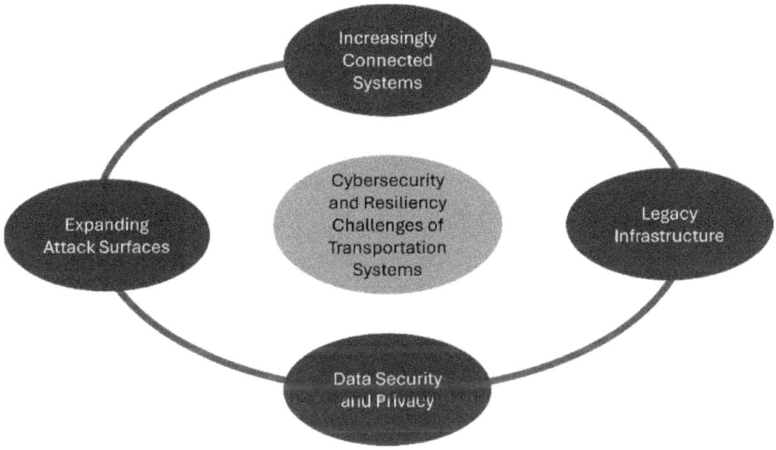

Fig. 11.3. Key cybersecurity and resiliency challenges of transportation systems.

to modern mobility but has simultaneously introduced unprecedented challenges for cybersecurity and resilience. ITS leverage Vehicle-to-Everything (V2X) communication, including Vehicle-to-Vehicle (V2V), Vehicle-to-Infrastructure (V2I), Vehicle-to-Network (V2N), and Vehicle-to-Pedestrian (V2P) interfaces, to enable real-time exchange of heterogeneous and high-volume data and cooperative decision-making among vehicles, infrastructure, and traffic management systems (Abdo *et al.*, 2024). This connectivity promises safer and more efficient mobility by facilitating coordinated lane merging, adaptive routing, and collision avoidance.

However, as ITS evolve toward highly connected ecosystems, their dependence on cyber-physical infrastructures significantly increases system complexity, exposing new vulnerabilities that challenge cybersecurity and resiliency of ITS. Modern CAVs are equipped with numerous sensors, cameras, LiDAR, radar, and control modules, all connected to cloud-based platforms for fleet coordination and traffic optimization (Beck, 2017). While these features enable advanced functionalities, they also open multiple vectors for adversarial exploitation. The inherent interdependence among vehicles, infrastructure, and communication systems means that an attack or disruption on a single component can propagate across the entire ecosystem, making ITS increasingly prone to cascading failures.

Moreover, the integration of ITS within smart city frameworks, where transportation networks are linked with energy grids, IoT-based surveillance, and cloud-enabled analytics, amplifies the challenge of maintaining security and operational resilience (Beck, 2017). In this context, cybersecurity is no longer confined to protecting isolated vehicles or subsystems but has become a systemic concern requiring holistic and adaptive strategies.

11.3.1.2 *Expanding Attack Surfaces*

The shift toward connected and automated mobility has dramatically expanded the attack surface of ITS, making cybersecurity a central concern for operational safety and resilience. The introduction of V2X communication channels, roadside IoT devices, and cloud-based management systems creates multiple entry points for malicious actors. One of the most prominent threats is a spoofing attack, where adversaries inject falsified Basic Safety Messages (BSMs) into V2V networks, misleading vehicles about traffic conditions and inducing unsafe maneuvers (Khattak, 2024). Another example, replay attacks exploit valid but outdated communication packets to cause inconsistencies in vehicle coordination, while denial-of-service (DoS) attacks overload V2I channels, preventing the delivery of safety-critical data.

Recent studies demonstrate the significant operational consequences of such attacks. Under controlled scenarios, DoS and replay attacks have been shown to increase acceleration-deceleration volatility by 35–40% and greenhouse gas emissions by 20%, highlighting the cascading environmental and safety impacts of even localized disruptions (Khattak, 2024). Beyond V2X vulnerabilities, modern CAV architectures with their million lines of code distributed across Electronic Control Units (ECUs), perception systems, and over-the-air update platforms, offer numerous points for exploitation (Abdo *et al.*, 2024). Attackers can remotely manipulate vehicle functions, such as braking and steering, through ECU compromises. Additionally, as ITS integrate with cloud platforms for fleet coordination and traffic management, a breach at the cloud layer could simultaneously disrupt thousands of vehicles, traffic signals, and logistics operations, exacerbating the overall challenge of maintaining resilience (Ahmed and Dey, 2020).

11.3.1.3 *Data Security and Privacy*

The extensive data collection capabilities of ITS and CAVs introduce significant challenges to data security and privacy. Vehicles continuously collect sensitive information, including geolocation, driving behavior, biometric identifiers, and real-time vehicle telemetry. These datasets are often transmitted across multiple networks, stored in cloud repositories, and shared among stakeholders, making them highly attractive targets for cybercriminals. Unsecured communication channels in roadside units and IoT-enabled infrastructure have already facilitated sensor spoofing and unauthorized manipulation of traffic control systems in real-world incidents (Beck, 2017).

Additionally, the integration of ITS into smart city ecosystems introduces further complexities. For instance, mobility marketplaces that handle financial transactions and personal identifiers are particularly vulnerable to breaches. An intrusion into such platforms can expose sensitive user data, including payment details and movement patterns, enabling profiling and targeted exploitation (Beck, 2017). Open data initiatives designed to improve traffic analytics and urban planning also heighten risks when insufficient anonymization protocols are in place. The resulting exposure undermines public trust in ITS deployments and raises ethical concerns regarding surveillance, user consent, and data governance. Ensuring robust end-to-end encryption, secure storage, and privacy-preserving analytics remain critical for maintaining confidence in ITS while enabling their data-driven benefits.

11.3.1.4 *Legacy Infrastructure*

One of the most persistent challenges in ITS security lies in the integration of advanced connected systems with legacy transportation infrastructure. Many existing assets, such as traditional traffic lights, roadside sensors, and communication backbones, were not designed with cybersecurity in mind. These components often lack built-in authentication mechanisms, secure firmware, and encryption protocols, making them easy targets for adversaries when integrated into modern ITS frameworks (Ahmed and Dey, 2020).

As ITS deployments expand, hybrid environments emerge where modern connected technologies coexist with outdated subsystems, creating inconsistencies in cybersecurity posture across the network. For example, while V2X-enabled intersections may feature encrypted communications and redundant monitoring systems, adjacent legacy-controlled segments might remain vulnerable to simple denial-of-service or physical tampering attacks. This uneven security landscape not only increases exposure but also complicates system-wide resilience planning.

11.3.2 *Future Threats of Quantum Computing in Transportation Systems*

Having discussed the cybersecurity and resiliency challenges of modern transportation systems, it is imperative to address a looming paradigm shift that could radically disrupt these infrastructures: quantum computing. As digital components in ITS continue to proliferate, the cryptographic protocols that secure these environments face an existential threat from quantum technologies. While quantum computing promises advances in AI and optimization, its potential to undermine foundational encryption methods cannot be overstated.

The core issue lies in the vulnerability of current asymmetric cryptographic systems to quantum algorithms, particularly Shor's algorithm. The Digital Signature Algorithm (DSA), which is widely used in V2X communications and authentication mechanisms, relies on the computational difficulty of factoring large numbers or solving discrete logarithm problems. A sufficiently advanced quantum computer would reduce these once-intractable problems to trivial ones, enabling adversaries to break encryption that currently protects sensitive vehicular and infrastructure data (Kirsch and Chow, 2015). Research indicates that cryptographically relevant quantum computers may emerge by the late 2020s, elevating the urgency for preparedness (Scholten *et al.*, 2024).

Quantum threats are particularly concerning in the context of ITS and CAVs. These systems depend on secure, low-latency communication between vehicles, roadside units, and cloud platforms. If encryption is compromised, attackers could eavesdrop on or manipulate data streams, impersonate legitimate actors, or issue malicious commands. This risk extends to AI models that process traffic patterns and sensor fusion data. Quantum-enhanced adversaries may exploit model vulnerabilities through accelerated optimization or reverse engineering, compromising safety-critical decisions (Kilber *et al.*, 2021).

Though the need for post-quantum cryptographic (PQC) migration is evident, significant challenges remain. Many transportation systems operate on embedded platforms with limited processing capacity, making them ill-equipped to adopt computationally intensive PQC algorithms. Moreover, the implementation of quantum-safe protocols is complicated by the need for industry-wide coordination across jurisdictions and stakeholders. Lattice-based schemes, hash-based signatures, and other PQC candidates selected by NIST offer promising alternatives, but may introduce latency or scalability concerns in real-time systems (Scholten *et al.*, 2024; NIST, 2017).

11.3.3 *Opportunities for Quantum Computing in Cybersecurity of Transportation Systems*

As transportation systems become more connected and increasingly autonomous, traditional cybersecurity mechanisms are proving inadequate to handle the complex, high-dimensional, and latency-sensitive nature of threats in this domain. Quantum computing presents novel opportunities to address these challenges. This subsection critically explores those opportunities, grounded entirely in recent scholarly work and empirical studies. The opportunities of quantum computing in cybersecurity of transportation systems are illustrated in Fig. 11.4.

11.3.3.1 *In-Vehicle Network Security*

In-vehicle networks such as the Controller Area Network (CAN) bus are foundational to modern vehicles but remain vulnerable due to their lack of built-in security mechanisms. Numerous studies have highlighted these vulnerabilities, including spoofing, replay, and message injection attacks. Traditional machine learning approaches for intrusion detection on CAN have achieved moderate success, but they often fail to generalize under real-world noisy data conditions. To address this limitation, hybrid quantum–classical intrusion detection frameworks are introduced. Among these, a classical convolutional neural network (CNN) is integrated with a quantum-restricted Boltzmann machine (QRBM) (Salek *et al.*, 2023). In this framework,

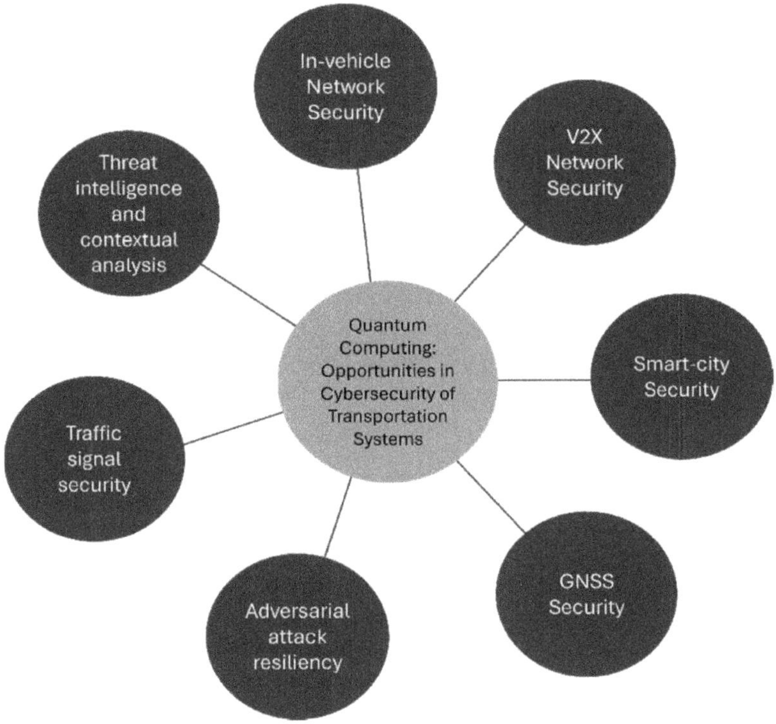

Fig. 11.4. Noteworthy opportunities for quantum computing in cybersecurity of transportation systems.

feature representations are first learned from CAN message images using a conventional CNN, after which a QRBM performs image reconstruction to enable the separation of normal and malicious traffic. Experimental results demonstrate that hybrid approaches can converge faster and yield higher detection rates compared to purely classical baselines.

Similarly, a hybrid quantum–classical neural network for cloud-supported in-vehicle intrusion detection is implemented, targeting amplitude-shift cyberattacks (Islam *et al.*, 2022). The hybrid approach demonstrated superior performance, reaching a detection accuracy of 94%, compared with 88% for a classical LSTM baseline and 62% for a fully quantum neural model. This highlights the potential of hybrid models to balance resource constraints while maintaining strong detection performance, especially in cloud or edge-deployed ITS security frameworks. The significance of this study lies not only in the detection accuracy but also in its architectural philosophy, where heavy pre-processing tasks are performed in the classical LSTM layers, while the quantum layer is invoked strategically for classification refinement. This design suggests a pathway for scaling quantum adoption in ITS security incrementally, allowing classical infrastructures to continue handling bulk data while leveraging quantum speedups and robustness in critical decision stages. Moreover, the research highlights that hybrid systems are not experimental curiosities; they can be deployed in operational ITS cloud settings to protect against specific attack vectors, such as amplitude-shift intrusions. The modular nature of the architecture means that similar designs could be extended to other attack types, forming a template for adaptive ITS defenses.

11.3.3.2 *V2X Network Security*

With the growing reliance on CAVs, ensuring secure, low-latency, and high-throughput communication between vehicles and roadside infrastructure is critical. A recent study developed a QKD-enabled framework for securing V2I communications within ITS (Stavdas *et al.*, 2024). The framework integrates Free-Space Optical (FSO)-based QKD with a Software-Defined Networking (SDN) architecture to establish quantum-secure symmetric keys for data encryption. The architecture utilizes QKD transceivers deployed between building-mounted Base Stations (BSs) and Roadside Units (RSUs), supported by Device Authentication Units (DAUs) that leverage Zero-Trust Authentication Protocols (ZAP) and Physical Unclonable Functions (PUFs) to provide strong, hardware-based identities and message-level authentication. A hierarchical SDN controller orchestrates both classical and quantum communication resources, enabling seamless multi-hop key relays and efficient end-to-end encryption across urban V2I networks. By integrating QKD, the framework achieves information-theoretic security, immune to attacks from both classical and quantum computers. It addresses challenges such as latency minimization, resilience against eavesdropping, and seamless handovers between RSUs and BSs. Additionally, the use of Quantum Key Pools (QKPs) ensures dynamic recovery from potential link disruptions and supports scalable, secure V2I deployments. This study demonstrates how combining quantum communication technologies with software-defined orchestration can enable a secure, adaptive, and resilient communication backbone for future ITS.

11.3.3.3 *Smart-City Security*

The idea of a hybrid quantum–classical neural network architecture can be extended to strengthen smart city security by integrating the power of QML with traditional Security Information and Event Management (SIEM) systems. As smart cities increasingly rely on interconnected IoT devices, sensors, and data-driven infrastructures, they face growing risks related to data privacy, IoT vulnerabilities, cyberattacks, and urban infrastructure security. To address these challenges, a recent study develops an architecture that leverages D-Wave's QBoost algorithm within the Leap Quantum Cloud alongside IBM QRadar SIEM to enable faster and more accurate threat detection (Barletta *et al.*, 2024a). The architecture is structured as a multi-layered system comprising several key components: a principal application for data processing, a quantum cloud service for high-performance classification, a dashboard for real-time monitoring, and a database for secure storage and retrieval. Incoming data from various smart city sources is analyzed simultaneously by the QBoost-based quantum component and QRadar's rule-based SIEM system. The results are presented through an integrated dashboard that enables security analysts to visualize attacks, assess threat severity, reconstruct attack chains, and respond effectively in real-time.

A major advantage of this hybrid approach is its ability to process large-scale, high-dimensional data rapidly while maintaining high detection accuracy. Traditional machine learning models can struggle with training time, scalability, and adversarial robustness. By contrast, quantum-enhanced classification accelerates decision-making and enhances the detection of complex and emerging cyber threats, such as distributed denial-of-service (DDoS) attacks, intrusion attempts, cross-site scripting (XSS) attacks, SQL injections, and reconnaissance scans. The study demonstrates that combining quantum algorithms with classical security frameworks provides a scalable and resilient defense strategy for smart cities. Rather than replacing existing systems, the architecture augments SIEM capabilities by introducing quantum-powered intelligence, reducing detection latency, and enabling proactive mitigation of threats (Barletta *et al.*, 2024a). This work highlights the growing importance of quantum-classical synergy in future urban cybersecurity frameworks, where real-time monitoring, intelligent automation, and adaptive responses are essential for safeguarding critical infrastructures and citizen data.

11.3.3.4 *GNSS Security*

Global Navigation Satellite Systems (GNSS) are crucial for transportation systems and critical infrastructure, providing precise positioning and timing solutions. Spoofing attacks against GNSS can mislead autonomous vehicles and disrupt traffic management systems. Quantum AI has recently demonstrated promising performance compared to classical AI in GNSS spoofing detection. A recent work presents a Hybrid Quantum-Classical Autoencoder (HQC-AE) framework for zero-day GNSS time-push spoofing detection, aiming to enhance the security and resilience of Positioning, Navigation, and Timing (PNT)-dependent infrastructures (Stavdas *et al.*, 2024). Unlike supervised approaches that require labeled spoofed data, this method is trained exclusively on authentic GNSS signals, enabling the detection of previously unseen spoofing attacks. The HQC-AE integrates a variational quantum circuit (VQC) within a classical autoencoder architecture to exploit the computational advantages of quantum-enhanced feature representation (Enan *et al.*, 2025).

During inference, the HQC-AE reconstructs inputs; deviations from expected reconstruction patterns produce anomaly scores that signal potential spoofing attempts. Using an interquartile range (IQR) thresholding strategy, the framework classifies spoofed signals without needing prior exposure to attack data. The results demonstrate significant improvements over classical autoencoders and traditional machine learning models, particularly against sophisticated and stealthy spoofing attacks. For challenging scenarios, HQC-AE achieves detection accuracies exceeding 98% while maintaining low false-negative and false-positive rates, thereby reducing both missed attacks and false alarms. Compared to state-of-the-art unsupervised methods, HQC-AE demonstrates superior robustness and generalization capabilities, making it well-suited for protecting critical GNSS-reliant infrastructures (Enan *et al.*, 2025).

11.3.3.5 *Adversarial Attack Resiliency*

Autonomous vehicles (AVs) rely heavily on the perception module for safe navigation; however, they are frequent targets of adversarial machine learning attacks, where slight perturbations in input images cause misclassifications. A recent study presents a Hybrid Classical-Quantum Deep Learning (HCQ-DL) framework to enhance the adversarial attack resilience of AV perception modules, specifically in traffic sign classification (Majumder *et al.*, 2025). Conventional deep learning (DL) models used in AV perception are highly vulnerable to adversarial attacks, where small, imperceptible perturbations cause misclassification of critical objects like stop signs, potentially leading to catastrophic safety consequences. The HCQ-DL models integrate quantum layers within classical convolutional neural network architectures, leveraging quantum entanglement, superposition, and a richer feature space to improve robustness against adversarial manipulations. Using classical CNN models, such as AlexNet and VGG-16 as feature extractors, the framework embeds quantum variational circuits between classical layers to exploit the expanded computational capabilities of quantum systems (Majumder *et al.*, 2025). The models were tested under three types of white-box adversarial attacks: projected gradient descent (PGD), fast gradient sign (FGSM), and gradient attacks (GA), without relying on conventional defenses such as image transformations or adversarial re-training.

Experimental results show that HCQ-DL models significantly outperform classical DL models in both no-attack and adversarial settings. While maintaining accuracy above 95% under normal conditions, HCQ-DL models sustain over 91% accuracy under GA and FGSA attacks, and 85% under PGD attacks, far exceeding classical models (Majumder *et al.*, 2025). These findings demonstrate that embedding quantum layers in AV perception pipelines can substantially enhance security, reliability, and robustness, making HCQ-DL a promising approach for safe autonomous navigation in adversarial environments.

11.3.3.6 *Traffic Signal Security*

Traffic signal controllers (TSCs) represent another critical point of vulnerability within ITS. Quantum AI contributes to TSC security in transportation systems. A recent study developed a Hybrid

QAI-based electromagnetic spectrum analysis framework aimed at strengthening the cybersecurity of TSC systems in CV environments (Vaidyan and Rimal, 2024). With the growing integration of CVs and intelligent transportation infrastructure, TSC systems are increasingly exposed to false data injection attacks and other adversarial threats. The framework introduces an air-gapped electromagnetic monitoring approach that isolates the monitoring system from the network, enhancing resilience and minimizing latency. The proposed solution leverages hybrid quantum-classical deep learning architectures to analyze electromagnetic emissions from TSCs, enabling the detection of malware, data spoofing, and falsified CV data attacks without interfering directly with the controller. Classical neural networks are combined with a quantum feature mapping layer to exploit quantum entanglement, superposition, and high-dimensional search spaces, thereby improving pattern recognition and anomaly detection. By converting electromagnetic traces into the frequency domain, the model identifies deviations from normal operational patterns, allowing real-time differentiation between legitimate signals and malicious manipulations.

Results demonstrate that the hybrid QAI framework achieves high detection accuracy under various scenarios, including multi-sensor fusion and controller interrupts, while significantly reducing false alarms. For instance, in low-penetration attack scenarios (1–10%), detection accuracy ranges from 76% to 84%, while increasing penetration levels improves accuracy up to 88% (Vaidyan and Rimal, 2024). The air-gapped design ensures robust isolation, preventing attackers from compromising the monitoring infrastructure itself, thus enhancing the overall security of connected vehicle ecosystems and ensuring the integrity and reliability of urban traffic operation.

11.3.3.7 *Threat Intelligence and Contextual Analysis*

Cybersecurity in transportation systems is not limited to real-time detection but also involves proactive threat intelligence. A quantum-based automotive threat intelligence and countermeasure framework was designed in a recent study to enhance the security of connected and autonomous vehicles (Barletta *et al.*, 2024b). With the rising integration of in-vehicle networks and electronic control units (ECUs), traditional intrusion detection techniques face significant challenges due to evolving cyberattack vectors on the CAN bus. The framework leverages QBoost, a quantum annealing-based ensemble algorithm, to perform multi-class intrusion detection using a one-vs-rest (OvR) strategy for attacks, such as flooding, fabrication, and malfunction (Barletta *et al.*, 2024b). Beyond detection, the framework introduces a threat intelligence support phase that integrates insights from the Automotive Attack Database (AAD) and the National Vulnerability Database (NVD). Once an attack is identified, the system automatically retrieves relevant information such as Common Vulnerabilities and Exposures (CVEs), Common Weakness Enumeration (CWE) categories, severity scores (CVSS), and remediation guidelines. This integration helps developers and cybersecurity specialists prioritize vulnerabilities and design countermeasures aligned with the ISO/SAE 21434 automotive cybersecurity standard (Barletta *et al.*, 2024b). The results demonstrate that quantum-enhanced IDS models can achieve high detection accuracy (above 97% for fabrication and malfunction attacks) with significantly reduced training and prediction times compared to classical methods (Barletta *et al.*, 2024b). Furthermore, the framework's knowledge-driven intelligence layer enables proactive risk assessment and adaptive security strategies for future automotive cybersecurity challenges.

11.4 Conclusions

Quantum computing is emerging as a transformative paradigm across diverse domains. Its unique computational capabilities, enabled by principles such as superposition, entanglement, and quantum parallelism, allow it to solve problems that are intractable for classical systems. In the context of security, quantum computing has introduced both significant challenges and remarkable opportunities. While its ability to undermine classical cryptographic schemes has raised concerns across industries, it simultaneously enables advanced techniques for threat detection, encryption, and resilient decision-making that were previously beyond reach. Transportation systems, as complex cyber-physical ecosystems, stand to benefit greatly from these emerging capabilities.

The power of quantum computing is particularly relevant for securing modern ITS and CAVs. As transportation evolves toward highly connected, data-driven ecosystems, cybersecurity has become a central concern. ITS rely on vast networks of V2X communication, cloud-based traffic management systems, edge computing nodes, and AI-driven decision frameworks to ensure safety, efficiency, and reduced emissions. However, this interconnectedness introduces unprecedented complexity and expands the attack surface dramatically. Vulnerabilities in V2X communication channels, roadside IoT infrastructure, and over-the-air software update mechanisms expose transportation networks to a variety of threats. Moreover, the integration of transportation networks within smart city ecosystems, where vehicle data interfaces with energy grids, IoT-based surveillance, and financial services, amplifies the stakes of cybersecurity breaches. Ensuring the resilience and trustworthiness of these systems is thus both technically challenging and operationally critical.

Against this backdrop, quantum computing offers a promising pathway to revolutionize cybersecurity in transportation systems. By leveraging quantum-enhanced machine learning, QKD and hybrid quantum–classical architectures, researchers and practitioners are developing advanced defenses tailored to the unique challenges of ITS. For instance, hybrid quantum–classical-based attack detection frameworks have demonstrated superior accuracy in detecting malicious activity within in-vehicle networks, zero-day GNSS spoofing, etc., overcoming the limitations of traditional machine learning approaches. Similarly, QKD-enabled architecture establishes information-theoretic security for V2X communications, ensuring that cryptographic keys remain secure even in the presence of quantum-capable adversaries. Moreover, the integration of quantum-enhanced threat intelligence and contextual analysis frameworks promises proactive, adaptive defenses capable of identifying, prioritizing, and mitigating threats in real time. Importantly, the potential of quantum computing in transportation cybersecurity extends beyond securing communication channels or detecting attacks. It enables resilient perception systems for autonomous vehicles by enhancing robustness against adversarial machine learning attacks, ensuring that critical decision-making processes remain reliable even under malicious perturbations. Similarly, quantum-assisted electromagnetic spectrum analysis strengthens the security of traffic signal controllers, preventing disruptions to urban mobility infrastructures. As transportation systems continue to integrate AI-driven decision pipelines, quantum computing provides the foundation for robustness guarantees, safeguarding both safety and operational continuity in highly adversarial environments.

In conclusion, quantum computing is poised to reshape the cybersecurity landscape of transportation systems. While transportation infrastructures face multifaceted challenges due to expanding connectivity, data proliferation, and evolving threat vectors, quantum technologies offer unprecedented opportunities to build secure, adaptive, and resilient systems. The convergence of quantum-enhanced detection frameworks, quantum-secure communication protocols, and AI-driven adaptive security marks a paradigm shift in defending critical mobility infrastructures. By proactively embracing these capabilities, stakeholders can transform transportation ecosystems into future-ready, quantum-resilient networks, ensuring trust, safety, and enhanced efficiency in the era of connected and autonomous mobility.

11.5 Future Directions

Looking ahead, the convergence of quantum technologies, AI-driven decision-making, and next-generation transportation infrastructures opens up promising research opportunities while presenting formidable challenges. Addressing these frontiers will be critical to ensuring the security, resilience, and scalability of future transportation systems.

11.5.1 *Post-Quantum Cryptography*

One of the most pressing priorities is migration toward post-quantum cryptography (PQC). As quantum computing capabilities mature, existing encryption standards will become obsolete. The development, deployment, and validation of quantum-safe cryptographic protocols, including lattice-based, hash-based, and code-based schemes, will require significant innovation to meet the low-latency, high-throughput requirements of ITS and CAV environments. Future research must focus on optimizing PQC for resource-constrained platforms such as embedded ECUs and IoT roadside devices while minimizing trade-offs in scalability and real-time responsiveness.

11.5.2 *Quantum-Secure Infrastructure Communication*

Another critical direction involves advancing quantum-secure communication infrastructures. The integration of QKD-enabled V2X frameworks with SDN and zero-trust authentication protocols demonstrates a promising foundation, but practical deployments remain limited. Future work must explore multi-hop QKD networks, quantum key pools, and dynamic routing strategies to support seamless key distribution in large-scale, heterogeneous transportation environments. Achieving operational efficiency will also require extensive field testing under real-world traffic and atmospheric conditions.

11.5.3 *Quantum-Enhanced Cybersecurity Analytics*

Additionally, the field is poised for rapid advancements in quantum-enhanced cybersecurity analytics. QAI adoption of hybrid quantum–classical architectures has already shown superior performance in intrusion detection, GNSS spoofing detection, and adversarial robustness for perception modules. Future research can expand these capabilities by developing federated QML frameworks that allow secure, collaborative learning across distributed ITS environments without compromising sensitive data. Integrating QML with real-time SIEM systems and threat intelligence platforms will enable transportation ecosystems to move from reactive defense toward predictive, adaptive security strategies.

11.5.4 *Quantum-Supported Privacy-Preserving Computation*

Moreover, addressing privacy preservation in data-driven transportation systems remains a key open problem. As mobility data becomes increasingly valuable for optimizing traffic flow, energy use, and urban planning, future research must explore quantum-assisted homomorphic encryption and privacy-preserving multi-party computation. These approaches have the potential to reconcile the competing demands of data utility and user privacy, particularly in smart city contexts where multi-stakeholder data sharing is unavoidable.

In conclusion, the coming decade will redefine the cybersecurity paradigm for transportation systems. Preparing for this future requires multidisciplinary collaboration across cybersecurity experts, quantum computing researchers, transportation engineers, policymakers, and industry stakeholders. By simultaneously mitigating quantum threats and harnessing quantum opportunities, the transportation sector can transition toward a secure, adaptive, and future-ready mobility ecosystem.

Bibliography

Abdo, A., Chen, H., Zhao, X., Wu, G., and Feng, Y. (2024). Cybersecurity on connected and automated transportation systems: A survey, *IEEE Transactions on Intelligent Vehicles* **9**, 1, pp. 1382–1401, https://doi.org/10.1109/TIV.2023.3326736.

Ahmed, S. and Dey, K. (2020). Resilience modeling concepts in transportation systems: A comprehensive review based on mode, and modeling techniques, *Journal of Infrastructure Preservation and Resilience* **1**, 1, p. 8, https://doi.org/10.1186/s43065-020-00008-9.

Barletta, V. S., Caivano, D., Catalano, C., and De Vincentiis, M. (2024b). Quantum-based automotive threat intelligence and countermeasures, in *Proceedings of the 28th International Conference on Evaluation and Assessment in Software Engineering* (ACM, Salerno Italy), pp. 548–554, https://doi.org/10.1145/3661167.3661278.

Barletta, V. S., Caivano, D., De Vincentiis, M., Pal, A., and Scalera, M. (2024a). Hybrid quantum architecture for smart city security, *Journal of Systems and Software* **217**, p. 112161, https://doi.org/10.1016/j.jss.2024.112161.

Beck, K. (2017). Smart security?: Evaluating security resiliency in the U.S. Department of Transportation's smart city challenge, *Transportation Research Record: Journal of the Transportation Research Board* **2604**, 1, pp. 37–43, https://doi.org/10.3141/2604-05.

Chatterjee, R. and Yu, T. (2017). Generalized coherent states, reproducing kernels, and quantum support vector machines, *Quantum Information & Computation* **17**, 15–16, pp. 1292–1306.

Cooper, C. H. V. (2022). Exploring potential applications of quantum computing in transportation modelling, *IEEE Transactions on Intelligent Transportation Systems* **23**, 9, pp. 14712–14720, https://doi.org/10.1109/TITS.2021.3132161.

Enan, A., Chowdhury, M., Dasgupta, S., and Rahman, M. (2025). Quantum-classical hybrid framework for zero-day time-push GNSS spoofing detection, https://doi.org/10.48550/ARXIV.2508.18085.

Enan, A., Salek, M. S., Chowdhury, M., Comert, G., Khan, S. M., and Majumder, R. (2024). Quantum annealing-enhanced virtual traffic lights and its evaluation using a quantum-in-the-loop simulation testbed, https://doi.org/10.48550/ARXIV.2412.18776.

Farhi, E., Goldstone, J., and Gutmann, S. (2014). A quantum approximate optimization algorithm, https://doi.org/10.48550/ARXIV.1411.4028.

Flöther, F. F. (2023). The state of quantum computing applications in health and medicine, *Research Directions: Quantum Technologies*, pp. 1–21, https://doi.org/10.1017/qut.2023.4.

Glover, F., Kochenberger, G., and Du, Y. (2018). A tutorial on formulating and using QUBO models, https://doi.org/10.48550/ARXIV.1811.11538.

Herman, D., Googin, C., Liu, X., Sun, Y., Galda, A., Safro, I., Pistoia, M., and Alexeev, Y. (2023). Quantum computing for finance, *Nature Reviews Physics* **5**, 8, pp. 450–465, https://doi.org/10.1038/s42254-023-00603-1.

Huang, H.-Y., Kueng, R., and Preskill, J. (2021). Information-theoretic bounds on quantum advantage in machine learning, *Physical Review Letters* **126**, 19, p. 190505, https://doi.org/10.1103/PhysRevLett.126.190505.

Islam, M., Chowdhury, M., Khan, Z., and Khan, S. M. (2022). Hybrid quantum-classical neural network for cloud-supported in-vehicle cyberattack detection, *IEEE Sensors Letters* **6**, 4, pp. 1–4, https://doi.org/10.1109/LSENS.2022.3153931.

Khan, Z., Tine, J. M., Khan, S. M., Majumdar, R., Comert, A. T., Rice, D., Comert, G., Michalaka, D., Mwakalonge, J., and Chowdhury, M. (2023). Hybrid quantum-classical neural network for incident detection, in *2023 26th International Conference on Information Fusion (FUSION)* (IEEE, Charleston, SC, USA), pp. 1–8, https://doi.org/10.23919/FUSION52260.2023.10224090.

Khattak, Z. H. (2024). Cybersecurity vulnerability and resilience of cooperative driving automation for energy efficiency and flow stability in smart cities, *Sustainable Cities and Society* **106**, p. 105368, https://doi.org/10.1016/j.scs.2024.105368.

Kilber, N., Kaestle, D., and Wagner, S. (2021). Cybersecurity for quantum computing, *arXiv preprint* arXiv:2110.14701.

Kirsch, Z. and Chow, M. (2015). Quantum Computing: The Risk to Existing Encryption Methods, http://www.cs.tufts.edu/comp/116/archive/fall2015/zkirsch.pdf.

Lucas, A. (2014). Ising formulations of many NP problems, *Frontiers in Physics* **2**, https://doi.org/10.3389/fphy.2014.00005.

Majumder, R., Chowdhury, M., Khan, S. M., Khan, Z., Ahmad, F., Ngeni, F., Comert, G., Mwakalonge, J., and Michalaka, D. (2025). Quantum computing supported adversarial attack-resilient autonomous vehicle perception module for traffic sign classification, https://doi.org/10.48550/ARXIV.2504.12644.

Marchetti, L., Nifosì, R., Martelli, P. L., Da Pozzo, E., Cappello, V., Banterle, F., Trincavelli, M. L., Martini, C., and D'Elia, M. (2022). Quantum computing algorithms: Getting closer to critical problems in computational biology, *Briefings in Bioinformatics* **23**, 6, p. bbac437, https://doi.org/10.1093/bib/bbac437.

McGeoch, C. C. (2020). Theory versus practice in annealing-based quantum computing, *Theoretical Computer Science* **816**, pp. 169–183, https://doi.org/10.1016/j.tcs.2020.01.024.

Munoz-Bauza, H. and Lidar, D. (2025). Scaling advantage in approximate optimization with quantum annealing, *Physical Review Letters* **134**, 16, p. 160601, https://doi.org/10.1103/PhysRevLett.134.160601.

National Institute of Standards and Technology (NIST) (2017). Post-Quantum Cryptography (PQC).

Paudel, H. P., Syamlal, M., Crawford, S. E., Lee, Y.-L., Shugayev, R. A., Lu, P., Ohodnicki, P. R., Mollot, D., and Duan, Y. (2022). Quantum computing and simulations for energy applications: Review and perspective, *ACS Engineering Au* **2**, 3, pp. 151–196.

Rawat, B., Mehra, N., Bist, A. S., Yusup, M., and Sanjaya, Y. P. A. (2022). Quantum computing and AI: Impacts & possibilities, *ADI Journal on Recent Innovation (AJRI)* **3**, 2, pp. 202–207, https://doi.org/10.34306/ajri.v3i2.656.

Salek, M. S., Biswas, P. K., Pollard, J., Hales, J., Shen, Z., Dixit, V., Chowdhury, M., Khan, S. M., and Wang, Y. (2023). A novel hybrid quantum-classical framework for an in-vehicle controller area network intrusion detection, *IEEE Access* **11**, pp. 96081–96092, https://doi.org/10.1109/ACCESS.2023.3304331.

Scholten, T. L., Williams, C. J., Moody, D., Mosca, M., Hurley, W., Zeng, W. J., Troyer, M., and Gambetta, J. M. (2024). Assessing the benefits and risks of quantum computers, *arXiv preprint* arXiv:2401.16317.

Schuld, M., Bergholm, V., Gogolin, C., Izaac, J., and Killoran, N. (2019). Evaluating analytic gradients on quantum hardware, *Physical Review A* **99**, 3, p. 032331, https://doi.org/10.1103/PhysRevA.99.032331.

Schuld, M. and Killoran, N. (2019). Quantum machine learning in feature Hilbert spaces, *Physical Review Letters* **122**, 4, p. 040504, https://doi.org/10.1103/PhysRevLett.122.040504.

Somvanshi, S., Islam, M. M., Polock, S. B. B., Chhetri, G., Anderson, D., Dutta, A., and Das, S. (2025). Quantum computing in transportation engineering: A survey, *SSRN* 5141686.

Stavdas, A., Kosmatos, E., Maple, C., Hugues-Salas, E., Epiphaniou, G., Fowler, D. S., Razak, S. A., Matrakidis, C., Yuan, H., and Lord, A. (2024). Quantum key distribution for V2I communications with software-defined networking, *IET Quantum Communication* **5**, 1, pp. 38–45, https://doi.org/10.1049/qtc2.12070.

Vaidyan, V. M. and Rimal, B. P. (2024). Hybrid quantum artificial intelligence electromagnetic spectrum analysis framework for transportation system security, *Journal of Hardware and Systems Security* **8**, 1, pp. 1–11, https://doi.org/10.1007/s41635-023-00142-2.

Valdez, F. and Melin, P. (2023). A review on quantum computing and deep learning algorithms and their applications, *Soft Computing* **27**, 18, pp. 13217–13236, https://link.springer.com/article/10.1007/s00500-026-11351-6.

Wierichs, D., Izaac, J., Wang, C., and Lin, C. Y.-Y. (2022). General parameter-shift rules for quantum gradients, *Quantum* **6**, p. 677, https://doi.org/10.22331/q-2022-03-30-677.

https://doi.org/10.1142/9789819831548_0012

Chapter 12

Navigating the Legal Terrain of Cybersecurity in Transportation: US and Global Frameworks for a Connected Future

Trayce Hockstad[*], **Evelyn Watson**[†], **and Colin Lawson**[†]

[*]*Alabama Transportation Institute, Tuscaloosa, Alabama, USA*
[†]*University of Alabama School of Law, Tuscaloosa, Alabama, USA*

Abstract

This article examines the rapidly evolving legal and regulatory landscape surrounding cybersecurity in the transportation sector, with particular emphasis on connected and autonomous vehicle systems. It argues that transportation infrastructure is increasingly dependent on digital networks that generate, transmit, and store vast quantities of sensitive data, thereby creating new legal obligations and cybersecurity risks for both public and private actors. The article is structured around three interconnected themes: data privacy and protection in connected transportation systems; cybersecurity preparedness and incident response; and the emerging legal and ethical challenges posed by autonomous and artificial intelligence-driven transportation technologies.

First, the article analyzes competing domestic and international approaches to transportation data privacy. It contrasts the fragmented United States model, which relies on sector-specific laws and Federal Trade Commission enforcement, with the European Union's comprehensive framework under the General Data Protection Regulation (GDPR). The discussion highlights how connected vehicles increasingly function as mobile data platforms, collecting information such as geolocation data, biometrics, driving behavior, and telemetry. The article also examines emerging international approaches in jurisdictions such as Japan and China, emphasizing the growing importance of privacy-by-design principles and cross-border compliance obligations.

Second, the article evaluates regulatory approaches to cybersecurity preparedness and incident response. It examines the United States' evolving framework, including NHTSA guidance, TSA cybersecurity directives, and the Cyber Incident Reporting for Critical Infrastructure Act of 2022 (CIRCIA), while comparing these approaches with stricter international regulatory models such as the EU's NIS2 Directive and UNECE cybersecurity regulations for vehicles. The article argues that transportation cybersecurity governance is shifting from voluntary best practices toward mandatory compliance obligations tied to operational resilience and market access.

Finally, the article explores the legal and ethical implications of autonomous and AI-powered transportation systems. It discusses the fragmented state-by-state regulatory framework governing autonomous vehicles in the United States and compares it with more centralized approaches in Europe. Particular attention is devoted to questions of liability, algorithmic accountability, safety regulation, and the EU Artificial Intelligence Act's risk-based regulatory model. The article concludes that transportation cybersecurity, data privacy, and AI governance are increasingly inseparable issues requiring integrated legal and technical solutions to ensure both innovation and public safety in the connected transportation environment.

Keywords: Law, Policy, Regulation, Data privacy, Data governance, Incident response.

12.1 Introduction

Modern transportation systems are no longer defined by engines and pavement alone. The rise of connected and automated technologies has turned vehicles into data platforms and infrastructure into targets. Transit fleets now run on software. Roadways communicate. Every sensor, signal, and system transmits information. In this environment, cybersecurity and data governance are not back-end technical issues but primary conditions for trust, functionality, and safety. This chapter traces the legal terrain that governs that reality. It focuses on the frameworks that regulate data privacy, structure cybersecurity obligations, and shape the deployment of artificial intelligence in transportation.

In the United States, the regulatory picture is layered and jurisdictionally complex. No single statute governs transportation cybersecurity across all modes. Instead, a latticework of federal and state laws, agency guidance, consumer protection norms, and sector-specific directives define the legal landscape. While this fragmentation poses coordination challenges, it also reflects the institutional pathways through which transportation governance has historically evolved. Internationally, the landscape looks different. Legal regimes like the European Union's General Data Protection Regulation (GDPR), the NIS2 Directive, and the United Nations Economic Commission for Europe (UNECE) cybersecurity rules offer a more unified, top-down approach. For multinational operators, these regimes increasingly set the pace.

This chapter adopts a three-part analytical structure focused on data privacy, cybersecurity preparedness and incident response, and autonomous and AI-enabled transportation because these domains reflect the principal legal vectors through which transportation cybersecurity is currently governed. Across federal, state, and international systems, most existing laws and regulatory initiatives cluster around these three areas. Data privacy regimes address the collection, use, and protection of personal and operational information generated by connected transportation systems. Cybersecurity regulation, by contrast, has emerged primarily through critical infrastructure protection, national security, and crime control frameworks that focus on resilience, reporting, and system integrity. Finally, the regulation of autonomous and AI-enabled transport represents the legal system's response to emerging technologies that reallocate decision-making authority from human actors to software-driven systems. Taken together, these perspectives capture the most consequential points at which law and policy presently intersect with transportation cybersecurity. They also reflect the institutional pathways through which regulatory authority has been exercised, contested, and operationalized.

Section 12.2 of this chapter outlines the legal standards that apply to data privacy in connected vehicles. It examines how personal and inferred data (ranging from geolocation and driving behavior to in-cabin biometrics) is treated under US law. It also considers how international models like the GDPR have influenced industry practices, particularly in engineering for compliance. Privacy, once peripheral to transportation law, now shapes design choices and operational protocols.

Section 12.3 turns to cybersecurity incident regulation. Here, the chapter traces regulatory developments across different modes, from NHTSA's best practice guidance in the automotive sector to TSA's security directives for pipelines and rail. It situates these efforts within a broader federal shift toward mandatory incident reporting, especially under the Cyber Incident Reporting for Critical Infrastructure Act (CIRCIA). At the global level, the section discusses how the EU and UNECE have imposed lifecycle governance, auditability, and technical obligations on transportation entities. Cybersecurity, in this framing, is not just a defensive measure. It is a precondition for access to markets, continuity of service, and compliance with law.

Section 12.4 addresses the legal and ethical dimensions of autonomous and AI-powered transport. These systems bring new forms of decision-making and risk into the legal frame. The section considers how liability is assigned when a vehicle acts on its own, how AI-driven systems are regulated as safety components, and how jurisdictions have begun to legislate around technical oversight, insurance design, and system transparency. It also highlights how traffic infrastructure, planning standards, and embedded rules shape not only vehicle behavior but the distribution of safety and harm.

Section 12.5 offers a focused examination of US case law and prosecutorial trends in cybersecurity and AI-enhanced cybercrime. It synthesizes recent appellate decisions on standing and injury in data breach litigation and examines the difficulties courts face when dealing with technologically mediated harm. It also explores how the rise of AI-enabled attacks like deepfakes, voice spoofing, and synthetic impersonation has created novel legal and evidentiary challenges, particularly in critical infrastructure contexts.

12.2 Legal Standards for Data Privacy and Protection in Connected Transportation Systems

Modern transportation systems are increasingly defined not by their mechanical features but by their capacity to collect, process, and transmit data. In the context of connected and automated vehicles (CAVs), this shift has transformed cars, trucks, and transit fleets into rolling data platforms. These systems generate continuous streams of information on geolocation, driver behavior, biometric identifiers, voice and video inputs, and even inferred psychological profiles. As these forms of data become integral to the operation, optimization, and monetization of transportation, they also raise urgent legal questions about how such data is regulated, who owns it, and what obligations attach to its collection and use. Data privacy, long peripheral to transportation law, is now central to understanding the obligations of transportation system operators, manufacturers, and service providers (Chander *et al.*, 2021).

In the United States, the legal framework that governs connected vehicle data has been characterized not by coherence but by fragmentation. There is no omnibus federal statute that regulates data privacy in transportation contexts. Instead, regulation has emerged piecemeal, through consumer protection doctrines, sector-specific statutes, agency guidance, and state-level data privacy laws (Hockstad *et al.*, 2024). The result is a legal terrain marked by jurisdictional ambiguity and uncertain enforcement mechanisms, particularly for automakers and telematics providers operating across state lines (Solove, 2022). While some federal oversight exists through the Federal Trade Commission (FTC), its authority hinges on broad principles of unfair or deceptive trade practices under Section 12.5 of the FTC Act, rather than any express statutory obligation to protect transportation-specific data (Evans, 2015).

A recent and illustrative example of the FTC's growing involvement in transportation privacy enforcement occurred in 2024, when the agency initiated action against a major automaker for the unauthorized sharing of driver behavior and location data with insurance companies. (FTC, 2025) The FTC characterized the data as "sensitive" and invoked a heightened duty of care under its evolving privacy doctrine (Peacher, 2020). This enforcement action, though grounded in general consumer protection law, marks a critical moment in federal recognition of vehicular data as a protected class of personal information. It also signals that regulatory pressure may intensify in the absence of congressional action.

At the state level, several jurisdictions have moved more aggressively. California's Consumer Privacy Act (CCPA), along with its successor the California Privacy Rights Act (CPRA), provides the most comprehensive statutory framework for data privacy in the US, applying to any entity that collects personal data from California residents (Buyuksagis, 2019). Importantly, this includes vehicle manufacturers, rental companies, and telematics service providers. The law establishes rights to access, delete, and opt-out of the sale of personal information. Other states including Colorado, Connecticut, and

Virginia have enacted similar statutes, creating a growing patchwork of compliance obligations (Solove, 2022).

The divergence among state laws presents acute challenges for transportation stakeholders (Hockstad *et al.*, 2025). For example, some states treat Event Data Recorder (EDR) information as the property of the vehicle owner or lessee, requiring explicit consent for access by third parties (Manny, 2018). Others have no such requirement. Even more complex is the legal treatment of inferred data like behavioral patterns, location trajectories, or biometric estimates that may not fall neatly within statutory definitions of "personal information" but are nonetheless capable of identifying individuals when cross-referenced with other data sets (Jones, 2016).

This fragmentation has led some legal scholars and privacy advocates to call for federal preemption or harmonization (Perumal, 2022). Yet such efforts have consistently stalled in Congress, often over disagreements about the scope of private rights of action and the extent to which preemption should override more protective state laws (Rustad and Koenig, 2019). The absence of uniformity has resulted in some companies adopting the highest applicable standard across their operations (usually modeled on the European Union's General Data Protection Regulation (GDPR)) as a means of managing legal risk.

The GDPR, by contrast, offers a comprehensive and deeply structured legal framework for regulating personal data in transportation systems. It applies to any entity that processes data of EU residents, regardless of the entity's geographic location. In practice, this means that non-European automakers and technology companies operating connected vehicles or digital services in the EU are bound by the regulation's provisions. GDPR defines personal data broadly to include any information related to an identifiable individual, including location data, driving patterns, and in-vehicle sensor outputs (Park, 2020).

Core to the GDPR's architecture are principles of data minimization, purpose limitation, and lawfulness of processing. Data minimization requires that only data necessary for the stated purpose be collected (Jones and Kaminski, 2021). Purpose limitation demands that data not be reused for incompatible purposes. Lawfulness of processing, in turn, requires that data collection occur pursuant to one of six lawful bases, including consent, contractual necessity, or legitimate interest. In the context of connected vehicles, this often translates into user-facing disclosures and opt-in mechanisms at the point of data collection, such as when a driver initiates infotainment services or enables location tracking (Wang, 2020).

Importantly, the GDPR also introduces concepts of "privacy by design" and "privacy by default," which require that data protection be integrated into the technical architecture of systems from the outset. This has led to substantial engineering changes in the design of connected car systems, particularly in Europe, where manufacturers now routinely anonymize data sets, segregate identifiers, and implement dynamic consent dashboards in vehicles (Wang, 2020). Enforcement of these principles falls to national data protection authorities, with the European Data Protection Board offering sector-specific guidance, like its recommendations on connected vehicles, which clarify that telematics, biometric access systems, and in-cabin audio sensors all qualify as personal data processing activities subject to strict controls (Park, 2020).

Other jurisdictions have adopted hybrid models that reflect varying degrees of alignment with the GDPR. Japan, for instance, amended its Act on the Protection of Personal Information (APPI) to secure an adequacy determination from the European Commission. The Japanese regime mirrors many GDPR principles, including transparency, data subject rights, and cross-border data transfer rules, but differs in its lighter enforcement mechanisms and broader reliance on industry self-regulation (Wang, 2020). Automotive companies operating in Japan must comply with local requirements for user notification and data security, though the government has encouraged voluntary certification frameworks to supplement legal mandates.

China, by contrast, has taken a national security-oriented approach to vehicle data privacy. The 2021 Provisions on the Management of Automotive Data set forth stringent requirements for local data storage and government security assessments before any vehicle data may be exported abroad. The law designates certain categories of information (such as route information in sensitive areas and biometric identification data) as "important data" or "core data," requiring high-level protections and subjecting companies to audits by the Cyberspace Administration of China (Pernot-Leplay, 2020). This model reflects a fundamentally different orientation: where GDPR views privacy as a human right, China frames it as a matter of state interest and regime stability.

The divergence in global privacy regimes has led multinational automakers to adopt a layered compliance approach. Rather than design separate systems for each jurisdiction, many companies have moved toward a "highest standard" model that defaults to GDPR-level protections globally, both to simplify operations and to future-proof their legal compliance (Rustad and Koenig, 2019). However, this strategy is not without cost. In the absence of harmonized standards, transportation firms must continuously adapt to regulatory developments, often retrofitting products or services to align with shifting interpretations of law.

In the US, this complexity is compounded by the potential liability that arises from the unauthorized use or disclosure of transportation data. Courts have only recently begun to confront cases involving the misuse of telematics data or overcollection of biometric information in vehicle cabins. While tort claims based on intrusion upon seclusion or breach of contract are possible, plaintiffs often face difficulty demonstrating concrete harm or tracing data usage to a specific defendant. The lack of a robust federal privacy law leaves individuals with few remedies and companies with little clarity on how to design legally compliant systems across jurisdictions (Peacher, 2020).

This regulatory uncertainty has spurred interest in the concept of "privacy engineering," whereby technical design decisions are made with explicit attention to legal compliance and ethical standards. In the context of connected transportation, this may include anonymizing location data after trip completion, limiting data retention periods, embedding local computation to reduce cloud dependency, and offering granular user controls over data sharing. Some manufacturers have begun incorporating consent screens into dashboard interfaces or mobile apps that pair with vehicles, offering opt-in features for location sharing, usage tracking, or infotainment personalization (Solove, 2022).

The integration of privacy engineering into vehicle design also opens up deeper questions about the role of engineers as regulatory actors. Traditionally, privacy law has treated legal compliance as the domain of attorneys and compliance officers. Yet in a transportation system mediated by software and sensors, the most consequential decisions about what data is collected, how it is stored, and to whom it is transmitted often lie with system designers, network architects, and embedded engineers. This shift underscores the need for interdisciplinary collaboration and for transportation professionals to be conversant not only in technical specifications but in the evolving legal obligations that accompany them (Corley, 2016).

As connected and autonomous vehicles become central to the mobility landscape, the regulation of personal data in these systems will continue to evolve. International models like the GDPR offer coherent frameworks but come with their own tradeoffs in terms of compliance burden and enforcement rigor (Colletti, 2024). Meanwhile, the US legal system faces its own challenge of making regulation and compliance consistent across federal and state jurisdictions. What is increasingly clear is that data privacy in transportation can no longer be treated as an ancillary issue. It is foundational to the legitimacy, safety, and public trust in next-generation mobility systems.

12.3 Regulatory Approaches to Cybersecurity Preparedness and Incident Response

As the cyber-physical dimensions of transportation infrastructure deepen, the imperative for cybersecurity regulation becomes both more urgent and more contested. Transportation systems increasingly depend on software-controlled components, networked sensors, and digital communication protocols that leave vehicles and infrastructure vulnerable to cyber intrusion. These vulnerabilities are not theoretical: ransomware attacks, data breaches, and system-level exploits have already disrupted pipeline operations, rail networks, and airline scheduling. Yet the regulatory response in the United States has been fragmented, inconsistent, and heavily shaped by sectoral jurisdiction. Rather than a comprehensive cybersecurity law governing transportation systems, what exists is a mosaic of voluntary guidance, agency-specific directives, and evolving statutory obligations that vary depending on the mode of transport, the public-private divide, and the perceived national security significance of the asset in question (Hockstad *et al.*, 2024).

In the automotive domain, the National Highway Traffic Safety Administration (NHTSA) has taken the lead, but its approach has favored industry-led standards over mandatory compliance. NHTSA's Cybersecurity Best Practices for the Safety of Modern Vehicles, first issued in 2016 and updated in 2022, encourage vehicle manufacturers to adopt a layered, risk-based approach to cybersecurity. The guidance calls for integration of the ISO/SAE 21434 standard, which formalizes cybersecurity engineering processes for road vehicles, and emphasizes organizational measures such as incident response plans and supply chain risk management. However, these documents are advisory rather than binding. NHTSA has repeatedly stated that it prefers to support innovation through non-prescriptive frameworks and has no current rulemaking that mandates specific technical safeguards (Hockstad *et al.*, 2024).

This voluntary posture has left a significant compliance gap, particularly as vehicles become more reliant on connectivity. Automakers may self-certify compliance with broad safety standards under the Federal Motor Vehicle Safety Standards (FMVSS), but cybersecurity is not currently enumerated as a performance requirement. This means that software vulnerabilities, including over-the-air update systems, infotainment modules, or vehicle-to-everything (V2X) communication layers, are effectively unregulated unless they can be shown to result in an unreasonable safety risk. Even then, NHTSA's recall authority only applies reactively and requires a high threshold of evidence that a defect exists and poses a safety hazard (USGAO, 2017).

In contrast to the automotive sector, other parts of the transportation ecosystem have seen more aggressive federal intervention. The Transportation Security Administration (TSA), which holds security oversight for aviation, rail, and pipelines, has issued binding directives following high-profile cyber incidents. In the wake of the Colonial Pipeline ransomware attack in 2021, TSA issued Security Directives 2021-01 and 2021-02, requiring pipeline operators to designate cybersecurity coordinators, report incidents within 24 hours, implement mitigation measures, and undergo annual cybersecurity assessments. Subsequent directives have extended similar requirements to surface transportation modes, including freight and passenger rail.

These directives are legally binding and represent a notable departure from TSA's prior reliance on voluntary frameworks. They reflect a shift toward treating cybersecurity not merely as an operational risk, but as a core component of national security. Still, the reach of these directives is limited by TSA's jurisdictional boundaries and enforcement capacities. Compliance auditing remains opaque, and the agency has struggled to balance prescriptive requirements with the operational complexity of large private operators who resist one-size-fits-all mandates (ITS JPO, 2025).

A pivotal development in the broader regulatory architecture came with the enactment of CIRCIA. This statute, administered by the Cybersecurity and Infrastructure Security Agency (CISA), creates a uniform federal obligation for designated critical infrastructure entities to report substantial cyber incidents within 72 hours and ransomware payments within 24 hours. While CIRCIA does not target transportation exclusively, it applies across all 16 critical infrastructure sectors identified by the Department of Homeland Security, including transportation systems, pipelines, and public transit networks (Evans, 2015).

CIRCIA's significance lies in its potential to harmonize incident reporting obligations that were previously scattered across agencies and states. Once fully implemented through CISA's forthcoming rulemaking, transportation operators covered under the statute will have to establish internal detection capabilities, prepare reporting protocols, and cooperate with federal investigators in near real time. This represents a shift away from the prior norm in which cybersecurity disclosures were largely voluntary or delayed, often obscured by reputational risk concerns. Moreover, by tying incident reporting to a centralized federal clearinghouse, CIRCIA aims to create a feedback loop in which aggregated data on threats can inform proactive risk mitigation across sectors (Cybersecurity Laws, 2023).

Parallel developments are underway internationally. The European Union's Network and Information Security Directive, recast in 2022 as the NIS2 Directive, expands cybersecurity obligations for operators of essential services and digital infrastructure. Under NIS2, medium and large transportation entities including airports, seaports, rail operators, and road authorities must implement technical and organizational security measures, conduct risk assessments, and report incidents to national authorities within 24 hours (Park, 2020). The Directive mandates board-level responsibility for cybersecurity and introduces substantial fines for non-compliance, aligning its enforcement structure more closely with that of the GDPR.

NIS2 represents a step toward deeper institutionalization of cybersecurity within transportation governance. Unlike the US approach, which often frames cybersecurity as a technical add-on or compliance burden, the EU model embeds it into the operational core of transport systems. The requirement that executive management be accountable for cybersecurity performance marks a cultural shift: cybersecurity is no longer delegated to IT departments but must be addressed as part of fiduciary oversight and risk governance (Park, 2020).

In the automotive context, this regulatory evolution is further extended by the UNECE, which has promulgated two binding regulations: UN Regulation No. 155 (on Cybersecurity Management Systems) and UN Regulation No. 156 (on Software Updates). These have been adopted in the EU, Japan, South Korea, and other participating countries, and are now required for new vehicle type approvals. As of July 2024, all new vehicles sold in the EU must comply with both regulations (Park, 2020).

UN Regulation No. 155 requires that automakers establish a certified Cybersecurity Management System (CSMS) that covers the entire vehicle lifecycle. This includes threat modeling, risk assessment, incident detection, mitigation strategies, and post-sale monitoring. Notably, compliance is not self-declared but subject to third-party audit as part of the vehicle type approval process. The regulation introduces a level of technical specificity that exceeds most national statutes and makes cybersecurity a condition for market entry. Regulation No. 156 complements this by mandating secure and traceable software update mechanisms. It establishes legal requirements for over-the-air update integrity, version control, rollback procedures, and user notification. These provisions reflect the growing recognition that vehicle software is no longer a fixed product but a dynamic system requiring continuous governance. Together, these UNECE regulations form the most detailed international regime governing vehicle cybersecurity to date.

Despite their technical complexity, these international standards are reshaping industry practice well beyond Europe. Global automakers seeking to sell in UNECE-aligned jurisdictions must now integrate cybersecurity compliance into their engineering workflows, supplier agreements, and product documentation. This has catalyzed the emergence of specialized cybersecurity assurance consultancies and embedded compliance modules in vehicle development pipelines. Unlike the US, where compliance risk is often managed post hoc through litigation or regulatory investigation, the UNECE approach builds compliance into the design process itself (Park, 2020).

This divergence between US and international approaches has practical consequences for transportation system operators and manufacturers. US-based companies with global operations must navigate conflicting timelines, enforcement philosophies, and technical specifications. For example, a company that treats ISO/SAE 21434 as voluntary guidance in the US must treat it as mandatory when seeking type approval in the EU. Moreover, the absence of a strong federal compliance framework in the US leaves companies exposed to state-level liability and insurance complications when cybersecurity failures occur (Hockstad *et al.*, 2024).

To manage this complexity, many firms are adopting multi-layered compliance strategies. This may include harmonizing internal standards with the most stringent international requirements, establishing global incident response playbooks, and building centralized security operations centers that can adapt to divergent legal expectations. Yet such strategies require significant investment and technical capacity, which may be beyond the reach of smaller suppliers or regional operators. The current regulatory landscape, in this sense, favors large incumbents with the resources to navigate its complexity (Goodyear, 2021).

Recent cyberattacks across transportation and related infrastructure sectors underscore the urgency of legal and regulatory reform. In 2025, the Texas Department of Transportation suffered a breach after a contractor's credentials were compromised, exposing nearly 300,000 crash reports containing sensitive information such as driver names, birthdates, and license numbers (Domiguez, 2025). While not confirmed to involve AI, the breach highlighted vulnerabilities in third-party access control and data compartmentalization. Rail systems have also long presented critical exposure points. Unencrypted wireless communications between train cars and control systems have been exploitable for over a decade, raising concerns about the potential for remote disruptions of braking systems and scheduling operations (CISA, 2025). Although no confirmed attack has exploited these weaknesses to date, their persistence reflects a failure to implement even baseline cybersecurity architecture. In aviation, the Scattered Spider group used AI-generated voice deepfakes to impersonate IT personnel, breaching the call center infrastructure of Qantas Airways and compromising more than five million passenger records (Williams, 2025). These incidents illustrate the porousness of existing defenses and the inadequacy of relying on voluntary technical guidance in systems where service continuity, public trust, and human safety are at stake.

The threat landscape is further complicated by the rapid deployment of AI-enabled attack tools. According to some reports, groups like FunkSec have employed generative models to produce polymorphic malware capable of adapting its signature to evade detection by traditional antivirus software (Hidalgo, 2025). These tools are increasingly offered as ransomware-as-a-service packages through dark web platforms, lowering the technical threshold for participation in complex cybercrime ecosystems. Other attacks have used large language models to produce persuasive phishing emails tailored to individual targets in multiple languages, significantly increasing breach success rates (ReliaQuest, 2024). In some cases, attackers have deployed prompt injection tactics that manipulate AI-powered defense tools into ignoring malicious files, thereby undermining detection systems designed to safeguard critical infrastructure.

These developments pose acute legal challenges. When a cybersecurity tool powered by machine learning is itself deceived, responsibility becomes difficult to localize. As these systems gain complexity, existing liability doctrines will be strained by questions of foreseeability, design responsibility, and causation. The increasing convergence of AI risk with critical infrastructure reliability suggests that transportation law must evolve to include formal accountability frameworks, enforceable standards, and lifecycle governance mechanisms robust enough to withstand adversarial innovation.

While the need for regulatory coherence has become increasingly apparent, attempts to unify cybersecurity law in the US have thus far stalled. Jurisdictional fragmentation, political polarization, and industry resistance have hindered comprehensive reform. Instead, governance is emerging through what might be called a functional convergence: voluntary standards, agency guidance, binding sectoral rules, and international obligations are gradually aligning around core principles of risk management, disclosure, lifecycle governance, and third-party auditing. While no single instrument provides full coverage, the interaction of these frameworks is beginning to shape a *de facto* standard for what constitutes adequate cybersecurity in transportation systems (Evans, 2015).

Ultimately, the success of this regulatory architecture depends not only on legal clarity, but on operational uptake. Transportation systems are among the most complex socio-technical systems in modern society. Cybersecurity in this domain cannot be reduced to technical fixes or compliance checklists. It requires an institutional capacity to anticipate threats, absorb shocks, and recover functionality without cascading failures. The legal frameworks discussed in this section, fragmented though they may be, represent the scaffolding upon which such capacities are being built. The challenge ahead lies in ensuring that this scaffolding supports not only regulatory compliance, but the broader goal of resilient and trustworthy transportation networks.

12.4 Emerging Legal and Ethical Challenges in Autonomous and AI-Powered Transport

The advent of autonomous vehicles (AVs) and AI-driven transportation systems has introduced a new frontier in legal regulation. While prior sections examined the fragmented governance of privacy and cybersecurity in connected vehicles, this section turns to the more speculative and contested domain of autonomy itself. Here, software supplants the human driver not just as a tool, but as the decision-making core of the vehicle. This transformation recasts long-standing assumptions about liability, agency, and the ethics of movement. Yet the law has not evolved with the same velocity. In the United States and abroad, the legal frameworks governing autonomous systems remain unsettled, revealing deep tensions between innovation and accountability, national interests and global standards, public safety and private control (Jones, 2016).

In the US, the regulatory regime for autonomous vehicles remains strikingly incomplete. No federal statute squarely governs the operation or safety certification of self-driving cars. Instead, oversight is divided between federal authorities, who retain jurisdiction over vehicle safety standards and interstate commerce, and state governments, who regulate licensing, liability, and roadway access. This division has produced a legal patchwork that complicates deployment across jurisdictions. As of 2025, over thirty states have enacted legislation or issued executive orders addressing some facet of autonomous vehicle operation, ranging from permitting regimes and insurance mandates to reporting requirements and definitions of autonomy (Jones, 2016).

NHTSA has issued several iterations of guidance on Automated Driving Systems (ADS), including the 2017 "Vision for Safety" document and its successors. These publications offer frameworks for voluntary safety assessments, encouraging manufacturers to disclose design choices, testing procedures, and ethical considerations. Yet the agency has declined to mandate such disclosures or impose binding performance standards specific to autonomy. This hands-off approach reflects a broader policy tendency to defer to industry in the hope that flexible regulation will foster innovation. However, this deferential stance also leaves critical issues unresolved, including how to evaluate the safety of machine decision-making and what threshold of reliability suffices for public deployment (USGAO, 2017).

Absent comprehensive federal law, states have taken divergent approaches. California, for instance, requires companies testing AVs to obtain permits, submit disengagement reports, and comply with specific operational restrictions. By contrast, Arizona has allowed AV testing and deployment under broad executive discretion, with limited public oversight. These inconsistencies create operational uncertainty for developers and raise broader concerns about forum-shopping and regulatory arbitrage. A system deemed roadworthy in one state may be unlawful in another, not because of technological differences, but because of institutional variance (Jones, 2016).

From a liability perspective, the introduction of autonomy disrupts traditional fault-based frameworks. If a vehicle operating in autonomous mode causes a crash, should liability fall on the manufacturer, the software provider, the user, or the system itself? Tort law presumes a human agent capable of negligence. Yet with Level 4 and Level 5 autonomy, the user may no longer play a meaningful role in driving decisions. In such contexts, attributing fault to the human operator may lack both moral and legal coherence (Rachlinksi and Wistrich, 2021).

Some jurisdictions have begun to legislate around this problem and others related to privacy issues of electric vehicle technology (Rogers, 2024). The United Kingdom's Automated and Electric

Vehicles Act of 2018 creates a first-party insurance scheme for AVs, treating the insurer as the initial payer in the event of an accident involving an autonomous vehicle. The insurer can then seek recovery from the manufacturer if the vehicle was operating improperly. This approach preserves compensation for victims while avoiding the need to immediately resolve complex questions of fault. It also treats the vehicle itself as the operative unit of analysis, rather than its driver or owner (UK, 2018).

Germany has taken a more expansive approach. Amendments to the Road Traffic Act allow for the operation of Level 4 vehicles in defined operational domains, provided that a technical oversight system is in place. This model emphasizes layered accountability, distributing responsibility between the human in charge of oversight, the automated system, and the manufacturer. The regime reflects a recognition that autonomy entails a spectrum of control and that responsibility may shift depending on the specific allocation of functions (Eastman *et al.*, 2023).

Beyond issues of legal liability, autonomous transport raises fundamental ethical questions about risk allocation, moral agency, and the trade-offs inherent in algorithmic decision-making. Scholars have debated whether the introduction of AVs creates morally troubling scenarios in which lives are implicitly traded (for instance, saving many while sacrificing a few, or protecting passengers at the expense of pedestrians). These debates often invoke the so-called trolley problem, repurposed in code. But others have criticized this framing as artificial, arguing that it obscures more pressing concerns about structural risk, systemic bias, and unequal exposure to harm (Rachlinksi and Wistrich, 2021).

The work of Alexander Hevelke and Julian Nida-Rümelin addresses this dilemma by challenging the presumption that responsibility can be meaningfully assigned to users of highly automated systems. If a human operator is not expected to intervene (and may in fact be incapable of doing so) it becomes ethically suspect to attribute blame for failing to prevent harm. Moreover, if autonomous systems reduce overall fatalities but shift risk toward different populations, they may still raise concerns about distributive justice. It is not only the number of lives saved that matters, but whose lives are at risk and who decides (Hevelke and Nida-Rümelin, 2014).

The ethical stakes of autonomy extend further in AI-powered decision-making, where opaque algorithms and machine learning systems increasingly guide navigation, traffic prediction, and collision avoidance. The opacity of these systems complicates accountability. Developers may not fully understand how a neural network processes input or arrives at decisions. In such cases, traditional notions of foreseeability and proximate cause—cornerstones of tort law—begin to erode. Legal scholars have debated whether such systems should be treated as agents, products, or something else entirely. Each framing carries distinct implications for liability, regulation, and public trust (Rachlinksi and Wistrich, 2021).

European regulators have responded to these challenges through both sector-specific rules and cross-cutting AI legislation. The EU's General Safety Regulation (Regulation 2019/2144), which came into effect in July 2022, requires that new vehicles be equipped with advanced safety features, including intelligent speed assistance, lane keeping systems, and emergency stop signals. These features lay the groundwork for higher levels of automation by ensuring a baseline of driver assistance. More broadly, the EU's Artificial Intelligence Act adopts a risk-based framework for regulating AI systems. Under the Act, AI used in safety-critical applications such as autonomous driving is classified as high-risk and subject to rigorous conformity assessments, transparency obligations, and human oversight requirements (EU, Regulation 2019/2144).

This dual approach of embedding functional safety requirements while also regulating AI as a class of technology illustrates a more integrated model of governance. It treats autonomy not as an exception to existing regimes, but as an extension of them, requiring new tools without abandoning existing principles. The AI Act's provisions on transparency and explainability are especially important in the context of transportation, where public trust is critical and the consequences of failure can be fatal (Park, 2020).

Meanwhile, non-governmental organizations and planning bodies have proposed alternative frameworks that center equity, safety, and public accountability. The National Association of City Transportation Officials (NACTO) has released successive editions of its Blueprint for Autonomous Urbanism, calling for AV deployment that prioritizes vulnerable road users, reduces vehicle miles traveled, and supports public transit (NACTO, 2021). NACTO's model stands in contrast to industry-led visions centered on individual convenience and commercial optimization. It reframes autonomy as a governance problem rather than a technological one (NACTO, 2020).

Legal scholars have also turned attention to the embedded standards that shape roadway safety at a structural level. The Manual on Uniform Traffic Control Devices (MUTCD), a foundational document governing traffic signage and infrastructure design in the United States, has been criticized for embedding a car-centric worldview that privileges speed over safety. In their critique, Bronin and Shill argue that revising this manual is essential to making streets safe for all users, especially in a world where AVs will rely on standardized infrastructure to make decisions (Bronin and Shill, 2021). Others have questioned whether the MUTCD's provisions are compatible with AV deployment at all, given that many of its assumptions presume a human driver capable of perception and discretion (FHWA, 2023).

These critiques underscore a broader point: the legal and ethical challenges of autonomous transport are not confined to the vehicle itself. They implicate land use, public investment, urban form, and the distribution of risk in the built environment. They require regulatory systems that are not only technologically literate but attuned to questions of justice, power, and institutional design.

Finally, the introduction of AVs and AI systems intersects with geopolitical concerns about data sovereignty, cross-border liability, and technological standards. China's regulatory regime, for instance, reflects a model in which AV data is treated as a national security asset, subject to localization requirements and state access. The Provisions on the Management of Automotive Data impose strict constraints on the export of data and require companies to undergo security assessments before transmitting data abroad (Pernot-Leplay, 2020). This model stands in sharp contrast to the EU's rights-based framework or the US market-driven model, highlighting the strategic stakes of AV regulation in a multipolar world.

Multinational firms navigating this landscape must not only comply with divergent legal requirements but must also anticipate how those requirements reflect broader political values. A design decision that satisfies GDPR's emphasis on transparency may run afoul of Chinese data localization rules. Conversely, a system optimized for rapid deployment in permissive US jurisdictions may lack the controls necessary for European market entry. The result is a global regulatory environment that is not merely fragmented, but ideologically plural.

In sum, the emergence of autonomous and AI-powered transport challenges the legal system at every level. It disrupts doctrines of liability, tests the limits of agency, and demands a rethinking of regulatory design. While existing laws provide some scaffolding, they are often ill-suited to address the distributed, opaque, and evolving nature of AI systems in motion. What is needed is not only statutory reform, but a jurisprudence of autonomy and an account of responsibility, fairness, and governance suited to a world where vehicles think, learn, and act (Rachum-Twaig and Perl, 2025). This section has sought to outline the contours of that challenge, drawing on comparative models, ethical analysis, and institutional critique. As transportation systems become increasingly intelligent, the law must do more than keep up. It must decide what kind of mobility future it is willing to authorize.

12.5 A Note on US Case Law and the Prosecution Challenges of Cybersecurity and AI-Enabled Cybercrime

As cyberattacks grow in frequency, sophistication, and societal impact, US courts are beginning to confront the legal challenges presented by cybersecurity breaches and data misuse. The landscape of case law remains in flux, but recent federal appellate decisions suggest a developing judicial framework for standing, liability, and remedies in cybersecurity-related litigation. These cases often turn on the issue of injury: whether a plaintiff who has experienced a data breach but has not yet suffered economic harm may sue under constitutional standing requirements. In parallel, the rise of AI-enabled cybercrime introduces new prosecutorial and evidentiary burdens that are only beginning to be addressed by courts and regulators. Together, these developments reflect a system grappling with the doctrinal limits of traditional tort and privacy law in the face of technologically distributed harm.

One of the central challenges for plaintiffs in cybersecurity cases has been demonstrating concrete injury for purposes of Article III standing. Courts have long held that speculative harms are insufficient, but the application of this principle to data breaches has varied. In *Bohnak v. Marsh & McLennan Cos.*, the Second Circuit reversed dismissal and found standing where a former employee's Social Security number had been accessed during a targeted breach. Although there was no confirmed misuse, the court held that the risk of identity theft, combined with emotional distress and mitigation costs, constituted a concrete injury. It also emphasized the nature of the data involved: immutable, sensitive, and exposed through a breach of confidence (Bohnak v. Marsh & McLennan Cos., 2023).

The Third Circuit reached a similar conclusion in *Clemens v. ExecuPharm Inc.*, where a hacking group accessed a pharmaceutical company's servers and published employee data on the dark web. The court ruled that the plaintiff suffered a concrete injury based on the public dissemination of her personal information and the resulting anxiety and out-of-pocket expenses. Importantly, the court treated the unauthorized publication of private information as analogous to harms traditionally recognized at common law, such as public disclosure of private facts or breach of confidentiality (Clemens v. ExecuPharm Inc., 2022).

By contrast, the Eleventh Circuit in *Green-Cooper v. Brinker Int'l, Inc.* drew a distinction between plaintiffs who experienced actual misuse of their data (such as fraudulent charges) and those who merely alleged exposure. The court permitted claims to proceed for the former but held that speculative future harm did not establish standing for others. It also found the class definition overbroad, underscoring the difficulty of establishing commonality in data breach litigation (Green-Cooper v. Brinker Int'l, Inc., 2023).

The Supreme Court's decision in *TransUnion LLC v. Ramirez* continues to shape the standing analysis in these cases. There, the Court held that only individuals whose inaccurate credit files were shared with third parties had standing to recover damages under the Fair Credit Reporting Act. Those whose information was never disseminated, despite the statutory violation, lacked standing absent a concrete, real-world injury. This decision narrowed the path to standing in data-related cases and emphasized the need for plaintiffs to tie statutory violations to recognizable harms (TransUnion LLC v. Ramirez, 2021).

Lower courts have attempted to apply this reasoning with varying results. In *Webb v. Injured Workers Pharmacy, LLC*, the First Circuit held that both actual identity theft and credible risk of future harm could establish standing, especially when accompanied by emotional distress or mitigation expenses. The case involved a breach of sensitive health-related data, and the court took a nuanced approach by affirming standing for damages while rejecting standing for injunctive relief (Webb v. Injured Workers Pharmacy, LLC, 2023).

While many data breach cases arise in tort or contract contexts, regulatory enforcement has also expanded. In *SEC v. SolarWinds Corp.*, the agency brought a civil action under the Securities Exchange Act, alleging that the company misled investors about the adequacy of its cybersecurity measures before a massive breach. The court allowed claims based on materially misleading pre-breach statements to proceed, highlighting the intersection between cybersecurity governance and corporate disclosure obligations. Although the court dismissed some of the SEC's internal controls claims as too vague, the case marks a shift toward viewing cyber risk as a material concern for securities regulation (SEC v. SolarWinds Corp., 2024).

These cases illustrate both the promise and limits of existing legal doctrines in responding to cyber threats. Standing requirements remain a central hurdle, particularly for plaintiffs who have not yet experienced economic losses. Courts have gradually recognized emotional distress, mitigation costs, and the nature of exposed data as relevant to injury analysis, but standards vary by circuit. Meanwhile, enforcement agencies are beginning to pursue cybersecurity failures as regulatory violations, signaling a broader transformation in how cyber risk is conceptualized across domains.

Layered atop these doctrinal challenges is the growing influence of AI-enabled cybercrime. As outlined in recent research, AI tools are rapidly transforming the threat landscape by enhancing the speed, scale, and deception capabilities of cyberattacks. Language models can be used to automate phishing emails, craft deepfake communications, and tailor attacks to specific psychological profiles (Akbar *et al.*, 2025). Criminal groups now offer "deepfake-as-a-service" or use generative tools to mimic voices and writing styles of trusted individuals, bypassing traditional social engineering defenses

(Hockstad *et al.*, 2026). These developments exacerbate the evidentiary burdens in litigation and prosecution. If an employee is tricked by an ultra-realistic synthetic voice or forged biometric credential, assigning blame becomes harder. The human factors that once served as firewalls like intuition, familiarity, and skepticism can now be simulated. This creates a legal gray zone in which negligence is difficult to prove, and responsibility for the breach may become dispersed across system design, employee conduct, and criminal ingenuity.

Moreover, AI-enhanced tools are enabling less technically skilled actors to participate in complex cybercrime ecosystems. This democratization of threat capabilities creates a broader base of risk while making attribution more difficult. Law enforcement must contend not only with anonymized actors operating across borders but with the rapid evolution of tools that outpace traditional investigative methods. Legal doctrines built for discrete actors and definable harms struggle to adapt to this environment, where harm is often latent, distributed, and anticipatory.

From a national defense and public safety perspective, this confluence of cyber vulnerability and AI-driven amplification raises acute concerns. Critical infrastructure systems, including transportation networks, are particularly exposed. A manipulated rail scheduling system, spoofed V2X communication, or falsified remote update could have catastrophic consequences. Prosecuting such acts requires both technical sophistication and a jurisprudence capable of assigning fault within complex sociotechnical systems. The boundary between civilian harm and national security threat is increasingly blurred, requiring legal frameworks that can accommodate both domains (Thomas *et al.*, 2025).

As the case law continues to evolve, and as AI deepens the reach and opacity of cybercrime, courts and legislatures must refine the tools available for redress and deterrence. This includes clarifying standing requirements, expanding definitions of injury to reflect psychological and risk-based harms, and adapting liability frameworks to systems in which human and machine agency intertwine. The rise of AI-enabled cybercrime does not merely intensify the risk. It demands a recalibration of the legal imagination to account for threats that are probabilistic, systemic, and technologically mediated. The cases surveyed here mark the early stages of that recalibration.

12.6 Conclusion

The legal framework surrounding cybersecurity and data governance in transportation is no longer a speculative or niche concern. It is a central feature of modern mobility. As vehicles collect more data, rely on more code, and operate across increasingly complex networks, the legal structures that govern their operation have expanded in kind. What once seemed like separate domains of data privacy, incident response, and artificial intelligence now converge in the everyday decisions of engineers, manufacturers, and regulators.

This chapter has outlined the current shape of that convergence. Data privacy laws define what can be collected, how it must be secured, and when users must be informed. Cybersecurity rules increasingly tie operational readiness to legal obligation. Artificial intelligence systems, once peripheral, now sit at the core of decision-making in automated fleets. Across each of these areas, the law is evolving. Sometimes it moves ahead of the technology, other times it lags behind. But in every case, it now plays an active role in how transportation systems are built and maintained. There is no single blueprint for compliance. Domestic and international obligations overlap. Agencies issue guidance, legislatures pass statutes, and courts develop doctrine through litigation. Transportation professionals must learn to read this terrain as carefully as they do technical specifications. In many cases, the design of a vehicle or system will reflect legal constraints just as much as engineering ones.

As cyber threats grow more sophisticated and AI systems more capable, the boundaries between technical risk and legal exposure will continue to blur. Security failures may trigger regulatory penalties, tort claims, or breach of contract actions. Data misuse can draw attention from both state attorneys general and federal agencies. And failures in AI logic or performance may raise questions not only of safety but of liability, bias, and fairness. In the face of these challenges, the most effective approach is one that treats cybersecurity and data governance not as checklists or regulatory hurdles, but as core elements of system design. That means engaging legal standards at the front end, integrating compliance into product development, and remaining alert to the broader institutional frameworks that structure transportation technology.

This chapter is necessarily limited by the pace at which both technology and law continue to evolve. Cybersecurity threats in the transportation sector are dynamic, with new vulnerabilities emerging as systems become more interconnected and increasingly reliant on machine learning and automated control. Legal responses, by contrast, tend to lag behind, and many existing statutes and regulatory instruments were not designed with AI-enabled infrastructure or synthetic attack vectors in mind. Jurisdictional variability further complicates analysis. Differences across states, sectors, and international regimes limit the possibility of drawing uniform conclusions about regulatory effectiveness or compliance strategy. Empirical data on enforcement, breach disclosure, and regulatory audits also remains scarce, particularly in private-sector transportation networks.

These constraints suggest several directions for future research. Scholars and practitioners may examine the operational impact of regulatory mandates across jurisdictions, assess the performance of sector-specific cybersecurity regimes over time, or explore how engineering-based risk models can be systematically integrated into legal compliance frameworks. A deeper empirical foundation will be essential for measuring whether current legal approaches promote resilience, allocate responsibility coherently, and adapt effectively to the realities of digital transportation systems. The legal landscape presented here is not static. It will continue to evolve with each new breach, court decision, or technological leap. But the contours are now visible enough to support active engagement. For engineers and transportation professionals, that engagement is not optional. It is the ground on which connected and automated mobility will succeed or fail.

Bibliography

Akbar, K. A., Uddin, N., Khan, L., Hockstad, T., Rahman, M., Chowdhury, M., and Thuraisingham, B. (2025). Retrieval-augmented generation–based large language models for bridging transportation cybersecurity legal knowledge gaps, in *Presented at the 103rd Annual Meeting of the Transportation Research Board*, Washington, D.C., January 2025, https://arxiv.org/abs/2505.18426.

Bohnak v. Marsh & McLennan Companies, Inc., 79 F.4th 276 (2d Cir. 2023).

Bronin, S. C. and Shill, G. H. (2021). Rewriting our nation's deadly traffic manual (B. withdraw the proposed chapter concerning automated vehicles), *Journal of Law and Mobility* (2021 Special Edition), pp. 54–62, https://doi.org/10.36635/jlm.2021.special.54.

Buyuksagis, E. (2019). Towards a transatlantic concept of data privacy, *Fordham Intellectual Property, Media & Entertainment Law Journal* **30**, pp. 139–196.

Chander, A., Abraham, M., Chandy, S., Fang, Y., Park, D., and Yu, I. (2021). Achieving privacy, *SMU Law Review* **74**, pp. 607–642.

Clemens v. ExecuPharm Inc., 48 F.4th 146 (3d Cir. 2022).

Colletti, M. (2024). Privacy law: A global comparison of data protection regulation, *Suffolk Transactional Law Review* **47**, pp. 262–285.

Corley, M. A. (2016). The need for an international convention on data privacy: Taking a cue from the CISG, *Brooklyn Journal of International Law* **41**, pp. 721–752.

Cybersecurity Laws and Regulations USA (2023). https://iclg.com/practice-areas/cybersecurity-laws-and-regulations/usa, accessed August 31, 2023.

Cybersecurity and Infrastructure Security Agency (2025). ICSA-25-191-10: End-of-train and head-of-train remote linking protocol (Update C), July 9, 2025, https://www.cisa.gov/news-events/ics-advisories/icsa-25-191-10.

Dominguez, C. (2025). TxDOT notifying drivers after breach results in 300K crash reports downloaded with personal info, *Houston Chronicle*, July 8, 2025, https://www.houstonchronicle.com/news/houston-texas/trending/article/txdot-crash-report-breach-20760518.php.

Eastman, B., Collins, S., Jones, R., Martin, J. J., Blumenthal, M. S., and Stanley, K. D. (2023). A comparative look at various countries' legal regimes governing automated vehicles, *Journal of Law and Mobility* **2023**, pp. 1–58.

European Parliament and Council of the European Union (EU) (2019). Regulation (EU) 2019/2144 of the European Parliament and of the Council of 27 November 2019 on type-approval requirements for motor vehicles and their trailers, and systems, components and separate technical units intended for such vehicles, as regards their general safety and the protection of vehicle occupants and vulnerable road users, and repealing regulation (EC) No 78/2009, *Official Journal of the European Union* **L 325**, pp. 1–40.

Evans, G. J. (2015). Regulating data practices: How state laws can shore up the FTC's authority to regulate data breaches, privacy and more, *Administrative Law Review* **67**, pp. 187–233.

Federal Highway Administration (FHWA) (2023). Manual on Uniform Traffic Control Devices (MUTCD), 11th Edition – Part 5: Traffic Control Device Considerations for Automated Vehicles, U.S. Department of Transportation, December 2023, https://mutcd.fhwa.dot.gov/pdfs/11th_Edition/part5.pdf.

Federal Trade Commission (FTC) (2025). General Motors and OnStar, LLC; Analysis of proposed consent order to aid public comment, Federal Register [File No. 242 3052].

Goodyear, M. P. (2021). A rising tide lifts all consumers: Penumbras of foreign data protection laws in the United States, *Richmond Journal of Law & Technology* **27**, pp. 1–37.

Green-Cooper v. Brinker Int'l, Inc., 73 F.4th 883 (11th Cir. 2023).

Hevelke, A. and Nida-Rümelin, J. (2014). Responsibility for crashes of autonomous vehicles: An ethical analysis, *Science and Engineering Ethics* **21**, 3, pp. 619–630.

Hidalgo, Á. (2025). Smarter threats, smarter defenses: The AI arms race in cybersecurity, *CyberProof*, April 4, 2025, https://www.cyberproof.com/blog/smarter-threats-smarter-defenses-the-ai-arms-race-in-cybersecurity/.

Hockstad, T., Rahman, M., Chowdhury, M., Akbar, K. A., Uddin, M. N., and Khan, L. (2025). Data security and privacy regulation in the U.S.A: A 50-state legislative survey, *Transportation Research Record*, https://doi.org/10.1177/03611981251342785.

Hockstad, T., Rahman, M., Jones, S., and Chowdhury, M. (2024). A regulatory gap analysis in transportation cybersecurity and data privacy, *Transportation Journal* **64**, 1, p. e12036, https://doi.org/10.1002/tjo3.12036.

Hockstad, T., Walker, R., Dasgupta, S., Uddin, M. N., Thomas, O., and Chowdhury, M. (2026). Cybersecurity at the crossroads: Addressing AI threats to U.S. Transportation, Transportation Research Board, Forthcoming.

Intelligent Transportation Systems Joint Program Office (ITS JPO) (2025). ITS Cybersecurity Research Program, https://www.its.dot.gov/.

Jones, A. (2016). Autonomous cars: Navigating the patchwork of data privacy laws that could impact the industry, *Catholic University Journal of Law and Technology* **25**, pp. 180–210.

Jones, M. L. and Kaminski, M. E. (2021). An American's guide to the GDPR, *Denver Law Review* **98**, pp. 93–142.

Manny, C. (2018). Driven data: Connected cars and privacy law, *Business Law Review* **51**, pp. 35–64.

National Association of City Transportation Officials (NACTO) (2020). Blueprint for Autonomous Urbanism, Version 1.3: "Local Action in the Face of Uncertainty" (NACTO, New York).

National Association of City Transportation Officials (NACTO) (2021). Recommendations for Reforms to NHTSA Safety Regulations, March 2021, https://nacto.org/wp-content/uploads/NACTO-NHTSA-Safety-Reforms.pdf.

Park, G. (2020). The changing wind of data privacy law: A comparative study of the European Union's General Data Protection Regulation and the 2018 California Consumer Privacy Act, *UC Irvine Law Review* **10**, pp. 1455–1490.

Peacher, H. B. (2020). Regulating data privacy of connected vehicles: How automotive giants can protect themselves and their golden goose, *Albany Law Journal of Science and Technology* **30**, pp. 74–112.

Pernot-Leplay, E. (2020). China's approach on data privacy law: A third way between the U.S. and the E.U.? *Penn State Journal of Law & International Affairs* **8**, pp. 49–98.

Perumal, V. (2022). The future of U.S. data privacy: Lessons from the GDPR and state legislation, *Notre Dame Journal of International & Comparative Law* **12**, 1, pp. 99–123.

Rachlinski, J. J. and Wistrich, A. J. (2021). Judging autonomous vehicles, *DePaul Law Review* **70**, 3, pp. 899–938.

Rachum-Twaig, O. and Perl, G. (2025). Human liability in semi-autonomous vehicles, *Journal of Law and Mobility* **2025**, pp. 1–25.

ReliaQuest Threat Research Team (2024). Introducing "AI-Powered Cybercrime" Report, *ReliaQuest Blog*, July 11, 2024, https://reliaquest.com/blog/introducing-ai-powered-cybercrime-report/.

Rogers, P. (2024). Leading the charge: How increased adoption of electric vehicles renews calls for data privacy protection in the United States, *William & Mary Environmental Law and Policy Review* **49**, pp. 277–306.

Rustad, M. L. and Koenig, T. H. (2019). Towards a global data privacy standard, *Florida Law Review* **71**, pp. 365–426.

SEC v. SolarWinds Corp., 741 F. Supp. 3d 37 (S.D.N.Y. 2024).

Solove, D. and Schwartz, P. M. (2022). ALI data privacy: Overview and black letter text, *UCLA Law Review* **68**, pp. 1252–1317.

Thomas, O., Salek, S., Tine, J.-M., Rahman, M., Hockstad, T., and Chowdhury, M. (2025). Cybersecurity in transportation systems: Policies and technology directions, in *Presented at the 104th Annual Meeting of the Transportation Research Board*, Washington, D.C., January 5–9, 2025, *arXiv preprint* arXiv:2501.05356.

TransUnion LLC v. Ramirez, 594 U.S. 413 (2021).

United Kingdom (UK) (2018). *Electric and Automated Vehicles Act 2018*, c. 18 (Her Majesty's Stationery Office, London).

U.S. Government Accountability Office (USGAO) (2017). Vehicle Data Privacy: Industry and Federal Efforts Under Way, but NHTSA Needs to Define Its Role, GAO-17-656.

Wang, F. Y. (2020). Cooperative data privacy: The Japanese model of data privacy and the EU-Japan GDPR adequacy agreement, *Harvard Journal of Law & Technology* **33**, pp. 661–701.

Williams, T. (2025). Did the Qantas hackers use AI voice deepfakes? *Information Age*. Australian Computer Society, July 17, 2025, https://ia.acs.org.au/article/2025/did-the-qantas-hackers-use-ai-voice-deepfakes-.html.

Webb v. Injured Workers Pharmacy, LLC, 72 F.4th 365 (1st Cir. 2023).

https://doi.org/10.1142/9789819831548_bmatter

Index

C

D

www.ingramcontent.com/pod-product-compliance
Ingram Content Group UK Ltd.
Pitfield, Milton Keynes, MK11 3LW, UK
UKHW061000120926
8556IPUK00032B/893

9 789819 83153